Cannabis

Cannabis

Neuropsychiatry and Its Effects on Brain and Behavior

Editors

Marco Colizzi
Sagnik Bhattacharyya

MDPI • Basel • Beijing • Wuhan • Barcelona • Belgrade • Manchester • Tokyo • Cluj • Tianjin

Editors
Marco Colizzi
Section of Psychiatry,
Department of Neurosciences,
Biomedicine and Movement
Sciences, University of Verona
Italy

Sagnik Bhattacharyya
Department of Psychosis
Studies, Institute of Psychiatry,
Psychology & Neuroscience,
King's College London
UK

Editorial Office
MDPI
St. Alban-Anlage 66
4052 Basel, Switzerland

This is a reprint of articles from the Special Issue published online in the open access journal *Brain Sciences* (ISSN 2076-3425) (available at: https://www.mdpi.com/journal/brainsci/special_issues/Cannabis_Neuropsychiatry_Brain_Behavior).

For citation purposes, cite each article independently as indicated on the article page online and as indicated below:

LastName, A.A.; LastName, B.B.; LastName, C.C. Article Title. *Journal Name* **Year**, *Volume Number*, Page Range.

ISBN 978-3-03943-995-9 (Hbk)
ISBN 978-3-03943-996-6 (PDF)

Cover image courtesy of Venere Rotelli.

Contents

About the Editors

Marco Colizzi, MD, PhD, is a Clinical Research Fellow in Psychiatry at the University of Verona, Italy. He is also a Visiting Researcher at King's College London where he received his PhD in Neuroscience for study of the neurocognitive and neurochemical effects of cannabinoids on the human brain. Dr Colizzi's research focuses on the neuropsychopharmacology and neurocognitive function of psychosis and neurodevelopmental conditions, with a focus on prevention and early intervention strategies in mental health To date, he has published over 50 scientific articles in refereed international journals, also receiving awards and recognition of several prestigious institutions, including the Royal College of Psychiatry, the European College of Neuropsychopharmacology, the Schizophrenia International Research Society, and the Royal Society of Medicine.

Sagnik Bhattacharyya, MBBS, MD, PhD, is a Professor of Translational Neuroscience and Psychiatry at King's College London and Consultant Psychiatrist the South London and Maudsley NHS foundation trust. The overarching goal of his research program is to advance understanding about the pathophysiological underpinnings of psychosis with and without comorbid substance misuse and their clinical correlates, so as to enable the translation of knowledge into practice, either by helping develop novel treatments or enabling prediction of the risk of adverse outcomes and their prevention. He heads the Translational Neuroscience and Psychiatry research group and leads multicentre clinical trials involving cannabinoids in various conditions.

Editorial

Cannabis: Neuropsychiatry and Its Effects on Brain and Behavior

Marco Colizzi [1,2,*] and Sagnik Bhattacharyya [2]

[1] Section of Psychiatry, Department of Neurosciences, Biomedicine and Movement Sciences, University of Verona, 37134 Verona, Italy

[2] Department of Psychosis Studies, Institute of Psychiatry, Psychology and Neuroscience, King's College London, London SE5 8AF, UK; sagnik.2.bhattacharyya@kcl.ac.uk

* Correspondence: marco.colizzi@univr.it; Tel.: +39-045-812-6832

Received: 30 October 2020; Accepted: 9 November 2020; Published: 10 November 2020

Possibly orally transmitted from before circa 2000 B.C., the first written evidence of a role of cannabis in health and disease dates back to Chinese medicine texts of the first to second century B.C. [1]. Between the 12th and the 13th century C.E., the detrimental effects of cannabis on mental health were first reported by the physician Iban Beitar [2]. Later in 1845, the French psychiatrist Jacque-Joseph Moreau described such effects as "acute psychotic reactions, generally lasting but a few hours, but occasionally as long as a week; the reaction seemed dose-related and its main features included paranoid ideation, illusions, hallucinations, delusions, depersonalization, confusion, restlessness, and excitement. There can be delirium, disorientation, and marked clouding of consciousness" [3]. Such evidence suggested a potential role of cannabis in the pathophysiology of psychosis and other mental disorders, as later corroborated by research studies performed over the last 50 years [4].

Now, in the 21st century, while the medicinal properties of cannabis are also under scrutiny through appropriate clinical development, testing and approval process, we are bombarded by claims about cannabis products that are sold over-the-counter with the promise to cure or prevent disease, improve health, and restore functioning. This has led to the question whether and why a drug like cannabis could be both a poison and an antidote. Much of the debate has been on the detrimental and potentially therapeutic effects of cannabis on brain and related behavior, with implications for a number of neuropsychiatric disorders [5].

Despite such apparent discrepancy, in recent times we have seen a considerable progress in our understanding of the role of specific cannabis ingredients and patterns of use for brain function, its neurobiology, and related behavior [6–10]. The chapters in this volume are but a sampling of the latest research evidence on the role of cannabis and its compounds in brain function and dysfunction as well as normal and aberrant behavior. Attention is also given to studies investigating how cannabis compounds may accelerate or prevent and even treat neuropsychiatric disorders.

Cognitive dysfunction as a consequence of cannabis use has been one of the hypotheses mostly investigated, even in this Special Issue, but also one of those mostly debated, due to conflicting evidence in both health and disease. Blest-Hopley et al. performed a systematic review of human studies investigating whether cannabis users and non-users differ in terms of memory-related brain functioning and related task performance. The authors found that cannabis use tends to be associated with poorer performance possibly underpinned by altered functioning of a wide network of brain substrates. However, they suggest that such evidence is far from unequivocal, due to difficulties in drawing conclusions from highly heterogeneous studies in terms of level and type of cannabis exposure, use during developmentally sensitive periods such as adolescence, and duration of abstinence, if any [11]. In order to clarify the effects of problematic cannabis use among young adults from both the neurophysiological and neurocognitive point of view, Imperatori et al. investigated triple-network electroencephalographic (EEG) functional connectivity in a case-control study. Results revealed an increased delta connectivity between the salience network and central executive network in the context of problematic cannabis use, specifically between the dorsal anterior cingulate cortex and right posterior parietal cortex. Such alteration, which is thought to regulate the general access to cognitive functions and to explain the development of psychopathological symptoms across multiple mental disorders, correlated with the severity of problematic cannabis use after controlling for the confounding effect of age, sex, educational level, tobacco use, problematic alcohol use, and general psychopathology [12]. In another case-control study among young adults, Shevorykin et al. investigated whether frontal alpha asymmetry (FAA), which is a measure of approach bias and inhibitory control, differs between cannabis users and healthy controls. Electroencephalographic measures revealed different patterns between the two groups, with healthy controls exhibiting greater relative right activity, that is associated with withdrawal-related tendencies, when exposed to cannabis cues during the filtering task. In contrast, cannabis users exhibited greater relative left frontal activity, which is associated with approach-related tendencies, independent of the cue. According to the authors, such a difference in using the behavioral inhibition system (BIS) and the behavioral activation system (BAS) may reflect a different organization of cognitive resources among cannabis users, with implication for emotions and behavior [13]. In another study, Sullivan et al. investigated structural brain abnormalities in the context of adolescent and adult cannabis use, finding larger cuneus surface area (SA). However, when clustering by gender, male cannabis users exhibited smaller SA and less complex local gyrification index (LGI) in frontal, cingulate and parietal regions, while female cannabis users tended to present with the opposite pattern. Moreover, independent of cannabis use, increased aerobic fitness was associated with more complex LGI and larger SA across different brain regions, possibly reflecting a superior cognitive functioning as a consequence of aerobic exercise which may mitigate the negative impact of chronic cannabis use on neurocognition [14]. Complementing this work, based on the evidence of a role of the endocannabinoid system in memory function as well as of an exercise–memory relationship, Loprinzi et al. proposed a model in which the endocannabinoid system may, at least in part, subserve the effects of exercise on memory function, through a number of endocannabinoid signaling mechanisms related to long-term potentiation, production of neurotrophic factors, and cellular neurogenesis. Its potential mechanistic paradigm, for instance, whether the site of cannabinoid receptor type 1 activation (e.g., gamma-aminobutyric acid (GABA)-ergic, glutamatergic) moderates the exercise–memory relationship, remains to be investigated [15]. Colizzi et al. discussed the importance of interpreting different lines of research evidence on cannabis and cognition altogether, including preclinical versus clinical evidence, acute versus long-term effects, occasional versus regular exposure, and organic versus synthetic cannabinoids, as a strategy to overcome the risks of interpreting the phenomenon based only on partial data. Their reappraisal concludes that earlier age of use, high-frequency and high-potency cannabis use, as well as sustained use over time and use of synthetic cannabinoids, are all correlated with a higher likelihood of developing potentially severe and persistent executive function impairments, as also corroborated by additional evidence from both structural and functional brain alterations associated with cannabis use. The authors call for attention regarding the effects that cannabis use may have in patients with neuropsychiatric conditions, whose cognitive function may already be less proficient as consequence of the underlying pathology [16].

Another recurring question in the field of cannabis and neuropsychiatry is whether the association between cannabis use and psychosis observed in many studied should be interpreted as cannabis use being a causal component in the development of psychosis [4]. Two studies published in this Special Issue advanced our understanding of the phenomenon. Colizzi et al. performed a double-blind, randomized, placebo-controlled crossover study where healthy young adults with modest previous cannabis exposure were acutely exposed to cannabis' key psychoactive ingredient, delta-9-tetrahydrocannabinol (Δ9-THC). Under such controlled experimental conditions, Δ9-THC elicited symptomatic manifestations that resembled those observed in psychosis in most of the participants, with one in five presenting with moderate to severe symptoms. Symptoms tended to quickly self-resolve; however, nearly one-third of the volunteers experienced mild symptomatic effects that lasted for at least 2.5 h [17]. van der Steur et al. performed a systematic review of the factors that may increase the risk of psychosis among cannabis users. They found that frequent cannabis use, especially on a daily basis, and the consumption of high-potency varieties, with high concentrations of Δ9-THC, are both associated with a higher risk of developing psychosis. Moreover, a common genetic background resulted to predispose to psychotic disorders as well as cannabis use, especially genetic variations in dopamine signaling. Finally, cannabis use was reported to be associated with an earlier onset of psychosis and to increase the risk of transition in individuals at clinical high risk of psychosis, thus potentially accelerating the cascade of neurobiological events leading to the manifestation of the disorder [18].

Another line of research is interested in investigating the psychobiological reasons for continuing using cannabis despite the potential experience of detrimental effects [19]. May et al. investigated the role of negative reinforcement by using the Cue Breathing fMRI paradigm which pairs a cue reactivity task with anticipation and experience of an unpleasant interoceptive stimulus, an inspiratory breathing load. Adolescents whose cannabis use reflected a substance use disorder experienced the aversive breathing load differently than experimental users and controls. However, instead of exhibiting an exaggerated activation in brain regions implicated in interoception and emotion regulation, as expected by the authors, the experience of the aversive interoceptive probe resulted in a greater deactivation across such regions. Moreover, findings did not support the hypothesis that cannabis use would be driven by negative reinforcement, as viewing substance images did not dampen uncomfortable sensations. On the contrary, results pointed in the direction of a positive reinforcement, such as increased sensation-seeking and reward responsivity, at least in adolescence [20]. A further study performed among African Americans in economically challenged areas found that current use of cannabis is more common in younger, healthier, less obese, and less educated African American older adults. In particular, findings suggest that African American older adults do not use cannabis to alleviate chronic disease, pain, or depression, and its use does not necessarily co-occur with cigarette smoking and alcohol drinking [21]. While these studies add to the increasing evidence against a self-medication hypothesis of cannabis use among both young and older people, the debate is still open.

Last but not least, cannabis use has seen a huge increase in its licit production, growing from 1.4 tons in 2000, mainly for purposes of scientific research, to 211.3 tons by 2016, due to the increasing implementation of medicinal programs with cannabis-related medicinal products for a wide range of neuropsychiatric conditions [5]. Aviram et al. reported the results of a cross-sectional questionnaire-based study aimed to investigate the impact of treatment with medical cannabis in people suffering from migraine. Medical cannabis resulted in long-term reduction of migraine frequency in >60% of treated patients, also reducing migraine disability severity and migraine analgesics consumption. Based on treatment response, indexed as a decrease in monthly migraine attacks frequency ≥50%, authors were able to identify a specific strain with potential benefits, containing higher doses of the phytocannabinoid ms_373_15c and lower doses of the phytocannabinoid ms_331_18d. As stated by the authors themselves, the anti-migraine effect of such phytocannabinoids and whether they are biological active will have to be elucidated in future studies [22]. This Special Issue also hosts the study protocol of a randomized controlled trial aiming to evaluate the efficacy of adjunctive dronabinol (licensed form of Δ9-THC) at the doses of 5 to 30 mg/die versus control (systemic analgesics only) for reducing opioid consumption in adults aged 18–65 years with traumatic injury [23].

We hope that the topics addressed in this Special Issue will result in new studies that will help further understanding the increasing role of cannabis and its components in neuropsychiatric health and disease. Thanks to such studies, we believe that in the near future we will witness important and exciting advances in the field of cannabis-related pharmacological treatments. Stay tuned.

Conflicts of Interest: The authors declare no conflict of interest.

References

1. Brand, E.J.; Zhao, Z. Cannabis in Chinese medicine: Are some traditional indications referenced in ancient literature related to cannabinoids? *Front. Pharm.* **2017**, *8*, 108. [CrossRef]
2. Dhunjibhoy, J.E. A brief resume of the types of insanity commonly met with in India with a full description of "Indian hemp insanity" peculiar to the country. *J. Ment. Sci.* **1930**, *76*, 254–264. [CrossRef]
3. Moreau, J.J. *Du Hachisch et de L'aliénation Mentale: Études Psychologiques*; Fortin Masson: Paris, France, 1845.
4. Colizzi, M.; Bhattacharyya, S. Is there sufficient evidence that cannabis use is a risk factor for psychosis? In *Risk Factors for Psychosis: Paradigms, Mechanisms, and Prevention*; Thompson, A.D., Broome, M.R., Eds.; Academic Press: Cambridge, MA, USA, 2020; pp. 305–331.
5. Colizzi, M.; Ruggeri, M.; Bhattacharyya, S. Unraveling the intoxicating and therapeutic effects of cannabis ingredients on psychosis and cognition. *Front. Psychol.* **2020**, *11*, 833. [CrossRef] [PubMed]
6. O'Neill, A.; Wilson, R.; Blest-Hopley, G.; Annibale, L.; Colizzi, M.; Brammer, M.; Giampietro, V.; Bhattacharyya, S. Normalization of mediotemporal and prefrontal activity, and mediotemporal-striatal connectivity, may underlie antipsychotic effects of cannabidiol in psychosis. *Psychol. Med.* **2020**, 1–11. [CrossRef] [PubMed]
7. Appiah-Kusi, E.; Petros, N.; Wilson, R.; Colizzi, M.; Bossong, M.G.; Valmaggia, L.; Mondelli, V.; McGuire, P.; Bhattacharyya, S. Effects of short-term cannabidiol treatment on response to social stress in subjects at clinical high risk of developing psychosis. *Psychopharmacology (Berl)* **2020**, *237*, 1121–1130. [CrossRef] [PubMed]
8. Colizzi, M.; Weltens, N.; McGuire, P.; Lythgoe, D.; Williams, S.; Van Oudenhove, L.; Bhattacharyya, S. Delta-9-tetrahydrocannabinol increases striatal glutamate levels in healthy individuals: Implications for psychosis. *Mol. Psychiatry* **2019**. [CrossRef]
9. Colizzi, M.; Bhattacharyya, S. Neurocognitive effects of cannabis: Lessons learned from human experimental studies. *Prog. Brain Res.* **2018**, *242*, 179–216. [CrossRef]
10. Schoeler, T.; Petros, N.; Di Forti, M.; Klamerus, E.; Foglia, E.; Ajnakina, O.; Gayer-Anderson, C.; Colizzi, M.; Quattrone, D.; Behlke, I.; et al. Effects of continuation, frequency, and type of cannabis use on relapse in the first 2 years after onset of psychosis: An observational study. *Lancet Psychiatry* **2016**, *3*, 947–953. [CrossRef]
11. Blest-Hopley, G.; Giampietro, V.; Bhattacharyya, S. A systematic review of human neuroimaging evidence of memory-related functional alterations associated with cannabis use complemented with preclinical and human evidence of memory performance alterations. *Brain Sci.* **2020**, *10*, 102. [CrossRef]
12. Imperatori, C.; Massullo, C.; Carbone, G.A.; Panno, A.; Giacchini, M.; Capriotti, C.; Lucarini, E.; Ramella Zampa, B.; Murillo-Rodríguez, E.; Machado, S.; et al. Increased resting state triple network functional connectivity in undergraduate problematic cannabis users: A preliminary EEG coherence study. *Brain Sci.* **2020**, *10*, 136. [CrossRef]
13. Shevorykin, A.; Ruglass, L.M.; Melara, R.D. Frontal alpha asymmetry and inhibitory control among individuals with cannabis use disorders. *Brain Sci.* **2019**, *9*, 219. [CrossRef]
14. Sullivan, R.M.; Wallace, A.L.; Wade, N.E.; Swartz, A.M.; Lisdahl, K.M. Assessing the role of cannabis use on cortical surface structure in adolescents and young adults: Exploring gender and aerobic fitness as potential moderators. *Brain Sci.* **2020**, *10*, 117. [CrossRef] [PubMed]
15. Loprinzi, P.D.; Zou, L.; Li, H. The endocannabinoid system as a potential mechanism through which exercise influences episodic memory function. *Brain Sci.* **2019**, *9*, 112. [CrossRef] [PubMed]
16. Colizzi, M.; Tosato, S.; Ruggeri, M. Cannabis and cognition: Connecting the dots towards the understanding of the relationship. *Brain Sci.* **2020**, *10*, 133. [CrossRef] [PubMed]
17. Colizzi, M.; Weltens, N.; McGuire, P.; Van Oudenhove, L.; Bhattacharyya, S. Descriptive psychopathology of the acute effects of intravenous delta-9-tetrahydrocannabinol administration in humans. *Brain Sci.* **2019**, *9*, 93. [CrossRef] [PubMed]
18. Van der Steur, S.J.; Batalla, A.; Bossong, M.G. Factors moderating the association between cannabis use and psychosis risk: A systematic review. *Brain Sci.* **2020**, *10*, 97. [CrossRef] [PubMed]
19. Bianconi, F.; Bonomo, M.; Marconi, A.; Kolliakou, A.; Stilo, S.A.; Iyegbe, C.; Gurillo Munoz, P.; Homayoun, S.; Mondelli, V.; Luzi, S.; et al. Differences in cannabis-related experiences between patients with a first episode of psychosis and controls. *Psychol. Med.* **2016**, *46*, 995–1003. [CrossRef] [PubMed]

20. May, A.C.; Jacobus, J.; Stewart, J.L.; Simmons, A.N.; Paulus, M.P.; Tapert, S.F. Do adolescents use substances to relieve uncomfortable sensations? A preliminary examination of negative reinforcement among adolescent cannabis and alcohol users. *Brain Sci.* **2020**, *10*, 214. [CrossRef]
21. Cobb, S.; Bazargan, M.; Smith, J.; Del Pino, H.E.; Dorrah, K.; Assari, S. Marijuana use among African American older adults in economically challenged areas of south Los Angeles. *Brain Sci.* **2019**, *9*, 166. [CrossRef]
22. Aviram, J.; Vysotski, Y.; Berman, P.; Lewitus, G.M.; Eisenberg, E.; Meiri, D. Migraine frequency decrease following prolonged medical cannabis treatment: A cross-sectional study. *Brain Sci.* **2020**, *10*, 360. [CrossRef]
23. Swartwood, C.; Salottolo, K.; Madayag, R.; Bar-Or, D. Efficacy of dronabinol for acute pain management in adults with traumatic injury: Study protocol of a randomized controlled Trial. *Brain Sci.* **2020**, *10*, 161. [CrossRef] [PubMed]

Review

A Systematic Review of Human Neuroimaging Evidence of Memory-Related Functional Alterations Associated with Cannabis Use Complemented with Preclinical and Human Evidence of Memory Performance Alterations

Grace Blest-Hopley [1], Vincent Giampietro [2] and Sagnik Bhattacharyya [1,3,*]

[1] Department of Psychosis Studies, Institute of Psychiatry, Psychology & Neuroscience, King's College London, London SE5 8AB, UK; k1334159@kcl.ac.uk

[2] Department of Neuroimaging, Centre for Neuroimaging Sciences, PO Box 089, Institute of Psychiatry, Psychology & Neuroscience, King's College London, London SE5 8AB, UK; vincent.giampietro@kcl.ac.uk

[3] South London and Maudsley NHS Foundation Trust, Denmark Hill, Camberwell, London SE5 8AB, UK

* Correspondence: sagnik.2.bhattacharyya@kcl.ac.uk; Tel.: +44-207-848-0955; Fax: +44-207-848-0976

Received: 8 January 2020; Accepted: 10 February 2020; Published: 13 February 2020

Abstract: Cannabis has been associated with deficits in memory performance. However, the neural correlates that may underpin impairments remain unclear. We carried out a systematic review of functional magnetic resonance imaging (fMRI) studies investigating brain functional alterations in cannabis users (CU) compared to nonusing controls while performing memory tasks, complemented with focused narrative reviews of relevant preclinical and human studies. Twelve studies employing fMRI were identified finding functional brain activation during memory tasks altered in CU. Memory performance studies showed CU performed worse particularly during verbal memory tasks. Longitudinal studies suggest that cannabis use may have a causal role in memory deficits. Preclinical studies have not provided conclusive evidence of memory deficits following cannabinoid exposure, although they have shown evidence of cannabinoid-induced structural and histological alteration. Memory performance deficits may be related to cannabis use, with lower performance possibly underpinned by altered functional activation. Memory impairments may be associated with the level of cannabis exposure and use of cannabis during developmentally sensitive periods, with possible improvement following cessation of cannabis use.

Keywords: cannabis; memory; functional magnetic resonance imaging; THC; systematic review

---

1. Introduction

Cannabis is the most-used illicit drug worldwide [1], with many beginning to use it during their adolescent years [2,3]. Acute effects of the drug have been shown on cognitive performance, particularly in the domain of memory [4], with impairments being observed in all aspects of memory function, such as encoding, storage, and recall [5,6]. In addition to evidence about its acute effects, meta-analytic evidence has documented that long-term use of cannabis is associated with memory deficits [7].

Brain-structural alterations in cannabis users have been previously attributed to underlie deficits in memory performance. Reduced hippocampal volumes have been observed in cannabis users [8–10], with some studies showing evidence of a dose-dependant effect [11–13]. Along with this, cannabis users have shown volume reductions in the medial temporal cortex, particularly in the parahippocampal gyrus and temporal pole [14], as well as decreased cortical thickness in the orbital frontal cortex [14–17], frontal gyrus [17], and prefrontal cortex [18]. Other evidence suggests that structural alterations are not robust in the hippocampus [17,19–21], the orbitofrontal cortex [13,22,23], frontal gyrus [23], or prefrontal regions [17], or for overall grey matter volumes [19,24–28], even following meta-analysis [29]. Therefore, proposed cognitive deficits in cannabis users may be better explained by alterations in the functioning of relevant brain regions.

The cannabinoid 1 (CB1) receptor is the main central cannabinoid receptor through which the leading psychoactive component of cannabis, delta-9-tertahydrocanabinol (THC), exerts its effect. The CB1 receptors are expressed ubiquitously throughout the brain [30], although higher densities are observed in regions key for memory functioning, such as the hippocampus and related medial temporal lobe structures and the frontal cortex [31]. Cannabis use may alter functioning of the key neural substrates involved in the processing of memory by affecting the homeostatic role of the endocannabinoid system, particularly when exposure occurs during developmentally sensitive periods [32].

Memory is a multidimensional construct and may be classified based on temporal characteristics into short-term (e.g., working) and long-term memory (e.g., declarative memory, i.e., the memory of facts and events; or procedural memory, i.e., the memory of skills or habits); its content (e.g., into verbal, visual, or spatial memory) or stage (e.g., encoding, consolidation, or retrieval) [33]. Declarative memory may be further classified into episodic or associative memory (i.e., memory for events and associations) and semantic memory (i.e., memory for meanings and facts) [34]. In the context of cannabis use, human neuroimaging studies have typically used cognitive paradigms involving working, associative, or spatial memory or encoding and recall stages [35]. Working memory requires the involvement of the prefrontal cortex, inferior and ventral temporal cortex, and the hippocampus [36–38], while spatial memory requires input from the hippocampus and prefrontal cortex [39], particularly for encoding [40]. Encoding into associative memory requires input from the hippocampus, medial temporal cortex, frontal cortex. and cingulate cortex [41–46], while recall of information relies on activation of the medial temporal cortex, including the hippocampus and parahippocampus, as well as the posterior parietal cortex and prefrontal cortex [47–49]. The hippocampus is therefore important in the context of multiple domains of memory processing and in both encoding and retrieval of information [50].

Previous work has reviewed both the cognitive [5,7,51–54] and neurofunctional [55–58] effects of cannabis, both acutely and chronically, in the context of memory processing. Although a number of systematic reviews have summarised brain-structural alterations [56,59,60] as well brain-functional alterations [56,59–63] more broadly over a wide range of cognitive domains associated with cannabis use, functional alterations in the context of memory processing in cannabis users have not been systematically and comprehensively summarized to include up-to-date literature [54]. Therefore, in order to summarise the current literature, we have conducted a systematic review of studies that have employed functional magnetic resonance (fMRI) techniques in conjunction with cognitive activation paradigms that involve memory processing, to investigate memory-related brain-functional alterations in long-term cannabis users (CU) compared to nonusers (NU). In addition, we review relevant preclinical and human studies investigating memory-related cognitive impairments (both cross-sectional and longitudinal) in association with nonacute cannabis or cannabinoid exposure, as well as human studies employing imaging techniques other than fMRI, to provide a comprehensive summary of current evidence linking the effects of persistent cannabis use on memory performance and brain functioning during memory processing. Furthermore, as the period of adolescence is thought to be a period of greater vulnerability to the effects of cannabis and cannabinoids [64–67], we also discuss the role of participant age (adolescent or adult) and age of onset of cannabis use as potential factors that may influence the extent of harm from cannabis use evident in current literature. We also link existing evidence to the effects of abstinence from cannabis exposure, as previous literature has documented the importance of this as a factor influencing the persistence of functional alterations associated with cannabis use [62,68]. Meta-analytic evidence focusing on memory performance in otherwise healthy recreational cannabis users suggests that cannabis use is associated with alterations in several memory domains, including prospective memory, working memory, verbal or visual memory/ learning/ recognition except for visual working memory, and visual immediate recall [7], suggesting that review of neuroimaging evidence should point toward altered activation in brain regions sub-serving these particular domains in cannabis users.

2. Methods

2.1. Systematic Search of fMRI Studies

A systematic search of previous studies comparing brain functional differences in CU and NU and employing fMRI in conjunction with memory processing tasks as activation paradigms was completed using the PUBMED database following the Cochrane Handbook [69] and the MOOSE guidelines [70]. We employed two categories of search terms: (1) those related to cannabis—cannabis, marijuana, marihuana, THC, and tetrahydrocannabinol—and (2) those related to neuroimaging technique: fMRI, imaging, functional activation, BOLD. The search was limited to human studies and was assessed for suitability through an initial screening of the titles, then abstracts, and a final full article review. An initial PUBMED search was completed on 21/10/2015 and was then repeated on 22/1/2020. Reference lists were also screened from included manuscripts and published reviews. Only manuscripts meeting the following criteria were included, as shown in Figure 1:

- Original peer-reviewed data-based publication, reported in the English language.
- Compared habitual, otherwise healthy cannabis users (>50 occasions of self-reported lifetime cannabis use) with healthy controls (<50 occasions of self-reported lifetime cannabis use).
- Used fMRI in conjunction with a memory-based cognitive activation task.

Studies were excluded if they did not use a cognitive activation paradigm or did not include a memory-based task; did not clearly indicate the extent of cannabis use in the cannabis user group; or involved use less than or equal to 50 times in their lifetime in the cannabis user group; were non-English-language studies.

2.2. Review of Other Evidence of Effects of Persistent Cannabis Use on Memory Performance (Preclinical and Clinical Evidence) and on Memory-Related Brain-Function Alterations Using Neuroimaging Modalities Other than fMRI

Studies investigating memory performance in humans and animals and brain-function alterations related to memory processing using neuroimaging techniques other than fMRI in humans were identified through a bibliography search of previous systematic and narrative reviews [5,56,58,71,72]. To capture papers that have been published since the previous reviews, a search was carried out using the PUBMED database for relevant studies using the search terms "cannabis" or "marijuana" or "cannabinoid" and "memory", which was completed on the 7/6/2018. These further papers were screened initially through a search of titles, then abstract, and finally a full article review. For the purposes of this review, we included only studies that used memory processing tasks with group comparison between cannabis or cannabinoid-exposed groups and a non-exposed or non-using control groups. Other studies that employed study designs different to this, but focused on the topics of interest in this review, have been discussed in the text, although they are not included in the tables.

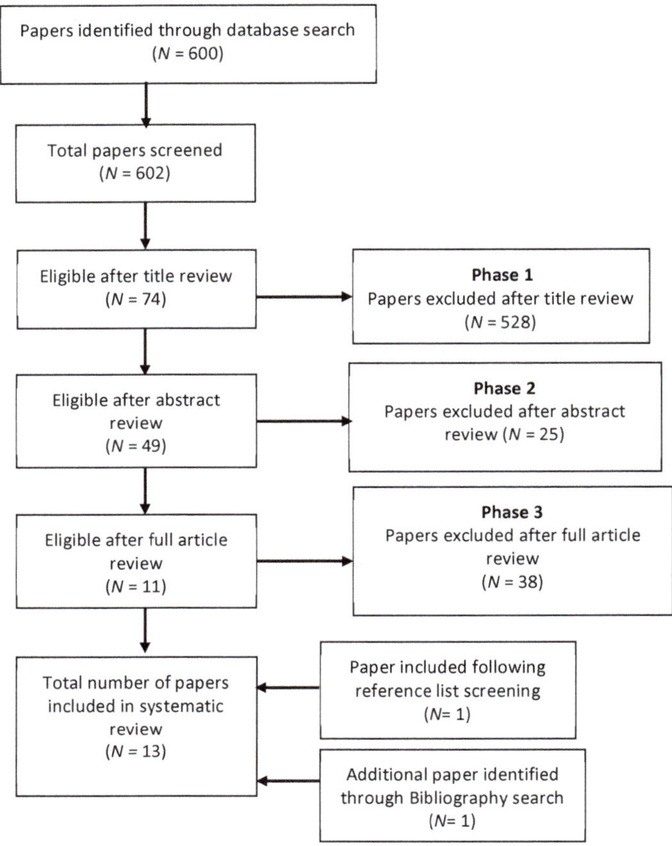

Figure 1. Identification of papers for systematic review.

3. Results

3.1. Systematic Review of Human fMRI Studies Investigating the Association between Cannabis Use and Memory-Related Brain Function

The initial search for fMRI studies comparing CU to NU while they performed memory-based cognitive activations tasks carried out in October 2015 identified 598 manuscripts. Of those, 10 met our inclusion criteria. Two further studies were identified by our final search on 22/1/2020 [73,74], and a further was identified from reference list screening [75]. Thirteen papers assessing memory-processing-related brain-activation differences between CU (n = 267) and NU (n = 261) using fMRI were identified in total. All included studies are reported in Table 1. Three of these papers involved only adolescents [76–78] CU (n = 72), NU (n = 79), while the remaining investigated adults CU (n = 195), NU (n = 182) with a group average age over 20 years. Four papers investigated spatial memory [73,78–80] and five associative memory [74,76,81–83], and four investigated working memory [76,84–86], while verbal learning [77] and false memory [75] were investigated by one paper each. Seven papers reported on group differences in whole-brain activation (WBA) [73,75,77,79–81,87] while eight investigated regions of interest (ROI) [76,77,79,81–85]. Four papers found CU to have performed worse than NU in the scanner-based memory task [75,79,81,83], nine found no significant performance difference [73,74,76,80,82,84–87].

Table 1. Systematic review of cannabis users and nonusing control comparisons, using functional magnetic resonance imaging.

Study	Task	No. NU and Age	No. CU and Age	Age of Cannabis Use Onset (Years)	Abstinence Before Scan	Cannabis Use Levels	Results		
							Whole Brain Analysis	Region of Interest	Task Performance
Adult studies									
Carey et al., 2015 [82]	Associative Memory	15 22.8 (2.9) (SD)	15 22.7 (4.2)	15.97 (0.42)	101.67 (37.45) h	6.43 (1.07) years 7341.40 (2340.80) lifetime uses	N/A	Decreased in the dACC and left hippocampus during processing of error-related and re-encoding or correct response in CU.	Decrease in error recall and correction rate in CU and poorer learning from errors.
Cousijn et al., 2014 [83]	Working Memory	41 22.0 (2.3) (SD)	32 21.4 (2.4)	18.9 (2.3) Onset of Heavy Use	24 h	2.5 (1.9) Years 1619.5 (1428.9) Lifetime uses	N/A	No significant difference found in the task-defined working-memory network.	No significant difference
Jager et al., 2006 [84]	Working Memory	10 23.3 (0.95) (SD)	10 22.4 (1.11)		7 days	7.1 (3.9) years 1300 [675–5400] lifetime uses	N/A	No significant difference found in learning or recall for both tasks.	No difference in task performance for both tasks.
Jager et al., 2007 [81]	Associative Memory	20 23.6 (3.9) (SD)	20 24.5 (5.2)		7 days	1900 [675–10150] lifetime uses	N/A	Decrease activation in bilateral parahippocampal regions and R DLPFC during learning for CU. Decreased activation in the right ACC in CU during recall.	No difference in task performance.
Kanayama et al., 2004 [79]	Spatial Working Memory Task	12 27.8 (7.9)	10 37.9 (7.4)		6–36 h	[5100–54000] lifetime use	Increased activation in; R SFG, IFG, STG, PCG, putamen; bilaterally, ACG, MFG, caudate. Decreased activation in bilateral MFC in CU.	N/A	No significant difference
Nestor et al., 2008 [80]	Associative Memory	14 24.1 ± 1.3 (SEM)	14 24.4 ± 1.4	16.5 ± 0.4	2–45 h	5.7 ± 0.6 years	Decreased activation in R STG, SFG, MFG, and L SFG during learning for CU. No difference in recall.	Increased in activation in R parahippocampal gyrus during learning for CU. No difference in recall.	Lower level of recall performance in CU.
Riba et al., 2015 [74]	False Memory Task	16 37.6 (11.8)	16 37.6 (10.8)	17 (12–20)	28 days	21 years [3–39] 42000 lifetime uses average	Decreased activation in CU in the R temporal cortex and precuneus; left, DLPFC, thalamus, caudate and medial temporal lobe, bilateral parietal cortex when recognising false memories over correct.	N/A	CU showed significantly more false memories.

Table 1. *Cont.*

Study	Task	No. NU and Age	No. CU and Age	Age of Cannabis Use Onset (Years)	Abstinence Before Scan	Cannabis Use Levels	Results		
							Whole Brain Analysis	Region of Interest	Task Performance
Sneider et al., 2013 [78]	Spatial Memory	18 22.8 (5.0) (SD)	10 20.3 (3.6)	15.6 (1.2)	12 hs	4 (2.4) years	Decreased activation in CU frontal pars triangularis, bilateral inferior frontal pars orbitalis, bilateral MFG, right pallidum and R putamen.	Decrease activation of R parahippocampal gyrus and cingulate gyrus in CU for recall motor control. No difference found in the hippocampus.	Similar performance, although CU showed more deficit in memory retrieval
Tervo-Clemmens et al., 2018 [72]	Spatial working memory	15 28.16 (0.71)	46 28.22 (0.72)	15.14 (2.27)	All THC negative	0.367 (0.683) mean joints per day—Only 15 used in the last year	No Significant Difference	N/A	No significant difference
Blest-Hopley et al., 2020 [73]	Associative Memory	21 24.24 (4.11)	22 24.95 (3.56)	14.67 (1.98)	12 h	6.19 (1.20) days per week 10.29 (3.10) years	Increased activation of bilateral; SFG, IFG, MFG, right medial FG in CU during encoding. No significant difference in Recall	N/A	No significant difference
Adolescent studies									
Jager et al., 2010 [75]	Working Memory and Associative Memory	24 16.8 (1.3) (SD)	21 17.2 (1.0) [15–19]	13.2 (2.3)	5.1 (4.2) weeks	4006 (7555) lifetime uses	N/A	Increased activation in the IFG, SPC and PCC/DLPFC of users during novel working-memory task. No group difference in associative memory task.	No difference in task performance.
Schweinsburg et al., 2008 [77]	Spatial working memory Task	17 17.9 (1.0) (SD)	15 18.1 (0.7) (SD)		28 days	480.7 (277.2) lifetime uses 4 (1.6) years	Increased activation R superior parietal lobe, decreased activation R DLPFC in CU.	N/A	No significant difference
Schweinsburg et al., 2011 [76]	Verbal Encoding Task	22/16 17.6 (0.8) 18.1 (0.7) (SD)	8/28 18.1 (0.9) 18.0 (1.0)	14.5 (2.5) 14.9 (3.4)	117.6 (153.9) 43.4 (37.1)	426.5 (280.1) 517.6 (451.3) lifetime uses	No significant difference	No significant activation in hippocampus.	No significant difference

NU = non-using cannabis group, CU = cannabis-using group, parentheses () used for SD, square brackets [] used for range.

3.1.1. Summary of Results—Adult Studies

Three studies investigated spatial memory in adults [73,79,80] using different types of tasks (water maze [79]; dot probe task [73,80]) and employed a whole-brain analysis approach. Opposite patterns of activation were identified in the superior and middle frontal gyri and putamen in two studies [79,80], while no difference was observed in a third study [73]. Tervo-Clemmens et al. included participants with low levels of cannabis use and long periods of abstinence, with only 15 of the 46 CU group having used in the previous year [73], which might explain the absence of difference between CU and NU in that study. Snieder et al. also employed an ROI analysis approach, finding only deceased activation in CU compared to NU in brain regions similar to their whole-brain analysis (WBA) approach, although they did not find any group difference in activation in the hippocampal ROI.

Associative memory in adults was assessed by four studies [74,81–83]. Three studies investigated activation during learning, with two finding that activation decreased in CU in the frontal and temporal regions, with one using both WBA and ROI [81] and another only using the ROI [82] approach, although Nestor et al. found an opposite direction of activation in the parahippocampal gyrus. Blest-Hopley et al. found CU to have increased activation in the inferior, superior, and middle frontal gyrus bilaterally and in the right medial frontal gyrus in a WBA. During recall of information, a decrease in activation was seen in two studies in the anterior cingulate cortex [82,83], but no group difference was found by another [74]. Carey et al. found activation decreased in other regions, including the hippocampus, using ROI analysis during a paired location number task, where CU had more repeated errors [83].

Two studies investigated working memory [84,85], where both studies employing ROI analysis found no difference in activation between CU and NU.

Using a task used to investigate brain activation associated with false memory, Riba et al. [75] found CU not only had more false memories but also decreased activation compared to NU in temporal, parietal, and frontal cortex, as well as thalamus, caudate, and precuneus, employing a whole-brain analysis approach.

Only one study found activation differences in the hippocampus [83] during the recall condition of an associative memory task, where CU had decreased activation compared to NU, whereas another found no significant differences using an ROI analysis approach [79] during a spatial memory task. Parahippocampal activation was, however, seen to be decreased in CU compared to NU during spatial and associative memory tasks [79,82], although another study found parahippocampal activation increased in CU compared to NU while performing an associative memory task [81]. The majority of studies reporting activation differences between groups found activation to be decreased in CU compared to NU in a variety of memory tasks [75,79,81–83]; however, some found regions of increased activation [74,80,81], with many regions overlapping with areas previously found as having decreased activation. Three studies, however, found no differences between CU and NU, using both whole-brain (WBA) and ROI analysis approaches [73,84,85].

Finally, a study not meeting our entry requirement for cannabis use levels compared 18- to 22-year-old cannabis users, based on their use over the previous 3 months, with those who had not used over that period. Using a visual memory task, no difference in activation was seen in the ROI of the IFG and hippocampus during the encoding condition; however, WBA found CU had decreased activation in the cerebellum (left), insula, basal ganglia, superior frontal gyrus, right precentral gyrus, and bilateral parahippocampal gyri. During the recognition condition of the task, ROI analysis showed CU had significant decreased activation in the hippocampus bilaterally and left IFG, while WBA revealed that CU had decreased activation in the cerebellum (bilateral), insula, basal ganglia and cingulate, and left posterior parietal cortices [88]. A longitudinal fMRI study of working memory from a baseline to 3 years in heavy cannabis users found that activation of the working memory network remained stable [89] over time despite continued moderate to heavy use of cannabis as well as nicotine, alcohol, and illegal substances.

3.1.2. Summary of Results—Adolescent Studies

Of the studies in adolescent cannabis users, one used a spatial working memory test and reported decreased activation in frontal and parietal regions in adolescent CU compared to NU [87]. Another study used an associative picture task, finding no significant difference in activation between adolescent CU and NU in ROI analysis [76]. Jager et al. also investigated working memory in adolescents using a letter recognition task and, using ROI analysis, found increased activation in CU compared to NU in frontal and parietal regions [76]. A third study of adolescent CU found no differences in brain activation during verbal encoding following both WBA and ROI analysis [77]. Of the two studies reporting activation differences between groups, both found activation in the superior parietal lobe to be increased in CU using different forms of working memory tasks, though opposite patterns of activation were seen in the dorsolateral prefrontal cortex by these studies [76,87].

3.2. Human Studies Investigating Memory-Related Brain Function Alterations Using Neuroimaging Modalities other than fMRI

Only two studies have employed neuroimaging techniques other than fMRI to investigate neurofunctional differences between CU and NU in the context of memory processing. Battisti et al. [90] investigated event-related potentials (ERP) during a verbal memory task wherein participants' responses were characterised based on whether they correctly recalled (CR) or did not recall (NR). In this study with 24 participants (CU = 24; NU = 24; average age of CU 36.4 (11.2) and NU 35.5 (11.5)), CU had an average of 17 years of near-daily use and had all used in the week prior to testing, with a minimum of 13 h between last use and testing. They identified attenuated latency in the frontal region of CU compared to NU in N4, a window around 350 ms, thought to originate in the hippocampus during encoding [91]. The amplitude of frontal and parietal zones was decreased in CU. The NR latency was attenuated in line with longer periods of cannabis use. Block et al. [90] investigated cerebral blood flow using positron emission technology (PET) during delayed and novel recall tasks in 18 CU who reported daily use of cannabis for over 2 years prior to recruitment and underwent 26 h of monitored abstinence and compared them with 13 NU. They found a decrease in frontal blood flow in CU compared to NU, which was most prominent whilst recalling newly presented words. Differences between CU and NU included the fact that language-based memory-related activity in the left hippocampus was observed to be higher in NU, with CU lacking this lateralization of hippocampal activation.

3.3. Human Studies Investigating Association between Cannabis Use and Memory Performance Alterations—Cross-Sectional Studies

Seventeen cross-sectional studies were identified that investigated the effects of cannabis use on memory performance by comparing CU and NU using various cognitive tasks engaging different domains of memory (Table 2). Twelve studies investigated adult cannabis users and five investigated adolescent participants.

Table 2. Memory performance studies comparing cannabis users to non-using controls.

Study	Task	No. NU and Age	No. CU and Age	Age of Cannabis Use Onset (years)	Abstinence Period	CU Levels	Results	Notes
Adult studies								
Gruber, Sagar, Dahlgren, Racine, and Lukas, 2012 [92]	Rey-Osterrieth Complex Gigure (visual memory)	28 24.32 (6.65)	34 22.76 (6.57)	15.53 (2.16)	12 h	7.24 (7.30) years 19.24 (19.58) smokes per week	No significant difference seen between CU and NU	Had an early- (<16 years) and late- (> 16 years) onset group. No significant difference was seen between early- and late-onset in either task
	California Verbal Learning Test						No significant difference seen between CU and NU	
Battisti et al., 2010 [90]	Verbal Memory Task	24 35.5 (11.5)	24 36.4 (11.2)	15 [12–25]	20 h (mean)	20.2 (9.7) years 30 [4–30] days per month	CU recalled significantly fewer words, which had a marginal correlation with the duration of use.	
Wadsworth, Moss, Simpson, and Smith, 2006 [93]	Immediate free recall task (episodic memory)	85 26.79 (4.64)	34 24.03 (5.28)			7.63 years [1–21]	No significant difference seen between CU and NU	This study was carried out in a workforce population
	Delayed free recall task (episodic memory)						No significant difference seen between CU and NU	
	Delayed recognition memory task (episodic memory)						No significant difference seen between CU and NU	
	Verbal reasoning task (working memory)						CU performed significantly worse; however, when cannabis use was considered for the last 24 h, this deficit was in line with the last use of cannabis	
	Semantic processing task						No significant difference seen between CU and NU	
Solowij et al., 2002 [94]	Rey Auditory Verbal Learning Test	33 34.8 (11.1)	ST—51 28.7 (5.5) LT—51 42.1 (5.2)	15.3 (2.6)	12 h	ST—10.2 [2.7–17] years 28.3 (5.2–30) days per month LT—23.9 [17.3–31.7] years (27.4 (3.5–30) days per month	LT—recalled fewer words and learned slower than both controls and ST—correlating with duration of use. ST- did not differ from controls	ST = Short term user LT = Long term user
	Rey Auditory Verbal Learning Test—long recall						All CU performed worse than controls overall; however, LT—recalled significantly less than before the delay time than ST—and NU	
Quednow et al., 2006 [95]	Rey Auditory Verbal Learning Test	19 23.42 (4.30)	19 21.42 (5.77)		3 days	6.55 (3.67) years 3.89 (4.72) times per week	No significant difference seen between CU and NU	This study mostly looking at an MDMA group. Did not have a high-using CU group

Table 2. *Cont.*

Study	Task	No. NU and Age	No. CU and Age	Age of Cannabis Use Onset (years)	Abstinence Period	CU Levels	Results	Notes
Pope, Gruber, Hudson, Huestis, and Yurgelun-Todd, 2002 [96]	Benton Visula Retention Test (visual memory)	87 40 [34–45]	77 36 [32–43]		0–28 days (month-long trial)		No significant difference seen between CU and NU on the test carried out on day 0, 7, and 28	
	Buschke's Selective Reminding Test (verbal memory)						CU had significantly poorer performance at day 0, 1, and 7. At day 28, these differences no longer met significance, except in the long-delay condition	
	Wechseler Memory Test				28 days		No significant difference seen between CU and NU	
Pope, Gruber, Hudson, Huestis, and Yurgelun-Todd, 2001 [97]	Benton Visula Retention Test (visual memory)	72 39.5 [34–44]	63 36 [32–41]		0–28 days (month long trial)	19 [15–24] years smoking >6 joints per week	No significant difference seen between CU and NU	A second former heavy CU group (n = 45) was recruited with < 12 times use in the last month
	Buschke's Selective Reminding Test (verbal memory)						CU had significantly poorer performance at day 0, 1, and 7. At day 28, these differences no longer met significance, except in the long-delay condition.	The former users showed no difference from controls on any task.
	Wechseler Memory Test						No significant difference seen between CU and NU	
Rodgers, 2000 [98]	Verbal Memory	15 32 [26–39]	15 30 [27–43]		1 month	4 days per week	CU performed significantly worse than NU	They did not test CU for abstinence
	Visual Memory						No Significant difference seen between CU and NU	
	General memory						CU performed significantly worse than NU	
	Delayed Recall						CU performed significantly worse than NU	
McKetin, Parasu, Cherbuin, Eramudugolla, and Anstey, 2016 [99]	Immediate Recall	4986 42.6 (1.5)	106 42.7 (1.4)			At least weekly	CU was related in a dose-related fashion to performance	
	Delayed Recall						CU was related in a dose-related fashion to performance	

Table 2. *Cont.*

Study	Task	No. NU and Age	No. CU and Age	Age of Cannabis Use Onset (years)	Abstinence Period	CU Levels	Results	Notes
Cengel et al., 2018 [100]	Immediate Memory	48 27.00 (6.19)	45 28.84 (6.37)	18.06 (3.95)	3 days	10.32 (6.12) years	No significant difference seen between CU and NU	
	Maximum Learning						CU scored significantly lower than NU	
	Number of repetitions						CU had significantly more/poorer performance than NU	
	Total Learning						CU performed significantly worse than NU	
	Recall Score						No significant difference seen between CU and NU	
	Recognition Scores						No significant difference seen between CU and NU	
	False Learning						CU had significantly more/poorer performance than NU	
	False Recall						CU had significantly more/poorer performance than NU	
Levar, Francis, Smith, Ho, and Gilman, 2018 [101]	California Verbal Learning Test	22 21.59 (1.94)	19 20.58 (2.52)	16.21 (1.69)	2.79 (3.10) days	4.37 (1.67) years	CU had worse performance, but only significant in the long-delay cued recall.	
Schuster et al. 2016 [102]	California Verbal Learning Test	48 21.5 (2)	27 19.6 (2.1)	15.1 (0.96)		2.9 (1.7) days per week 3.8 (2.1) years	CU performed significantly worse at encoding and recall than NU	Early-onset cannabis-using group (<16 years of age)
			21 21.2 (1.8)	17.8 (0.83)		2.9 (1.6) days per week 5.5 (1.7) years	No significant differences	Late-onset cannabis-using group (>16 years of age)
				Adolescent studies				
Ashtari et al., 2011 [12]	California Verbal Learning Test	14 18.5 (1.4)	14 19.3 (0.8)	13.1 [9–15]	6.7 months [3–11]	5.3 (2.1) years	No significant differences	
Solowij et al., 2011 [103]	Rey Auditory Verbal Learning Test	62 18.07 (0.48)	52 18.67 (0.82)	15 [10–17]	12 h	2.36 (1.17) years 13.87 [0.5–30] days per month	CU recalled significantly fewer words than NU	
	Rey Auditory Verbal Learning Test—long recall						CU recalled significantly fewer words than NU	
	Word Recognition Test						CU recognised significantly fewer words that NU	

Table 2. *Cont.*

Study	Task	No. NU and Age	No. CU and Age	Age of Cannabis Use Onset (years)	Abstinence Period	CU Levels	Results	Notes
Hanson et al., 2010 [104]	Hopkins Verbal Learning Test	21 17.4 (1.0)	19 18.1 (0.8)	15.6 (1.6) regular weekly use	3.3 (3.2)	16 (9.2) days past month 465 (294.5) life-time use episodes	CU performed significantly worse than NU	
	Verbal Working Memory						CU performed significantly worse than NU	
Medina et al., 2007 [105]	California Verbal Learning Test	34 17.86 (0.99)	31 18.07 (0.87)		30 days	2.91 (2.08) years of weekly cannabis use	CU performed at trend level ($p < 0.10$) worse than NU	
	Verbal Story Memory						CU performed at a trend level ($p < 0.10$) than NU. Performance correlated with cannabis use	
	Verbal List Learning						No significant difference seen between CU and NU	
	Visuo-spatial Memory						No significant difference seen between CU and NU	
Fried, Watkinson, and Gray, 2005 [106]	Immediate memory	59 17.7 (0.7)	19 current light CU < 5 joints a week 18.0 (1.2)	15.7 (1.7)		1.8 (2.0) years	Current light CU did not differ significantly from NU on all three memory tasks	Three groups of CU in the study, completed all three of the tasks.
	General Memory		19 current heavy CU > 5 joints a week 17.8 (0.8)	15.0 (1.5)		2.6 (1.3) years	Current heavy CU performed significantly worse to NU at general and immediate memory, but not working memory	
	Working Memory		16 former CU 17.9 (1.1)	14.3 (1.3)		2.2 (1.4) years	Former CU did not differ significantly from NU on all three memory tasks	Former users had no regular use for 3 months

NU = Nonusing cannabis group, CU = Cannabis using group, parentheses () used for SD, Square Brackets [] used for range.

3.3.1. Summary of Results—Adult Studies

Four studies employed a verbal learning task where stimuli were visually presented, finding that CU performed significantly worse at recall of words [93,96–98]. However, in the study by Wadsworth and colleagues, this was only observed in CU that had used in the 24 h prior to testing [96]. Pope et al. found in both studies that CU performed worse compared to NU at verbal memory test over the first week of examination following an abstinence of 0, 1, and 7 days, but by day 25, CU only performed worse on long-delay recall [97,98]. In contrast, no difference in performance was seen in a smaller former CU group compared to NU on the verbal memory test [98].

Auditory verbal learning tasks were used by six studies, where word lists were read out to the participants. CU had worse recall performance than NU [94], with higher performance deficits seen in those who had used for longer periods [99,107] or at a higher dose [95]. While Rodgers et al. tested participants after a month of self-reported abstinence from cannabis use, they did not carry out testing of urine or blood to confirm abstinence [94]. McKetin et al. did not report time since last use but interestingly found that abstinence did not improve performance at two waves of four-year retesting [95]. No difference between CU and NU was seen in recall performance in two studies [92,100]. However, Cengel et al. found that CU performed worse compared to NU on five of the eight conditions tested, including false recall and maximum and total learning [92] after three days' abstinence.

Three studies used the California verbal learning test [101–103], with two finding that CU performed significantly worse than NU [102,103], and the third study no significant difference [101]. Levar et al. only found a significant difference in the long-delay cued recall condition out of four tests of short and long delay free and cued recall with earlier-onset users performing worse than late-onset users [102], and Schuster et al. found that CU performed significantly worse at encoding and recall only in early-onset users, i.e., before the age of 16 [103]. It is unclear whether duration of abstinence or extent of cannabis use may have accounted for the difference in results in these three studies. While Levar et al. studied participants with an average abstinence of a few days, participants in the study by Schuster et al. were only required to be abstinent on the day of testing. In contrast, the study by Gruber et al. [92] required only a 12-hour abstinence period of their participants but failed to detect significant performance difference between users and nonusers, although their participants reported the highest mean years of cannabis use of these three studies.

Using the Wechsler memory scale, Pope et al. compared a set of heavy-using CU to NU at four time-points of abstinence, finding no significant difference between CU and NU after 25 days of abstinence [97], replicating findings of their previous smaller study [98]. Rodgers et al. also used a test for general memory, finding significant impairment in their CU group after abstinence for 1 month compared to NU [94].

None of the four studies investigating visual memory found a significant difference between CU and NU [94,97,98,101].

3.3.2. Summary of Results—Adolescent Studies

Two studies investigated auditory verbal learning in adolescent CU compared to NU [104,105]. Solowij et al. observed these deficits were in line with the quantity and frequency of cannabis used, as well as the age of onset of use, which remained even after controlling for premorbid intellectual ability [104]. In contrast, Hanson et al. found that performance in CU returned to a level comparable to NU after a 3-week period of abstinence, though users in this study had comparatively low levels of cannabis use [105].

The California verbal learning test was used by two studies with one finding some deficits in adolescent CU compared to NU following a 1 month abstinence [106], while the other found no significant difference following a 3- to 11-month period of abstinence [12]. Medina and colleagues found deficits were trend level in CU (with cannabis use ranging between 60 and 1800 times per lifetime) compared to NU in the California verbal learning test and Wechsler Memory scale Logical memory test of first recall, immediate, and delayed recall and recognition scores, while there were no impairments in verbal list learning and visuospatial memory [106].

Both immediate and general memory performance was tested by Fried et al. in current heavy (average 12.4 (9.8) joints per week), light (<5 joints per week), and former cannabis users (over 3 months abstinence), with all three groups compared to NU separately. Heavy CU performed worse in both immediate and general memory performance, whereas light CU and former CU had no significant difference in performance compared to NU [108].

Two studies investigated working memory in adolescent CU compared to NU [105,108], with one finding CU performed significantly worse initially, which was no longer evident after 3 weeks of abstinence [105]. These results were consistent with evidence from another study reporting significantly impaired immediate and delayed memory in current heavy cannabis users but not in light users or in former users [108].

One study investigated spatial learning performance following a 1 month abstinence in a group with cannabis use ranging between 60 and 1800 times per lifetime, with no significant difference in performance between CU and NU [106].

3.4. Human Studies Investigating Association between Cannabis Use and Memory—Longitudinal Studies

We identified six studies that used some form of longitudinal study design to investigate whether memory deficits seen in CU predated the use of the drug or developed following cannabis use. In one of the earliest reports, Fried et al. controlled for differences in cognitive performance prior to initiation of drug use and compared immediate, general, and working memory performance between heavy CU and NU. Heavy CU performed significantly worse in all memory domains compared to NU. Immediate and general memory impairments persisted after controlling for pre-drug-use performance, though working memory performance was no longer significantly impaired after controlling for pre-drug-use performance [108]. In an 8-year follow-up study, Tait et al. found that cessation of use in heavy cannabis users was associated with significant longitudinal improvement in immediate recall performance compared to continued heavy cannabis users [109]. In another cohort study, Meier et al. measured IQ at the age of 13 years old and used it to control for memory performance at a follow-up age of 38 years. After also controlling for years of education, cannabis use was found to be significantly associated with decline in memory performance [110]. A 25-year follow-up study by Auer et al. found that after excluding current CU and adjusting for potential confounders such as baseline memory performance cumulative lifetime exposure to cannabis was strongly associated with poorer performance subsequently in a verbal memory task in a dose-dependant manner [111]. However, another study employing a longitudinal design did not find any significant adverse effect of cannabis use on longitudinal change in performance in memory tasks at 4 and 8 years follow-up in an older (40–46 years) cohort of participants [95]. In contrast, in another study, Castellanos-Ryan et al. found a bidirectional relationship between cannabis use and cognitive performance such that poorer short-term memory and working memory performance at age 13 (prior to initiation of cannabis use) was associated with earlier age of onset of cannabis use, and earlier onset, and more frequent cannabis during adolescence, in turn, was associated with neurocognitive decline by age 20 [112]. However, a specific effect of cannabis use on subsequent memory performance was not reported in this study.

3.5. Preclinical Studies Investigating the Effect of Cannabis Use on Memory

A total of 18 animal studies were identified in our search (listed in Table 3). Exposure times to cannabinoids ranged from 14–180 days, while washout periods ranged from 0–116 days. All studies presented used rats. Thirteen studies investigated spatial memory, eleven investigated short-term memory, and four examined working memory. Six studies treated two separate groups of animals with cannabinoids during either adolescence and adulthood.

Table 3. Animal studies using memory tasks to investigate exposure to exogenous cannabinoids.

Study	Task Used	Age of Exposure	Washout Period	Animal Type	Drug Used	Results	Other Notes
Renard, Krebs, Jay, and Le Pen, 2013 [113]	Object Recognition	29–50 PND	28 days	Wister Rat	CP55,940	Drug-treated animals spent less time exploring novel objects and had significantly different times exploring familiar objects to control	Wister Rats had a larger effect of memory performance following drug exposure than Listerhooded Rats
	Object Recognition	29–50 PND	28 days	Listerhooded Rat	CP55,940	Drug-treated animals spent less time exploring novel objects and had significantly different times exploring familiar objects to control	
	Object Recognition	70–91 PND	28 days	Wister Rat	CP55,940	No difference in time exploring novel objects and familiar objects to control	
	Object Recognition	70–91 PND	28 days	Listerhooded Rat	CP55,940	No difference in time exploring novel objects and familiar objects to control	
	Object location	29–50 PND	28 days	Wister Rat	CP55,940	Drug-treated animals did not show a significant change to novel exploration time, where control did	
	Object location	29–50 PND	28 days	Listerhooded Rat	CP55,940	Drug-treated animals did not show a significant change to novel exploration time, where control did	
	Object location	70–91 PND	28 days	Wister Rat	CP55,940	Drug-treated animals showed no difference in behaviour to control	
	Object location	70–91 PND	28 days	Listerhooded Rat	CP55,940	Drug-treated animals showed no difference in behaviour to control	
Kirschmann, Pollock, Nagarajan, and Torregrossa, 2017 [114]	Object Recognition	34–54 PND	0 days	Sprague Dawlet Rats	WIN55,212-2	Drug-treated animals showed significant effect of drug on object recognition	
	Working Memory test	34–54 PND	17 days	Sprague Dawlet Rats	WIN55,212-2	No significant effect of drug on working memory performance	
	Object Location	34–54 PND	17 days	Sprague Dawlet Rats	WIN55,212-2	No significant effect of drug on object location test	
Harte and Dow-Edwards, 2010 [115]	Active Place Avoidance Testing	22–40 PND	33 days	Sprague Dawlet Rats	THC	Drug-treated animals performed worse than control	
	Active Place Avoidance Testing	41–60 PND	16 days	Sprague Dawlet Rats	THC	No significant effect of drug seen in performance	
Schneider and Koch, 2003 [116]	Object Recognition	40–65 PND	20 days	Wister Rat	WIN55,212-2	Drug-treated animals showed significantly impairment of recognition memory	
	Object Recognition	>70 PND	20–25 days	Wister Rat	WIN55,212-2	No significant effect of drug seen in performance	
O'Shea, Singh, McGregor, and Mallet, 2004 [117]	Object Recognition	30–51 PND		Wister Rat	CP55,940	Novel object recognition was significantly lower in drug-treated animals to controls; however, delay time had no significant effect between groups.	
	Object Recognition	56–77 PND		Wister Rat	CP55,940	No effect of treatment was seen between groups.	Only nine animals in the adult 56-77 PND-treated group

Table 3. *Cont.*

Study	Task Used	Age of Exposure	Washout Period	Animal Type	Drug Used	Results	Other Notes
Rubino et al., 2008 [118]	Elevated Plus-Maze	35–45 PND	30 days	Sprague Dawlet Rats	THC	No significant effect of drug seen in performance	
Schneider, Drews, and Koch, 2005 [119]	Object Recognition	15–40 PND	45 days	Wister Rat	WIN55,212-2	No significant effect of drug seen in performance	
	Progressive Ration/Operant learning	15–40 PND	35 days	Wister Rat	WIN55,212-2	No significant effect of drug seen in performance	
Abush and Akirav, 2012 [120]	Water Maze	45–60 PND	24 h	Sprague Dawlet Rats	WIN55,212-2	Drug-treated animals took longer to find the platform	
	Water Maze	45–60 PND	10 days	Sprague Dawlet Rats	WIN55,212-2	A significant difference was found compared to rats tested at 24 h abstinence.	
	Object Location	45–60 PND	24 h	Sprague Dawlet Rats	WIN55,212-2	Drug-treated animals showed impaired long-term memory to control animals	
	Object Location	45–60 PND	10 days	Sprague Dawlet Rats	WIN55,212-2	Drug-treated animals showed impaired long-term memory to control animals	
	Object Location	45–60 PND	30 days	Sprague Dawlet Rats	WIN55,212-2	Drug-treated animals showed impaired long-term memory to control animals	
	Object Recognition	45–60 PND	24 h	Sprague Dawlet Rats	WIN55,212-2	Drug-treated animals spent significantly less time exploring novel objects	
	Object Recognition	45–60 PND	10 days	Sprague Dawlet Rats	WIN55,212-2	Drug-treated animals spent significantly less time exploring novel objects	
	Object Recognition	45–60 PND	30 days	Sprague Dawlet Rats	WIN55,212-2	Drug-treated animals spent significantly less time exploring novel objects	
Fehr, Kalant, and LeBlanc, 1976 [112]	Closed Field Maze	14 day treatment	24 h	Rats	THC	Drug-treated animals performed worse than control animals	
	Closed Field Maze		25 days	Rats	THC	No significant difference was seen between the groups	
Hill, Froc, Fox, Gorzalka, and Christie, 2004 [121]	Water Maze	15 day treatment	During daily treatment	Long-Evans Rats	3-11-Δ8-THC	Drug-treated group took much longer to learn the task, showed similar performance	
	Water Maze, plus time delay				3-11-Δ8-THC	Drug-treated animals had significantly worse performance	
Mateos et al., 2011 [122]	Spontaneous Alternation (Short-term memory) Task	28–43 PND	24 h	Wister Rat	CP55,940	Drug-treated animals performed significantly worse than control	
	Object Location	28–43 PND	37 days	Wister Rat	CP55,940	No significant effect of drug seen in performance	
	Object Recognition	28–43 PND	43 days	Wister Rat	CP55,940	Drug-treated animals performed significantly worse than control	
Rubino et al., 2009 [123]	Passive Avoidance	35–45 PND	30 days	Sprague Dawlet Rats	THC	No significant effect of drug seen in performance	
	Radial Maze	35–45 PND	30 days	Sprague Dawlet Rats	THC	Drug-treated animals had significantly more errors and took significantly more time to learn the maze layout	

Table 3. *Cont.*

Study	Task Used	Age of Exposure	Washout Period	Animal Type	Drug Used	Results	Other Notes
Stiglick and Kalant, 1982 [124]	Radial Maze	180 days	30 days	Wister Rat	THC and CBN	Drug-treated animals made significantly more errors and less correct responses and took longer to learning the overall task	
Stiglick and Kalant, 1985 [125]	Radial Maze	90 days	31 days	Wister Rat	THC, CBN and CBD	No effect of drug was seen between groups	
	Avoidance test	90 days	116 days	Wister Rat	THC, CBN and CBD	Drug-treated animals performed worse than controls	
Cha, Jones, Kuhn, Wilson, and Swartzwelder, 2007 [126]	Water Maze	30–51 PND	28 days	Sprague Dawlet Rats	THC	No effect of drug was seen between groups	
	Water Maze	70–91 PND	28 days	Sprague Dawlet Rats	THC	No effect of drug was seen between groups	
Cha, White, Kuhn, Wilson, and Swartzwelder, 2006 [127]	Water Maze—Spatial task	34/36 PND + 21 day	28 days	Sprague Dawlet Rats	THC	No effect of drug was seen between groups	
	Water Maze—Non-Spatial Task	34/36 PND + 21 day	28 days	Sprague Dawlet Rats	THC	No effect of drug was seen between groups	
	Water Maze—Spatial task	69/74 PND + 21 days		Sprague Dawlet Rats	THC	No effect of drug was seen between groups	
	Water Maze—Non-Spatial Task	69/74 PND + 21 days		Sprague Dawlet Rats	THC	No effect of drug was seen between groups	
Higuera-Matas et al., 2009 [128]	Object Recognition	28–38 PND	63 days	Wister Rat	CP55,940	No effect of drug was seen between groups	
	Water Maze—Reference memory	28–38 PND	67 days	Wister Rat	CP55,940	No effect of drug was seen between groups	
	Water Maze—Spatial task	28–38 PND	67 days	Wister Rat	CP55,940	No effect of drug was seen between groups	

PND = Postnatal day, 3-11-Δ8-THC = 3-(1,1-dimethylheptyl)-(–)-11-hydroxy-Δ8-tetrahydrocannabinol, THC = Δ-9-tetrahydrocannabinol, CBN = cannabinol, CBD = cannabidiol.

Spatial memory, investigated using learning maze-based tasks, was been found to be impaired in rats following chronic exposure to THC or CB1 receptor agonist by five studies [120,121,123,124,126]. Five other studies using maze tasks found that previous exposure to cannabinoids did not have a significant effect on performance [113,118,125,127,128]. Object location tasks have also been used to assess spatial memory in rats, with impairments found in two studies following a prolonged cannabinoid exposure [122,123], but not in two others [114,115].

Impairments have also been observed in short-term memory in eight studies [113–117,119,122,123], although these findings have not been replicated in three other studies [121,128,129]. Two studies reported deficits in working memory performance following cannabinoid exposure [123,126], although two other studies did not find cannabinoid exposure having a significant negative effect in working memory performance [115,119]. No effect was seen on recognition or operant learning in rats following adolescent exposure [129].

Of the studies that exposed the animals during adolescence to adulthood, four found that cannabinoids had a negative effect on memory performance in the adolescent group, but not in the adult group in the same study [116,117,119,122]. Two studies, however, found no difference between the two age groups and did not find any deficit following chronic cannabinoid exposure at all [127,128].

Investigating memory performance after a period of washout is useful to disentangle the residual effects of cannabinoids from any acute effects. Of the four studies that included a memory test done within 24 h of the last drug administration, all reported deficits in the cannabinoid-treated animals [114,115,123,124]. Impairments in spatial memory appeared to reverse after an abstinence of a few days in two studies [123,126]; however, they were still present at 75 days abstinence in short-term memory tasks in one [123]. Rats who showed deficit following WIN 55,212-2 exposure in short-term memory after a 24 hr abstinence had no significant differences to controls after a 51 day washout period [115]. Similarly, those that had shown poorer performance after THC exposure in spatial memory at 24 h had no significant differences after 25 days [124]. Memory deficits remained following washout periods of 30 [120] and 116 [113] days.

4. Discussion

Our objective was to carry out a comprehensive review of alterations in brain functioning and performance in the domain of memory associated with persistent cannabis use, by drawing upon evidence from human and relevant preclinical studies. To this end, first, we carried out a systematic review of studies investigating brain-functional alterations in CU compared to NU using fMRI. We also reviewed published studies that have employed neuroimaging techniques other than fMRI to investigate memory-related functional alterations associated with cannabis use. Finally, to situate this understanding within the context of specific subdomains of memory affected, we performed two focused narrative reviews of studies investigating alterations in memory performance associated with persistent cannabis use employing a range of designs: studies in humans using cross-sectional and longitudinal designs to compare CU and NU and preclinical studies comparing cannabinoid exposed animals with nonexposed animals. These results are discussed under separate subsections below.

4.1. Systematic Review of fMRI Studies Using Memory Tasks

Our systematic review of functional brain activation during memory performance found CU to have altered brain activation, although no consistent pattern emerged either in terms of the direction of alteration in activation or the brain regions affected, although changes appeared to be mostly focused in the frontal and temporal regions. Altered activation in the hippocampus was found in some studies, particularly those employing ROI analysis approaches focusing on the hippocampus, but without a conclusive direction of change. Alteration of hippocampal activation was perhaps less frequently observed than one would have expected considering the central role of the hippocampal region in memory processing. Activation was altered more often during the encoding/learning stage than while recalling information, similar to previous evidence investigating memory performance [130].

Of the three studies reporting hippocampal activation differences between CU and NU during learning, there was no consistent direction of change [81–83], despite having similar ages of participants as well as ages of onset and levels of cannabis use in their participant groups. Inconsistencies in abstinence periods could have played a role in these differences, as cannabis users in the study by Nestor et al. had a self-reported abstinent range of 2–45 h, while those in the study by Carey et al. had an average of 101.67 h of self-reported abstinence, and Jager et al. required subjects to have tested negative for THC on urine screening indicating that the psychoactive substance was no longer present in their system. A further study investigating hippocampal activation with ROI analysis approach found no difference in brain activation between CU and NU during encoding but decreased activation in CU compared to NU during recognition. WBA found the parahippocampal gyrus to have significantly less activation during encoding in CU compared to NU [88].

The level of previous cannabis use might have been a potential confounder in studies that have reported no significant differences between CU and NU. Significant differences in activation between CU and NU were observed in studies using similar tasks investigating CU with high levels and prolonged periods of cannabis use [74,76,80,87] but not in those with lower levels and/or less extensive cannabis use [73,84]. Counter to this, however, was that continued cannabis use was not associated with altered activation patterns during working memory at 3-year follow up in another study [89]. Although a study with a wide range of cannabis use levels in the previous three months found higher levels of cannabis use was correlated with decreased hippocampal activation during recognition [88], it should be noted that no abstinence period prior to scanning was reported for cannabis use. Differences in task performance may also have contributed to the differences seen in activation; however, task difference during fMRI was only found in four of the studies [75,79,81,83].

The literature on brain activation differences measured by fMRI between CU and NU during memory tasks still lacks clarity, possibly due to heterogeneity of both the extent of cannabis use as well as the quantity, frequency, and age of onset of cannabis use, which have been found to correlate with alterations seen in cannabis users [88,104,131]. In addition, the wide range of tasks employed measuring different domains of memory function (e.g., spatial or working memory) mean it is difficult to perform a robust meta-analysis of their results or to draw consistent conclusions between studies. Subgroup analysis in a previous meta-analytic study focusing only on memory tasks in adult cannabis users found only decreased activation in the inferior frontal gyrus, pre- and post-central gyrus and precuneus, although the previously mentioned inconsistencies and the modest number of studies available for this analysis were major limitations [61]. In the fullness of time, meta-analysis of well-matched studies focusing on a particular domain of memory, such as verbal memory, spatial memory, etc., may better serve to reveal a consistent pattern of functional alterations in the context of memory processing associated with cannabis use.

Functional activation differences between CU and NU were more consistently seen in adult populations of cannabis users than in adolescents, although a previous meta-analysis of all cognitive domains has identified functional difference in CU compared to NU in both adults and adolescents [61,63]. Lack of differences in brain activation between adolescent CU and NU may be attributable to the design of some adolescent studies, with participants having used cannabis for a shorter period of time and therefore not having been exposed to the threshold at which functional differences become detectable using fMRI, and also simply the smaller subset of studies available. For example, one study in adolescent users that found no difference between CU and NU in whole-brain or ROI analysis during verbal learning had four groups of participants with and without alcohol abuse as well as with and without CU, meaning that comparisons of CU to NU groups without alcohol abuse involved a relatively small number of participants. CU participants also had a relatively long period of abstinence, and their cannabis use levels were relatively modest [77], indicating that certain functional differences may become less evident with longer periods of abstinence. This was further supported from a comparison of adolescent CU with recent cannabis use (2–7 days of abstinence) and a group of CU following a longer period of abstinence (27–60 days of abstinence) performing a spatial working memory task, suggesting that duration of abstinence may have an impact on alterations in functional activation associated with cannabis use [86]. More recent CU had greater activation compared to abstinent users in the bilateral insula and superior frontal gyrus; right—inferior gyrus; left—precentral gyrus, medial and middle frontal, and gyrus. In contrast, abstinent users only had greater activation compared to recent users in the right precentral gyrus, which may reflect a compensatory response in recent CU requiring recruitment of additional brain regions compared to abstinent CU in order to perform the memory task, as has been suggested previously [80].

Acute effects of THC confounding previous literature cannot be ruled out in some studies that have investigated participants who were not confirmed to have a negative result for THC on urine drug screening and/or studied participants after only short periods of abstinence from cannabis. Although a sustained period of abstinence may have alleviated some of the group differences in functional activation, abstinence periods did not consistently predict the detection of group differences in functional activation. While this may suggest that certain functional alterations are more robust than others and therefore detectable in CU even following a washout period of the drug, it is very likely that this also reflects the possibility that functional alterations observed cross-sectionally are not just attributable to the effects of drug exposure but an interplay with baseline differences between CU and NU that predate initiation of cannabis use.

4.2. Review of Cross-Sectional Human Studies Investigating Memory Task Performance

We reviewed studies of both adults and adolescents, comparing the performance of memory tasks between CU to NU. Verbal memory performance was negatively affected by cannabis use [93,94,96–98,102–106], with increased deficits associated with cannabis use levels [95,99,107]. This was also shown in a study of cannabis users ranging from light to heavy use after a month abstinence [132], in both adolescent and adult users, though deficits in performance were not reported in all studies [12,97,100,101]. Visual memory did not appear to have been affected in CU [94,97,98,101], although there were only a limited number of studies that investigated this paradigm, which were all conducted only in adults. General memory performance, as tested in adults and adolescents, was also negatively affected by cannabis use [94,108]. Working memory, in contrast, was shown to be affected in some studies, but these effects were not sustained following abstinence from the drug [96,108]. Spatial memory was only tested in one study in adolescents and was found to be unaffected by cannabis use [106].

Evidence from some studies suggests that longer duration of cannabis use may have an adverse impact on memory performance [93,107] with the amount of cannabis used correlating with performance in some studies [95,106], implying that high levels and longer use of cannabis are related to reduced memory performance. An increased likelihood of memory impairments was found to be associated with an earlier age of onset of cannabis use [103,133], although this association may also have been a result of the longer duration of exposure in those that started earlier and therefore the greater amount of cannabis that they had been exposed to and not necessarily an effect of earlier age of onset of use per se. This issue was investigated in one study that found significant deficits in early-onset users compared to late-onset users, despite late-onset users having used for longer periods [103], suggesting that there is very likely a developmentally sensitive period when the effects of cannabis exposure on memory performance are more prominent in humans, consistent with preclinical evidence [116,117,119,122]. However, another study did not identify differences between early- and late-onset CU, although they also failed to detect any significant difference between all CU and NU [101], perhaps due to a large range in years of cannabis use.

Abstinence from cannabis use reversed some memory deficits observed, with an earlier meta-analysis showing no significant effect of the drug after a 4-week abstinence on performance [68]. We also observed in our review, which included a number of studies published subsequent to that meta-analysis, that the interval period from last cannabis use to the assessment of task performance may have a bearing on the likelihood of studies reporting poorer memory performance in CU compared to NU. This may reflect two different factors influencing task performance in cannabis users in these studies: the residual acute effects of cannabis influencing memory task performance in studies involving shorter periods of abstinence, as well as the recovery of CB1 receptor density following an initial downregulation after prolonged exposure to cannabis, resulting in absence of detectable differences in memory performance in studies involving longer periods of abstinence [134]. Lack of significant difference in memory performance between CU and NU, as found in several studies with abstinence periods from as little as 3 weeks, may mean there is a reversal of the negative effects of cannabis use on memory performance.

It is of interest to note that the effort made during memory tasks has also been found to be negatively associated with frequency of cannabis use [135,136]. Therefore, it is possible that the effort made in completing the tasks may have also influenced the relationship between cannabis use frequency and performance deficits observed in learning and memory tasks. There is some evidence that the extent of impairments in task performance is not always perceived fully by cannabis-using participants [137], which may also adversely affect their effort during the task, thereby influencing their performance.

4.3. Review of Longitudinal Human Studies Investigating Memory Task Performance

While some longitudinal studies of memory performance suggest that CU may be associated with a deficit in memory performance following prolonged use, the evidence is not unequivocal, especially when baseline cognitive ability predating initiation of cannabis use is taken into consideration. McKetin conducted a follow-up study finding that CU performed the same irrespective of whether they had continued to use or had ceased use [95]. This may indicate that cannabis use does not contribute to memory decline in a linear fashion and continued use past a critical sensitive neurodevelopmental period may no longer be associated with continuing decline in memory performance, especially as participants in McKetin's study were middle-aged individuals. Of course, one cannot completely rule out the possibility that cannabis use does not have a direct causal effect on poor memory performance or indeed of a bidirectional effect.

Although poorer premorbid memory performance may be partly attributed in some users to their lower cognitive attainment in the domain of memory performance, improvement in performance observed following the cessation of cannabis use suggests that cannabis use may in fact have a direct deleterious effect. Further longitudinal studies are needed to tease apart the effects of cannabis use from other genetic and environmental effects. Other studies have aimed to address the question of the causal nature of the relationship between cannabis use and memory performance and have employed study designs that allowed them to account for genetic and other environmental confounders. A monozygotic twin study by Lyons et al. [138] investigated the effects of cannabis use on memory in 54 pairs of twins. They reported only a trend level decrement in performance in the CU group compared to nonusers during recall of a verbal learning task. During other memory tasks, no group differences were detected. However, participants in this study had a wide range of exposure to cannabis, with 37% of the CU using less than 52–300 times in total lifetime. Furthermore, CU participants included in the study were only required to have used cannabis regularly for one year in total with no restriction as to how long ago the period of use was. In many subjects, they had ceased to use regularly for an average of 27 years, with all subjects having at least a 1-month period of abstinence from cannabis when they were tested [138]. Another twin study by Meier et al. found that greater cannabis use by one twin was associated with poorer working memory performance compared to their nonusing twin, in the absence of any difference in their IQ at baseline. However, such a difference was not observed for the other memory (spatial memory) and executive (visual processing) tasks [139]. While cannabis use has been found to be associated with lower intelligence scores, another study employing a discordant twin design reported that family traits were more associated with intelligence performance [140]. Collectively, these studies provide some further evidence suggesting that cannabis use may have a causal role in memory deficits observed in cannabis users, although whether the deficits persist following abstinence from use remains debatable, and genetics may play a larger role in determining memory performance.

4.4. Review of Preclinical Studies Investigating Memory Task Performance

We found that animal studies have investigated the effect of not only extracts directly obtained from dried cannabis plants with varying potency in earlier experiments [113,120], but also analytical grade plant-derived cannabinoids such THC, or synthetic cannabinoids such as the CB1 receptor agonists WIN and CP55940 [141]. The discrepancy in results between studies may reflect difference in dose, duration of exposure, experimental conditions, animal species used, and the precise periods of exposure during adolescence or adulthood. Both age of onset as well as the washout period following exposure to cannabinoids appeared to influence whether deficits in memory performance were detected following chronic exposure to cannabinoids. However, in two studies using cannabis extracts in adult rats for a prolonged period of 90 and 180 days, deficits were still detectable even after a washout period of 116 and 30 days respectively [113,120]. This may suggest that, although adolescence acts as a critical period for exposure and abstinence periods may be associated with decrease in observed deficits and a possible reversal of residual effect, duration of exposure is also a critical factor that may influence the extent and persistence of deficits. Age of onset also appeared to be a critical factor as exposure to cannabinoids during the early adolescent phase produced a significant impairment in learning, compared to the late adolescent period, when no significant effect was seen [116,117,119,122].

The period of adolescence in rat models is short and so makes it difficult to administer drugs for long enough periods of time when targeting them in the developmental period. However, many studies did attempt to administer drugs for the majority of the adolescent period and saw significant results. The lengthier period of adolescence in humans allows for much longer exposure during neuronal maturation and possibly presents an extended period during which cannabis may have a greater detrimental effect on cognition. Preclinical studies with shorter periods of administration of drugs found fewer deficits in memory, but the period of washout from drug administration seemed to have a major influence on the outcome of studies.

The hippocampus is a key region for memory performance [142], and animal studies have been useful to investigate structural, functional, and histological effects of chronic cannabinoid exposure here. In rats, following exposures to THC and WIN, even after a washout period, alteration in dendrites of hippocampal pyramidal neurons has been found [143,144]. Interestingly after a washout period, synaptic density was not different in drug-treated animals, even though hippocampal volume and structure were found to be decreased in rats [143,145]. In monkeys administered cannabis and THC chronically, both structure and function in the hippocampus were found to have changed, and there were also synaptic changes [146], suggesting that alterations in hippocampal structure and function may underlie functional and performance differences.

Behavioral tests in animals have not provided conclusive evidence of memory deficits following cannabis use, although structural and histological investigations have shown robust evidence of cannabinoid-induced changes. Collectively, the body of research completed in animals investigating the effects of chronic cannabis use and its possible effects on memory function appears limited in comparison to that investigating the acute effects of cannabinoids.

5. Conclusions

To summarize, cannabis use has been shown in some studies to negatively affect memory-related brain functioning and task performance, particularly verbal memory and encoding in human studies, with preclinical evidence generally consistent with human evidence. Effects have also been observed in recall and working memory tasks, though these findings have been less robust. However, existing evidence regarding the effect of cannabis use on memory function is far from unequivocal, as evident from the heterogeneity in conclusions from the different studies.

From the evidence reviewed above, three clear factors emerge that may underlie differences in results seen in studies of memory function following cannabis use. Firstly is abstinence/washout period from last use, which seems to vary a lot in all the animal and human studies reviewed. Longer abstinence periods do appear to be associated with a less pronounced difference between CU and NU participants and cannabis-exposed and -unexposed animals, although not in all cases, particularly following a long exposure. Collectively, results from human studies reviewed here showing recent cannabis use being associated with alteration in memory-related functional activation, which becomes less prominent following periods of abstinence and longitudinal data [89], suggest that cannabis users may compensate for neurophysiological deficits associated with drug use by recruiting a network of additional brain regions.

Secondly, cannabis use parameters, such as use of higher quantities, longer periods of use, and more regular use, appear to increase the chances of detecting differences in memory-related outcomes. Some, but not all, studies do report linear relationships with time and quantity of cannabis used. Future studies should therefore aim to systematically investigate associations with these parameters of cannabis use and also incorporate other parameters, such as type of cannabis used, that have been shown to be associated with other health outcomes [147]. Interestingly, some studies over extended periods of cannabis use suggest a plateau in observed changes from cannabis use, where alteration may only be linear in the initial period of use or only during developmental periods.

Finally, the age at which cannabis use starts seems to be another important determinant, with animal literature in particular providing robust evidence for adolescence being a period of higher risk of brain alterations from cannabinoid exposure. From a biological perspective, this is a period of neuronal developmental processes, including brain development and altering binding affinity of CB1 receptors [64–67]. From a more social perspective, cannabis use during this period may result in poorer educational outcomes that may then in turn exacerbate memory performance impairments [148]. Cannabis use acutely impairs memory performance [149] and alters memory-related brain function [150] and may therefore adversely affect educational attainment. This is consistent with evidence showing that CU have an increased chance of leaving school earlier [151] and have poorer educational achievements [152] compared to NU. Pope et al. [133] also showed in their study comparing early- and late-onset users that there was a significant difference in the completion of a 4-year college course rates, with only 32% of early-onset CU compared to 60% of late-onset CU and 82% NU completing.

The detrimental effects of cannabis may be due to CB1-receptor-mediated disruption of hippocampal plasticity, a finding supported by animal histological investigations. Although there is a lack of evidence of significant effects on hippocampal activation from human studies, changes in hippocampal cerebral blood flow and ERPs [90,93] as well as structural differences [29] have been observed in CU compared to NU. Functional activation in other brain regions that express a high density of CB1 receptors were also found to be altered during memory processing tasks in cannabis users, possibly due to disruption of the normal functioning of the endocannabinoid system as suggested by overlap between brain regions with high CB1 receptor distribution [31] regions showing altered functioning in cannabis users (regions highlighted in [35]).

One limitation inherent to studies investigating chronic cannabis users, as in any studies investigating recreational drug users and that affects this systematic review as well, is the bias associated with retrospective recall of usage pattern and type of cannabis consumed over time. This is particularly important as the longer-term effects of cannabis use may depend on the specific usage pattern as well as type of cannabis, particularly the ratio of different cannabinoids, which may have often opposing effects on brain function and connectivity [153–155]. Future investigations should therefore focus on employing prospective designs in conjunction with more accurate ways of quantifying cannabis use, perhaps using a similar model to alcohol units, in order to improve the quality of future studies [156].

To conclude, evidence summarized here suggests that memory performance deficits may be related to cannabis use, with lower performance in memory tasks possibly underpinned by altered functioning of a wide network of brain substrates that may result from changes at the synaptic level. This review did not summarise how functional connectivity may be altered in cannabis users, as was discussed in a previous review [35], which reported altered connectivity in cannabis users during cognitive task performance. However, further studies of functional connectivity, particularly during memory processing tasks, are necessary in order to understand how the coordinated activity of brain networks may be affected rather than just brain regions. Further research is also necessary, taking into account baseline cognitive performance or ability prior to initiation of cannabis use in order to conclusively establish whether these changes persist or are indeed reversible following cessation of cannabis use and to fully understand the potential determinants of reversibility such as period of use and the quantity or length of time of exposure to cannabinoids. Such granular understanding is necessary to inform public health policy to help mitigate harm from cannabis use in those that are most vulnerable.

Funding: The views expressed are those of the authors and not necessarily those of the NHS, the NIHR, or the Department of Health. The funders had no role in the design and conduct of the study; collection, management, analysis, and interpretation of the data; preparation, review, or approval of the manuscript; or decision to submit the manuscript for publication. All authors have approved the final version of the paper.

Acknowledgments: S.B. has been funded by the National Institute for Health Research (NIHR), UK through a Clinician Scientist award (NIHR-CS-11-001) and also supported by the NIHR Biomedical Research Centre for Mental Health at the South London and Maudsley NHS Foundation Trust and Institute of Psychiatry, Psychology and Neuroscience, King's College London jointly funded by the Guy's and St Thomas' Trustees and the South London and Maudsley Trustees.

Conflicts of Interest: The authors declare none conflict of interest.

References

1. Unodc. Market Analysis of Plant-Based Drugs. Available online: http://www.unodc.org/wdr2016/en/cannabis.html (accessed on 3 July 2017).
2. Clark, T.T.; Doyle, O.; Clincy, A. Age of first cigarette, alcohol, and marijuana use among U.S. biracial/ethnic youth: A population-based study. *Addict. Behav.* **2013**, *38*, 2450–2454. [CrossRef]
3. Kann, L.; Kinchen, S.; Shanklin, S.L.; Flint, K.H.; Kawkins, J.; Harris, W.A.; Lowry, R.; Olsen, E.O.; McManus, T.; Chyen, D.; et al. Youth risk behavior surveillance—United States, 2013. *Mmwr. Suppl.* **2014**, *63*, 1–168.
4. Crean, R.D.; Crane, N.A.; Mason, B.J. An evidence based review of acute and long-term effects of cannabis use on executive cognitive functions. *J. Addict. Med.* **2011**, *5*, 1–8. [CrossRef]
5. Schoeler, T.; Bhattacharyya, S. The effect of cannabis use on memory function: An update. *Subst. Abus. Rehabil.* **2013**, *4*, 11–27. [CrossRef]
6. Bossong, M.G.; Jansma, J.M.; van Hell, H.H.; Jager, G.; Oudman, E.; Saliasi, E.; Kahn, R.S.; Ramsey, N.F. Effects of delta9-tetrahydrocannabinol on human working memory function. *Biol. Psychiatry* **2012**, *71*, 693–699. [CrossRef] [PubMed]
7. Schoeler, T.; Kambeitz, J.; Behlke, I.; Murray, R.; Bhattacharyya, S. The effects of cannabis on memory function in users with and without a psychotic disorder: Findings from a combined meta-analysis. *Psychol. Med.* **2016**, *46*, 177–188. [CrossRef] [PubMed]
8. Yucel, M.; Lorenzetti, V.; Suo, C.; Zalesky, A.; Fornito, A.; Takagi, M.J.; Lubman, D.I.; Solowij, N. Hippocampal harms, protection and recovery following regular cannabis use. *Transl. Psychiatry* **2016**, *6*, e710. [CrossRef]
9. Lorenzetti, V.; Solowij, N.; Yucel, M. The Role of Cannabinoids in Neuroanatomic Alterations in Cannabis Users. *Biol. Psychiatry* **2016**, *79*, e17–e31. [CrossRef] [PubMed]
10. Chye, Y.; Suo, C.; Yucel, M.; den Ouden, L.; Solowij, N.; Lorenzetti, V. Cannabis-related hippocampal volumetric abnormalities specific to subregions in dependent users. *Psychopharmacology (Berl.)* **2017**, *234*, 2149–2157. [CrossRef] [PubMed]
11. Yücel, M.; Solowij, N.; Respondek, C.; Whittle, S.; Fornito, A.; Pantelis, C.; Lubman, D.I. Regional brain abnormalities associated with long-term heavy cannabis use. *Arch. Gen. Psychiatry* **2008**, *65*, 694–701. [CrossRef] [PubMed]

12. Ashtari, M.; Avants, B.; Cyckowski, L.; Cervellione, K.L.; Roofeh, D.; Cook, P.; Gee, J.; Sevy, S.; Kumra, S. Medial temporal structures and memory functions in adolescents with heavy cannabis use. *J. Psychiatr. Res.* **2011**, *45*, 1055–1066. [CrossRef] [PubMed]
13. Cousijn, J.; Wiers, R.W.; Ridderinkhof, K.R.; van den Brink, W.; Veltman, D.J.; Goudriaan, A.E. Grey matter alterations associated with cannabis use: Results of a VBM study in heavy cannabis users and healthy controls. *Neuroimage* **2012**, *59*, 3845–3851. [CrossRef] [PubMed]
14. Battistella, G.; Fornari, E.; Annoni, J.M.; Chtioui, H.; Dao, K.; Fabritius, M.; Favrat, B.; Mall, J.F.; Maeder, P.; Giroud, C. Long-term effects of cannabis on brain structure. *Neuropsychopharmacol. Off. Publ. Am. Coll. Neuropsychopharmacol.* **2014**, *39*, 2041–2048. [CrossRef] [PubMed]
15. Filbey, F.M.; Aslan, S.; Calhoun, V.D.; Spence, J.S.; Damaraju, E.; Caprihan, A.; Segall, J. Long-term effects of marijuana use on the brain. *Proc. Natl. Acad. Sci. USA* **2014**, *111*, 16913–16918. [CrossRef] [PubMed]
16. Price, J.S.; McQueeny, T.; Shollenbarger, S.; Browning, E.L.; Wieser, J.; Lisdahl, K.M. Effects of marijuana use on prefrontal and parietal volumes and cognition in emerging adults. *Psychopharmacology (Berl.)* **2015**, *232*, 2939–2950. [CrossRef]
17. Mashhoon, Y.; Sava, S.; Sneider, J.T.; Nickerson, L.D.; Silveri, M.M. Cortical thinness and volume differences associated with marijuana abuse in emerging adults. *Drug Alcohol. Depend.* **2015**, *155*, 275–283. [CrossRef]
18. Shollenbarger, S.G.; Price, J.; Wieser, J.; Lisdahl, K. Impact of cannabis use on prefrontal and parietal cortex gyrification and surface area in adolescents and emerging adults. *Dev. Cogn. Neurosci.* **2015**, *16*, 46–53. [CrossRef]
19. Tzilos, G.K.; Cintron, C.B.; Wood, J.B.; Simpson, N.S.; Young, A.D.; Pope, H.G., Jr.; Yurgelun-Todd, D.A. Lack of hippocampal volume change in long-term heavy cannabis users. *Am. J. Addict.* **2005**, *14*, 64–72. [CrossRef]
20. Medina, K.L.; Nagel, B.J.; Park, A.; McQueeny, T.; Tapert, S.F. Depressive symptoms in adolescents: Associations with white matter volume and marijuana use. *J. Child. Psychol Psychiatry* **2007**, *48*, 592–600. [CrossRef]
21. Koenders, L.; Lorenzetti, V.; de Haan, L.; Suo, C.; Vingerhoets, W.; van den Brink, W.; Wiers, R.W.; Meijer, C.J.; Machielsen, M.; Goudriaan, A.E.; et al. Longitudinal study of hippocampal volumes in heavy cannabis users. *J. Psychopharmacol.* **2017**, *31*, 1027–1034. [CrossRef]
22. Lorenzetti, V.; Solowij, N.; Whittle, S.; Fornito, A.; Lubman, D.I.; Pantelis, C.; Yucel, M. Gross morphological brain changes with chronic, heavy cannabis use. *Br. J. Psychiatry* **2015**, *206*, 77–78. [CrossRef] [PubMed]
23. Orr, J.M.; Paschall, C.J.; Banich, M.T. Recreational marijuana use impacts white matter integrity and subcortical (but not cortical) morphometry. *Neuroimage Clin.* **2016**, *12*, 47–56. [CrossRef] [PubMed]
24. Mata, I.; Perez-Iglesias, R.; Roiz-Santianez, R.; Tordesillas-Gutierrez, D.; Pazos, A.; Gutierrez, A.; Vazquez-Barquero, J.L.; Crespo-Facorro, B. Gyrification brain abnormalities associated with adolescence and early-adulthood cannabis use. *Brain Res.* **2010**, *1317*, 297–304. [CrossRef] [PubMed]
25. Zalesky, A.; Solowij, N.; Yucel, M.; Lubman, D.I.; Takagi, M.; Harding, I.H.; Lorenzetti, V.; Wang, R.; Searle, K.; Pantelis, C.; et al. Effect of long-term cannabis use on axonal fibre connectivity. *Brain* **2012**, *135*, 2245–2255. [CrossRef]
26. Gilman, J.M.; Kuster, J.K.; Lee, S.; Lee, M.J.; Kim, B.W.; Makris, N.; van der Kouwe, A.; Blood, A.J.; Breiter, H.C. Cannabis use is quantitatively associated with nucleus accumbens and amygdala abnormalities in young adult recreational users. *J. Neurosci.* **2014**, *34*, 5529–5538. [CrossRef]
27. Weiland, B.J.; Thayer, R.E.; Depue, B.E.; Sabbineni, A.; Bryan, A.D.; Hutchison, K.E. Daily marijuana use is not associated with brain morphometric measures in adolescents or adults. *J. Neurosci.* **2015**, *35*, 1505–1512. [CrossRef]
28. Chye, Y.; Suo, C.; Lorenzetti, V.; Batalla, A.; Cousijn, J.; Goudriaan, A.E.; Martin-Santos, R.; Whittle, S.; Solowij, N.; Yucel, M. Cortical surface morphology in long-term cannabis users: A multi-site MRI study. *Eur. Neuropsychopharmacol.* **2018**. [CrossRef]
29. Rocchetti, M.; Crescini, A.; Borgwardt, S.; Caverzasi, E.; Politi, P.; Atakan, Z.; Fusar-Poli, P. Is cannabis neurotoxic for the healthy brain? A meta-analytical review of structural brain alterations in non-psychotic users. *Psychiatry Clin. Neurosci.* **2013**, *67*, 483–492. [CrossRef]
30. Iversen, L. Cannabis and the brain. *Brain* **2003**, *126*, 1252–1270. [CrossRef]
31. Herkenham, M. Characterization and localization of cannabinoid receptors in brain: An in vitro technique using slide-mounted tissue sections. *Nida Res. Monogr.* **1991**, *112*, 129–145.

32. Rubino, T.; Parolaro, D. The Impact of Exposure to Cannabinoids in Adolescence: Insights From Animal Models. *Biol. Psychiatry* **2016**, *79*, 578–585. [CrossRef] [PubMed]
33. Eustache, F.; Desgranges, B.; Laville, P.; Guillery, B.; Lalevee, C.; Schaeffer, S.; de la Sayette, V.; Iglesias, S.; Baron, J.C.; Viader, F. Episodic memory in transient global amnesia: Encoding, storage, or retrieval deficit? *J. Neurol. Neurosurg. Psychiatry* **1999**, *66*, 148–154. [CrossRef] [PubMed]
34. Riedel, W.J.; Blokland, A. Declarative memory. *Handb Exp. Pharm.* **2015**, *228*, 215–236. [CrossRef]
35. Weinstein, A.; Livny, A.; Weizman, A. Brain Imaging Studies on the Cognitive, Pharmacological and Neurobiological Effects of Cannabis in Humans: Evidence from Studies of Adult Users. *Curr. Pharm. Des.* **2016**, *22*, 6366–6379. [CrossRef] [PubMed]
36. Stokes, M.G.; Kusunoki, M.; Sigala, N.; Nili, H.; Gaffan, D.; Duncan, J. Dynamic coding for cognitive control in prefrontal cortex. *Neuron* **2013**, *78*, 364–375. [CrossRef] [PubMed]
37. Sugase-Miyamoto, Y.; Liu, Z.; Wiener, M.C.; Optican, L.M.; Richmond, B.J. Short-term memory trace in rapidly adapting synapses of inferior temporal cortex. *PLoS Comput. Biol.* **2008**, *4*, e1000073. [CrossRef] [PubMed]
38. Ranganath, C.; Yonelinas, A.P.; Cohen, M.X.; Dy, C.J.; Tom, S.M.; D'Esposito, M. Dissociable correlates of recollection and familiarity within the medial temporal lobes. *Neuropsychologia* **2004**, *42*, 2–13. [CrossRef]
39. Curtis, C.E.; D'Esposito, M. The effects of prefrontal lesions on working memory performance and theory. *Cogn. Affect. Behav. Neurosci.* **2004**, *4*, 528–539. [CrossRef]
40. Spellman, T.; Rigotti, M.; Ahmari, S.E.; Fusi, S.; Gogos, J.A.; Gordon, J.A. Hippocampal-prefrontal input supports spatial encoding in working memory. *Nature* **2015**, *522*, 309–314. [CrossRef]
41. Brasted, P.J.; Bussey, T.J.; Murray, E.A.; Wise, S.P. Role of the hippocampal system in associative learning beyond the spatial domain. *Brain* **2003**, *126*, 1202–1223. [CrossRef]
42. Law, J.R.; Flanery, M.A.; Wirth, S.; Yanike, M.; Smith, A.C.; Frank, L.M.; Suzuki, W.A.; Brown, E.N.; Stark, C.E. Functional magnetic resonance imaging activity during the gradual acquisition and expression of paired-associate memory. *J. Neurosci.* **2005**, *25*, 5720–5729. [CrossRef] [PubMed]
43. Suzuki, W.A. Making new memories: The role of the hippocampus in new associative learning. *Ann. N. Y. Acad. Sci.* **2007**, *1097*, 1–11. [CrossRef] [PubMed]
44. Wirth, S.; Yanike, M.; Frank, L.M.; Smith, A.C.; Brown, E.N.; Suzuki, W.A. Single neurons in the monkey hippocampus and learning of new associations. *Science* **2003**, *300*, 1578–1581. [CrossRef]
45. Pergola, G.; Suchan, B. Associative learning beyond the medial temporal lobe: Many actors on the memory stage. *Front. Behav. Neurosci.* **2013**, *7*, 162. [CrossRef] [PubMed]
46. Yonelinas, A.P.; Aly, M.; Wang, W.C.; Koen, J.D. Recollection and familiarity: Examining controversial assumptions and new directions. *Hippocampus* **2010**, *20*, 1178–1194. [CrossRef]
47. Elger, C.E.; Grunwald, T.; Lehnertz, K.; Kutas, M.; Helmstaedter, C.; Brockhaus, A.; Van Roost, D.; Heinze, H.J. Human temporal lobe potentials in verbal learning and memory processes. *Neuropsychologia* **1997**, *35*, 657–667. [CrossRef]
48. Rugg, M.D.; Vilberg, K.L. Brain networks underlying episodic memory retrieval. *Curr. Opin. Neurobiol.* **2013**, *23*, 255–260. [CrossRef]
49. Wagner, A.D.; Shannon, B.J.; Kahn, I.; Buckner, R.L. Parietal lobe contributions to episodic memory retrieval. *Trends Cogn. Sci.* **2005**, *9*, 445–453. [CrossRef]
50. Zeineh, M.M.; Engel, S.A.; Thompson, P.M.; Bookheimer, S.Y. Dynamics of the hippocampus during encoding and retrieval of face-name pairs. *Science* **2003**, *299*, 577–580. [CrossRef]
51. Ganzer, F.; Bröning, S.; Kraft, S.; Sack, P.M.; Thomasius, R. Weighing the Evidence: A Systematic Review on Long-Term Neurocognitive Effects of Cannabis Use in Abstinent Adolescents and Adults. *Neuropsychol. Rev.* **2016**, *26*, 186–222. [CrossRef]
52. Broyd, S.J.; van Hell, H.H.; Beale, C.; Yucel, M.; Solowij, N. Acute and Chronic Effects of Cannabinoids on Human Cognition-A Systematic Review. *Biol. Psychiatry* **2016**, *79*, 557–567. [CrossRef] [PubMed]
53. Crane, N.A.; Schuster, R.M.; Fusar-Poli, P.; Gonzalez, R. Effects of cannabis on neurocognitive functioning: Recent advances, neurodevelopmental influences, and sex differences. *Neuropsychol. Rev.* **2013**, *23*, 117–137. [CrossRef]
54. Solowij, N.; Battisti, R. The chronic effects of cannabis on memory in humans: A review. *Curr. Drug Abus. Rev.* **2008**, *1*, 81–98. [CrossRef] [PubMed]

55. Bossong, M.G.; Jager, G.; Bhattacharyya, S.; Allen, P. Acute and non-acute effects of cannabis on human memory function: A critical review of neuroimaging studies. *Curr. Pharm. Des.* **2014**, *20*, 2114–2125. [CrossRef] [PubMed]
56. Batalla, A.; Bhattacharyya, S.; Yucel, M.; Fusar-Poli, P.; Crippa, J.A.; Nogue, S.; Torrens, M.; Pujol, J.; Farre, M.; Martin-Santos, R. Structural and functional imaging studies in chronic cannabis users: A systematic review of adolescent and adult findings. *PLoS ONE* **2013**, *8*, e55821. [CrossRef]
57. Batalla, A.; Crippa, J.A.; Busatto, G.F.; Guimaraes, F.S.; Zuardi, A.W.; Valverde, O.; Atakan, Z.; McGuire, P.K.; Bhattacharyya, S.; Martin-Santos, R. Neuroimaging studies of acute effects of THC and CBD in humans and animals: A systematic review. *Curr. Pharm. Des.* **2014**, *20*, 2168–2185. [CrossRef]
58. Martin-Santos, R.; Fagundo, A.B.; Crippa, J.A.; Atakan, Z.; Bhattacharyya, S.; Allen, P.; Fusar-Poli, P.; Borgwardt, S.; Seal, M.; Busatto, G.F.; et al. Neuroimaging in cannabis use: A systematic review of the literature. *Psychol. Med.* **2010**, *40*, 383–398. [CrossRef] [PubMed]
59. Nader, D.A.; Sanchez, Z.M. Effects of regular cannabis use on neurocognition, brain structure, and function: A systematic review of findings in adults. *Am. J. Drug Alcohol. Abus.* **2018**, *44*, 4–18. [CrossRef]
60. Brumback, T.; Castro, N.; Jacobus, J.; Tapert, S. Effects of Marijuana Use on Brain Structure and Function: Neuroimaging Findings from a Neurodevelopmental Perspective. *Int. Rev. Neurobiol.* **2016**, *129*, 33–65. [CrossRef] [PubMed]
61. Blest-Hopley, G.; Giampietro, V.; Bhattacharyya, S. Residual effects of cannabis use in adolescent and adult brains—A meta-analysis of fMRI studies. *Neurosci. Biobehav. Rev.* **2018**, *88*, 26–41. [CrossRef]
62. Blest-Hopley, G.; Giampietro, V.; Bhattacharyya, S. Regular cannabis use is associated with altered activation of central executive and default mode networks even after prolonged abstinence in adolescent users: Results from a complementary meta-analysis. *Neurosci. Biobehav. Rev.* **2018**, *96*, 45–55. [CrossRef] [PubMed]
63. Yanes, J.A.; Riedel, M.C.; Ray, K.L.; Kirkland, A.E.; Bird, R.T.; Boeving, E.R.; Reid, M.A.; Gonzalez, R.; Robinson, J.L.; Laird, A.R.; et al. Neuroimaging meta-analysis of cannabis use studies reveals convergent functional alterations in brain regions supporting cognitive control and reward processing. *J. Psychopharmacol.* **2018**, *32*, 283–295. [CrossRef] [PubMed]
64. Andersen, S.L. Trajectories of brain development: Point of vulnerability or window of opportunity? *Neurosci. Biobehav. Rev.* **2003**, *27*, 3–18. [CrossRef]
65. Belue, R.C.; Howlett, A.C.; Westlake T.M.; Hutchings, D.E. The ontogeny of cannabinoid receptors in the brain of postnatal and aging rats. *Neurotoxicol. Teratol.* **1995**, *17*, 25–30. [CrossRef]
66. Rice, D.; Barone, S., Jr. Critical periods of vulnerability for the developing nervous system: Evidence from humans and animal models. *Env. Health Perspect* **2000**, *108*, 511–533.
67. Spear, L.P. Assessment of adolescent neurotoxicity: Rationale and methodological considerations. *Neurotoxicol. Teratol.* **2007**, *29*, 1–9. [CrossRef]
68. Schreiner, A.M.; Dunn, M.E. Residual effects of cannabis use on neurocognitive performance after prolonged abstinence: A meta-analysis. *Exp. Clin. Psychopharmacol.* **2012**, *20*, 420–429. [CrossRef]
69. Jpt, H. *Cochrane Handbook for Systematic Reviews of Interventions*; The Cochrane Collaboration, 2011. Available online: www.handbook.cochrane.org (accessed on 12 February 2020).
70. Stroup, D.F.; Berlin, J.A.; Morton, S.C.; Olkin, I.; Williamson, G.D.; Rennie, D.; Moher, D.; Becker, B.J.; Sipe, T.A.; Thacker, S.B. Meta-analysis of observational studies in epidemiology: A proposal for reporting. Meta-analysis Of Observational Studies in Epidemiology (MOOSE) group. *JAMA* **2000**, *283*, 2008–2012. [CrossRef]
71. Cohen, K.; Weinstein, A. The Effects of Cannabinoids on Executive Functions: Evidence from Cannabis and Synthetic Cannabinoids-A Systematic Review. *Brain Sci.* **2018**, *8*, 40. [CrossRef]
72. Figueiredo, P.R.; Tolomeo, S.; Steele, J.D.; Baldacchino, A. Neurocognitive consequences of chronic cannabis use: A systematic review and meta-analysis. *Neurosci. Biobehav. Rev.* **2020**, *108*, 358–369. [CrossRef]
73. Tervo-Clemmens, B.; Simmonds, D.; Calabro, F.J.; Day, N.L.; Richardson, G.A.; Luna, B. Adolescent cannabis use and brain systems supporting adult working memory encoding, maintenance, and retrieval. *Neuroimage* **2018**, *169*, 496–509. [CrossRef] [PubMed]
74. Blest-Hopley, G.; O'Neill, A.; Wilson, R.; Giampietro, V.; Bhattacharyya, S. Disrupted parahippocampal and midbrain function underlie slower verbal learning in adolescent-onset regular cannabis use. *Psychopharmacology (Berl.)* **2019**, *9*, 1–17. [CrossRef]

75. Riba, J.; Valle, M.; Sampedro, F.; Rodriguez-Pujadas, A.; Martinez-Horta, S.; Kulisevsky, J.; Rodriguez-Fornells, A. Telling true from false: Cannabis users show increased susceptibility to false memories. *Mol. Psychiatry* **2015**, *20*, 772–777. [CrossRef] [PubMed]
76. Jager, G.; Block, R.I.; Luijten, M.; Ramsey, N.F. Cannabis use and memory brain function in adolescent boys: A cross-sectional multicenter functional magnetic resonance imaging study. *J. Am. Acad. Child. Adolesc. Psychiatry* **2010**, *49*, 561–572. [CrossRef] [PubMed]
77. Schweinsburg, A.D.; Schweinsburg, B.C.; Nagel, B.J.; Eyler, L.T.; Tapert, S.F. Neural correlates of verbal learning in adolescent alcohol and marijuana users. *Addiction* **2011**, *106*, 564–573. [CrossRef] [PubMed]
78. Schweinsburg, A.D.; Brown, S.A.; Tapert, S.F. The influence of marijuana use on neurocognitive functioning in adolescents. *Curr. Drug Abus. Rev.* **2008**, *1*, 99–111.
79. Sneider, J.T.; Gruber, S.A.; Rogowska, J.; Silveri, M.M.; Yurgelun-Todd, D.A. A preliminary study of functional brain activation among marijuana users during performance of a virtual water maze task. *J. Addict.* **2013**, *2013*, 461029. [CrossRef]
80. Kanayama, G.; Rogowska, J.; Pope, H.G.; Gruber, S.A.; Yurgelun-Todd, D.A. Spatial working memory in heavy cannabis users: A functional magnetic resonance imaging study. *Psychopharmacology (Berl.)* **2004**, *176*, 239–247. [CrossRef]
81. Nestor, L.; Roberts, G.; Garavan, H.; Hester, R. Deficits in learning and memory: Parahippocampal hyperactivity and frontocortical hypoactivity in cannabis users. *Neuroimage* **2008**, *40*, 1328–1339. [CrossRef]
82. Jager, G.; Van Hell, H.H.; De Win, M.M.; Kahn, R.S.; Van Den Brink, W.; Van Ree, J.M.; Ramsey, N.F. Effects of frequent cannabis use on hippocampal activity during an associative memory task. *Eur. Neuropsychopharmacol.* **2007**, *17*, 289–297. [CrossRef]
83. Carey, S.E.; Nestor, L.; Jones, J.; Garavan, H.; Hester, R. Impaired learning from errors in cannabis users: Dorsal anterior cingulate cortex and hippocampus hypoactivity. *Drug Alcohol. Depend.* **2015**, *155*, 175–182. [CrossRef]
84. Cousijn, J.; Wiers, R.W.; Ridderinkhof, K.R.; van den Brink, W.; Veltman, D.J.; Goudriaan, A.E. Effect of baseline cannabis use and working-memory network function on changes in cannabis use in heavy cannabis users: A prospective fMRI study. *Hum. Brain Mapp.* **2014**, *35*, 2470–2482. [CrossRef] [PubMed]
85. Jager, G.; Kahn, R.S.; Van Den Brink, W.; Van Ree, J.M.; Ramsey, N.F. Long-term effects of frequent cannabis use on working memory and attention: An fMRI study. *Psychopharmacology (Berl.)* **2006**, *185*, 358–368. [CrossRef] [PubMed]
86. Schweinsburg, A.D.; Schweinsburg, B.C.; Medina, K.L.; McQueeny, T.; Brown, S.A.; Tapert, S.F. The influence of recency of use on fMRI response during spatial working memory in adolescent marijuana users. *J. Psychoact. Drugs* **2010**, *42*, 401–412. [CrossRef]
87. Schweinsburg, A.D.; Nagel, B.J.; Schweinsburg, B.C.; Park, A.; Theilmann, R.J.; Tapert, S.F. Abstinent adolescent marijuana users show altered fMRI response during spatial working memory. *Psychiatry Res.* **2008**, *163*, 40–51. [CrossRef] [PubMed]
88. Dager, A.D.; Tice, M.R.; Book, G.A.; Tennen, H.; Raskin, S.A.; Austad, C.S.; Wood, R.M.; Fallahi, C.R.; Hawkins, K.A.; Pearlson, G.D. Relationship between fMRI response during a nonverbal memory task and marijuana use in college students. *Drug Alcohol. Depend.* **2018**, *188*, 71–78. [CrossRef]
89. Cousijn, J.; Vingerhoets, W.A.; Koenders, L.; de Haan, L.; van den Brink, W.; Wiers, R.W.; Goudriaan, A.E. Relationship between working-memory network function and substance use: A 3-year longitudinal fMRI study in heavy cannabis users and controls. *Addict. Biol.* **2014**, *19*, 282–293. [CrossRef]
90. Block, R.I.; O'Leary, D.S.; Hichwa, R.D.; Augustinack, J.C.; Boles Ponto, L.L.; Ghoneim, M.M.; Arndt, S.; Hurtig, R.R.; Watkins, G.L.; Hall, J.A.; et al. Effects of frequent marijuana use on memory-related regional cerebral blood flow. *Pharm. Biochem Behav.* **2002**, *72*, 237–250. [CrossRef]
91. Friedman, D.; Johnson, R., Jr. Event-related potential (ERP) studies of memory encoding and retrieval: A selective review. *Microsc. Res. Tech.* **2000**, *51*, 6–28. [CrossRef]
92. Cengel, H.Y.; Bozkurt, M.; Evren, C.; Umut, G.; Keskinkilic, C.; Agachanli, R. Evaluation of cognitive functions in individuals with synthetic cannabinoid use disorder and comparison to individuals with cannabis use disorder. *Psychiatry Res.* **2018**, *262*, 46–54. [CrossRef]
93. Battisti, R.A.; Roodenrys, S.; Johnstone, S.J.; Respondek, C.; Hermens, D.F.; Solowij, N. Chronic use of cannabis and poor neural efficiency in verbal memory ability. *Psychopharmacology (Berl.)* **2010**, *209*, 319–330. [CrossRef] [PubMed]

94. Rodgers, J. Cognitive performance amongst recreational users of "ecstasy". *Psychopharmacology (Berl.)* **2000**, *151*, 19–24. [CrossRef] [PubMed]
95. McKetin, R.; Parasu, P.; Cherbuin, N.; Eramudugolla, R.; Anstey, K.J. A longitudinal examination of the relationship between cannabis use and cognitive function in mid-life adults. *Drug Alcohol. Depend.* **2016**, *169*, 134–140. [CrossRef] [PubMed]
96. Wadsworth, E.J.; Moss, S.C.; Simpson, S.A.; Smith, A.P. Cannabis use, cognitive performance and mood in a sample of workers. *J. Psychopharmacol.* **2006**, *20*, 14–23. [CrossRef] [PubMed]
97. Pope, H.G., Jr.; Gruber, A.J.; Hudson, J.I.; Huestis, M.A.; Yurgelun-Todd, D. Cognitive measures in long-term cannabis users. *J. Clin. Pharm.* **2002**, *42*, 41S–47S. [CrossRef] [PubMed]
98. Pope, H.G., Jr.; Gruber, A.J.; Hudson, J.I.; Huestis, M.A.; Yurgelun-Todd, D. Neuropsychological performance in long-term cannabis users. *Arch. Gen. Psychiatry* **2001**, *58*, 909–915. [CrossRef]
99. Messinis, L.; Kyprianidou, A.; Malefaki, S.; Papathanasopoulos, P. Neuropsychological deficits in long-term frequent cannabis users. *Neurology* **2006**, *66*, 737–739. [CrossRef]
100. Quednow, B.B.; Jessen, F.; Kuhn, K.U.; Maier, W.; Daum, I.; Wagner, M. Memory deficits in abstinent MDMA (ecstasy) users: Neuropsychological evidence of frontal dysfunction. *J. Psychopharmacol.* **2006**, *20*, 373–384. [CrossRef]
101. Gruber, S.A.; Sagar, K.A.; Dahlgren, M.K.; Racine, M.; Lukas, S.E. Age of onset of marijuana use and executive function. *Psychol. Addict. Behav.* **2012**, *26*, 496–506. [CrossRef]
102. Levar, N.; Francis, A.N.; Smith, M.J.; Ho, W.C.; Gilman, J.M. Verbal Memory Performance and Reduced Cortical Thickness of Brain Regions Along the Uncinate Fasciculus in Young Adult Cannabis Users. *Cannabis. Cannabinoid. Res.* **2018**, *3*, 56–65. [CrossRef]
103. Schuster, R.M.; Hoeppner, S.S.; Evins, A.E.; Gilman, J.M. Early onset marijuana use is associated with learning inefficiencies. *Neuropsychology* **2016**, *30*, 405–415. [CrossRef] [PubMed]
104. Solowij, N.; Jones, K.A.; Rozman, M.E.; Davis, S.M.; Ciarrochi, J.; Heaven, P.C.; Lubman, D.I.; Yucel, M. Verbal learning and memory in adolescent cannabis users, alcohol users and non-users. *Psychopharmacology (Berl.)* **2011**, *216*, 131–144. [CrossRef] [PubMed]
105. Hanson, K.L.; Winward, J.L.; Schweinsburg, A.D.; Medina, K.L.; Brown, S.A.; Tapert, S.F. Longitudinal study of cognition among adolescent marijuana users over three weeks of abstinence. *Addict. Behav.* **2010**, *35*, 970–976. [CrossRef] [PubMed]
106. Medina, K.L.; Hanson, K.L.; Schweinsburg, A.D.; Cohen-Zion, M.; Nagel, B.J.; Tapert, S.F. Neuropsychological functioning in adolescent marijuana users: Subtle deficits detectable after a month of abstinence. *J. Int. Neuropsychol. Soc.* **2007**, *13*, 807–820. [CrossRef]
107. Solowij, N.; Stephens, R.S.; Roffman, R.A.; Babor, T.; Kadden, R.; Miller, M.; Christiansen, K.; McRee, B.; Vendetti, J.; Marijuana Treatment Project Research, G. Cognitive functioning of long-term heavy cannabis users seeking treatment. *JAMA* **2002**, *287*, 1123–1131. [CrossRef]
108. Fried, P.A.; Watkinson, B.; Gray, R. Neurocognitive consequences of marihuana–a comparison with pre-drug performance. *Neurotoxicol. Teratol.* **2005**, *27*, 231–239. [CrossRef]
109. Tait, R.J.; Mackinnon, A.; Christensen, H. Cannabis use and cognitive function: 8-year trajectory in a young adult cohort. *Addiction* **2011**, *106*, 2195–2203. [CrossRef]
110. Meier, M.H.; Caspi, A.; Ambler, A.; Harrington, H.; Houts, R.; Keefe, R.S.; McDonald, K.; Ward, A.; Poulton, R.; Moffitt, T.E. Persistent cannabis users show neuropsychological decline from childhood to midlife. *Proc. Natl. Acad. Sci. USA* **2012**, *109*, E2657–E2664. [CrossRef]
111. Auer, R.; Vittinghoff, E.; Yaffe, K.; Kunzi, A.; Kertesz, S.G.; Levine, D.A.; Albanese, E.; Whitmer, R.A.; Jacobs, D.R., Jr.; Sidney, S.; et al. Association Between Lifetime Marijuana Use and Cognitive Function in Middle Age: The Coronary Artery Risk Development in Young Adults (CARDIA) Study. *JAMA Intern. Med.* **2016**, *176*, 352–361. [CrossRef]
112. Castellanos-Ryan, N.; Pingault, J.B.; Parent, S.; Vitaro, F.; Tremblay, R.E.; Seguin, J.R. Adolescent cannabis use, change in neurocognitive function, and high-school graduation: A longitudinal study from early adolescence to young adulthood. *Dev. Psychopathol.* **2017**, *29*, 1253–1266. [CrossRef]
113. Stiglick, A.; Kalant, H. Residual effects of chronic cannabis treatment on behavior in mature rats. *Psychopharmacology (Berl.)* **1985**, *85*, 436–439. [CrossRef] [PubMed]

114. Mateos, B.; Borcel, E.; Loriga, R.; Luesu, W.; Bini, V.; Llorente, R.; Castelli, M.P.; Viveros, M.P. Adolescent exposure to nicotine and/or the cannabinoid agonist CP 55,940 induces gender-dependent long-lasting memory impairments and changes in brain nicotinic and CB(1) cannabinoid receptors. *J. Psychopharmacol.* **2011**, *25*, 1676–1690. [CrossRef] [PubMed]
115. Kirschmann, E.K.; Pollock, M.W.; Nagarajan, V.; Torregrossa, M.M. Effects of Adolescent Cannabinoid Self-Administration in Rats on Addiction-Related Behaviors and Working Memory. *Neuropsychopharmacol. Off. Publ. Am. Coll. Neuropsychopharmacol.* **2017**, *42*, 989–1000. [CrossRef] [PubMed]
116. Harte, L.C.; Dow-Edwards, D. Sexually dimorphic alterations in locomotion and reversal learning after adolescent tetrahydrocannabinol exposure in the rat. *Neurotoxicol. Teratol.* **2010**, *32*, 515–524. [CrossRef] [PubMed]
117. Schneider, M.; Koch, M. Chronic pubertal, but not adult chronic cannabinoid treatment impairs sensorimotor gating, recognition memory, and the performance in a progressive ratio task in adult rats. *Neuropsychopharmacol. Off. Publ. Am. Coll. Neuropsychopharmacol.* **2003**, *28*, 1760–1769. [CrossRef]
118. Higuera-Matas, A.; Botreau, F.; Miguens, M.; Del Olmo, N.; Borcel, E.; Perez-Alvarez, L.; Garcia-Lecumberri, C.; Ambrosio, E. Chronic periadolescent cannabinoid treatment enhances adult hippocampal PSA-NCAM expression in male Wistar rats but only has marginal effects on anxiety, learning and memory. *Pharm. Biochem Behav.* **2009**, *93*, 482–490. [CrossRef]
119. O'Shea, M.; Singh, M.E.; McGregor, I.S.; Mallet, P.E. Chronic cannabinoid exposure produces lasting memory impairment and increased anxiety in adolescent but not adult rats. *J. Psychopharmacol.* **2004**, *18*, 502–508. [CrossRef]
120. Stiglick, A.; Kalant, H. Learning impairment in the radial-arm maze following prolonged cannabis treatment in rats. *Psychopharmacology (Berl.)* **1982**, *77*, 117–123. [CrossRef]
121. Rubino, T.; Realini, N.; Braida, D.; Guidi, S.; Capurro, V.; Vigano, D.; Guidali, C.; Pinter, M.; Sala, M.; Bartesaghi, R.; et al. Changes in hippocampal morphology and neuroplasticity induced by adolescent THC treatment are associated with cognitive impairment in adulthood. *Hippocampus* **2009**, *19*, 763–772. [CrossRef]
122. Renard, J.; Krebs, M.O.; Jay, T.M.; Le Pen, G. Long-term cognitive impairments induced by chronic cannabinoid exposure during adolescence in rats: A strain comparison. *Psychopharmacology (Berl.)* **2013**, *225*, 781–790. [CrossRef]
123. Abush, H.; Akirav, I. Short- and long-term cognitive effects of chronic cannabinoids administration in late-adolescence rats. *PLoS ONE* **2012**, *7*, e31731. [CrossRef] [PubMed]
124. Fehr, K.A.; Kalant, H.; LeBlanc, A.E. Residual learning deficit after heavy exposure to cannabis or alcohol in rats. *Science* **1976**, *192*, 1249–1251. [CrossRef] [PubMed]
125. Rubino, T.; Vigano, D.; Realini, N.; Guidali, C.; Braida, D.; Capurro, V.; Castiglioni, C.; Cherubino, F.; Romualdi, P.; Candeletti, S.; et al. Chronic delta 9-tetrahydrocannabinol during adolescence provokes sex-dependent changes in the emotional profile in adult rats: Behavioral and biochemical correlates. *Neuropsychopharmacol. Off. Publ. Am. Coll. Neuropsychopharmacol.* **2008**, *33*, 2760–2771. [CrossRef] [PubMed]
126. Hill, M.N.; Froc, D.J.; Fox, C.J.; Gorzalka, B.B.; Christie, B.R. Prolonged cannabinoid treatment results in spatial working memory deficits and impaired long-term potentiation in the CA1 region of the hippocampus in vivo. *Eur. J. Neurosci.* **2004**, *20*, 859–863. [CrossRef]
127. Cha, Y.M.; Jones, K.H.; Kuhn, C.M.; Wilson, W.A.; Swartzwelder, H.S. Sex differences in the effects of delta9-tetrahydrocannabinol on spatial learning in adolescent and adult rats. *Behav. Pharm.* **2007**, *18*, 563–569. [CrossRef]
128. Cha, Y.M.; White, A.M.; Kuhn, C.M.; Wilson, W.A.; Swartzwelder, H.S. Differential effects of delta9-THC on learning in adolescent and adult rats. *Pharm. Biochem Behav.* **2006**, *83*, 448–455. [CrossRef]
129. Schneider, M.; Drews, E.; Koch, M. Behavioral effects in adult rats of chronic prepubertal treatment with the cannabinoid receptor agonist WIN 55,212-2. *Behav. Pharm.* **2005**, *16*, 447–454. [CrossRef]
130. Ranganathan, M.; Radhakrishnan, R.; Addy, P.H.; Schnakenberg-Martin, A.M.; Williams, A.H.; Carbuto, M.; Elander, J.; Pittman, B.; Andrew Sewell, R.; Skosnik, P.D.; et al. Tetrahydrocannabinol (THC) impairs encoding but not retrieval of verbal information. *Prog Neuropsychopharmacol. Biol. Psychiatry* **2017**, *79*, 176–183. [CrossRef]
131. Harvey, M.A.; Sellman, J.D.; Porter, R.J.; Frampton, C.M. The relationship between non-acute adolescent cannabis use and cognition. *Drug Alcohol. Rev.* **2007**, *26*, 309–319. [CrossRef]

132. Bolla, K.I.; Brown, K.; Eldreth, D.; Tate, K.; Cadet, J.L. Dose-related neurocognitive effects of marijuana use. *Neurology* **2002**, *59*, 1337–1343. [CrossRef]
133. Pope, H.G., Jr.; Gruber, A.J.; Hudson, J.I.; Cohane, G.; Huestis, M.A.; Yurgelun-Todd, D. Early-onset cannabis use and cognitive deficits: What is the nature of the association? *Drug Alcohol. Depend.* **2003**, *69*, 303–310. [CrossRef]
134. D'Souza, D.C.; Cortes-Briones, J.A.; Ranganathan, M.; Thurnauer, H.; Creatura, G.; Surti, T.; Planeta, B.; Neumeister, A.; Pittman, B.; Normandin, M.; et al. Rapid Changes in CB1 Receptor Availability in Cannabis Dependent Males after Abstinence from Cannabis. *Biol. Psychiatry Cogn. Neurosci. Neuroimaging* **2016**, *1*, 60–67. [CrossRef] [PubMed]
135. Hirst, R.B.; Young, K.R.; Sodos, L.M.; Wickham, R.E.; Earleywine, M. Trying to remember: Effort mediates the relationship between frequency of cannabis use and memory performance. *J. Clin. Exp. Neuropsychol.* **2017**, *39*, 502–512. [CrossRef] [PubMed]
136. Silveira, M.M.; Adams, W.K.; Morena, M.; Hill, M.N.; Winstanley, C.A. Delta(9)-Tetrahydrocannabinol decreases willingness to exert cognitive effort in male rats. *J. Psychiatry Neurosci.* **2017**, *42*, 131–138. [CrossRef] [PubMed]
137. McClure, E.A.; Lydiard, J.B.; Goddard, S.D.; Gray, K.M. Objective and subjective memory ratings in cannabis-dependent adolescents. *Am. J. Addict.* **2015**, *24*, 47–52. [CrossRef]
138. Lyons, M.J.; Bar, J.L.; Panizzon, M.S.; Toomey, R.; Eisen, S.; Xian, H.; Tsuang, M.T. Neuropsychological consequences of regular marijuana use: A twin study. *Psychol. Med.* **2004**, *34*, 1239–1250. [CrossRef]
139. Meier, M.H.; Caspi, A.; Danese, A.; Fisher, H.L.; Houts, R.; Arseneault, L.; Moffitt, T.E. Associations between adolescent cannabis use and neuropsychological decline: A longitudinal co-twin control study. *Addiction* **2018**, *113*, 257–265. [CrossRef]
140. Jackson, N.J.; Isen, J.D.; Khoddam, R.; Irons, D.; Tuvblad, C.; Iacono, W.G.; McGue, M.; Raine, A.; Baker, L.A. Impact of adolescent marijuana use on intelligence: Results from two longitudinal twin studies. *Proc. Natl. Acad. Sci. USA* **2016**, *113*, E500–E508. [CrossRef]
141. Howlett, A.C.; Johnson, M.R.; Melvin, L.S.; Milne, G.M. Nonclassical cannabinoid analgetics inhibit adenylate cyclase: Development of a cannabinoid receptor model. *Mol. Pharm.* **1988**, *33*, 297–302.
142. Milner, B.; Squire, L.R.; Kandel, E.R. Cognitive neuroscience and the study of memory. *Neuron* **1998**, *20*, 445–468. [CrossRef]
143. Scallet, A.C.; Uemura, E.; Andrews, A.; Ali, S.F.; McMillan, D.E.; Paule, M.G.; Brown, R.M.; Slikker, W. Morphometric studies of the rat hippocampus following chronic delta-9-tetrahydrocannabinol (THC). *Brain Res.* **1987**, *436*, 193–198. [CrossRef]
144. Lawston, J.; Borella, A.; Robinson, J.K.; Whitaker-Azmitia, P.M. Changes in hippocampal morphology following chronic treatment with the synthetic cannabinoid WIN 55,212-2. *Brain Res.* **2000**, *877*, 407–410. [CrossRef]
145. Landfield, P.W.; Cadwallader, L.B.; Vinsant, S. Quantitative changes in hippocampal structure following long-term exposure to delta 9-tetrahydrocannabinol: Possible mediation by glucocorticoid systems. *Brain Res.* **1988**, *443*, 47–62. [CrossRef]
146. Heath, R.G.; Fitzjarrell, A.T.; Fontana, C.J.; Garey, R.E. Cannabis sativa: Effects on brain function and ultrastructure in rhesus monkeys. *Biol. Psychiatry* **1980**, *15*, 657–690.
147. Schoeler, T.; Petros, N.; Di Forti, M.; Klamerus, E.; Foglia, E.; Ajnakina, O.; Gayer-Anderson, C.; Colizzi, M.; Quattrone, D.; Behlke, I.; et al. Effects of continuation, frequency, and type of cannabis use on relapse in the first 2 years after onset of psychosis: An observational study. *Lancet. Psychiatry* **2016**, *3*, 947–953. [CrossRef]
148. Jager, G.; Ramsey, N.F. Long-term consequences of adolescent cannabis exposure on the development of cognition, brain structure and function: An overview of animal and human research. *Curr. Drug Abus. Rev.* **2008**, *1*, 114–123. [CrossRef]
149. Ranganathan, M.; D'Souza, D.C. The acute effects of cannabinoids on memory in humans: A review. *Psychopharmacology (Berl.)* **2006**, *188*, 425–444. [CrossRef]
150. Bhattacharyya, S.; Fusar-Poli, P.; Borgwardt, S.; Martin-Santos, R.; Nosarti, C.; O'Carroll, C.; Allen, P.; Seal, M.L.; Fletcher, P.C.; Crippa, J.A.; et al. Modulation of mediotemporal and ventrostriatal function in humans by Delta9-tetrahydrocannabinol: A neural basis for the effects of Cannabis sativa on learning and psychosis. *Arch. Gen. Psychiatry* **2009**, *66*, 442–451. [CrossRef]

151. Horwood, L.J.; Fergusson, D.M.; Hayatbakhsh, M.R.; Najman, J.M.; Coffey, C.; Patton, G.C.; Silins, E.; Hutchinson, D.M. Cannabis use and educational achievement: Findings from three Australasian cohort studies. *Drug Alcohol. Depend.* **2010**, *110*, 247–253. [CrossRef]
152. Lynskey, M.; Hall, W. The effects of adolescent cannabis use on educational attainment: A review. *Addiction* **2000**, *95*, 1621–1630. [CrossRef]
153. Bhattacharyya, S.; Falkenberg, I.; Martin-Santos, R.; Atakan, Z.; Crippa, J.A.; Giampietro, V.; Brammer, M.; McGuire, P. Cannabinoid modulation of functional connectivity within regions processing attentional salience. *Neuropsychopharmacol. Off. Publ. Am. Coll. Neuropsychopharmacol.* **2015**, *40*, 1343–1352. [CrossRef]
154. Bhattacharyya, S.; Morrison, P.D.; Fusar-Poli, P.; Martin-Santos, R.; Borgwardt, S.; Winton-Brown, T.; Nosarti, C.; CM, O.C.; Seal, M.; Allen, P.; et al. Opposite effects of delta-9-tetrahydrocannabinol and cannabidiol on human brain function and psychopathology. *Neuropsychopharmacol. Off. Publ. Am. Coll. Neuropsychopharmacol.* **2010**, *35*, 764–774. [CrossRef] [PubMed]
155. Bhattacharyya, S.; Crippa, J.A.; Allen, P.; Martin-Santos, R.; Borgwardt, S.; Fusar-Poli, P.; Rubia, K.; Kambeitz, J.; O'Carroll, C.; Seal, M.L.; et al. Induction of psychosis by Delta9-tetrahydrocannabinol reflects modulation of prefrontal and striatal function during attentional salience processing. *Arch. Gen. Psychiatry* **2012**, *69*, 27–36. [CrossRef] [PubMed]
156. Freeman, T.P.; Lorenzetti, V. 'Standard THC units': A proposal to standardize dose across all cannabis products and methods of administration. *Addiction* **2019**. [CrossRef] [PubMed]

Article

Increased Resting State Triple Network Functional Connectivity in Undergraduate Problematic Cannabis Users: A Preliminary EEG Coherence Study

Claudio Imperatori [1,*], Chiara Massullo [1], Giuseppe Alessio Carbone [1], Angelo Panno [1], Marta Giacchini [1], Cristina Capriotti [1], Elisa Lucarini [1], Benedetta Ramella Zampa [1], Eric Murillo-Rodríguez [2], Sérgio Machado [3] and Benedetto Farina [1]

1 Cognitive and Clinical Psychology Laboratory, Department of Human Science, European University of Rome, Italy, Via degli Aldobrandeschi 190, 00163 Roma, Italy; chiara_massullo@yahoo.it (C.M.); giuseppea.carbone@gmail.com (G.A.C.); angelo.panno@unier.it (A.P.); marta.giacchini@hotmail.com (M.G.); cristinacap1996@hotmail.it (C.C.); elisalucarini1941@gmail.com (E.L.); b.ramellazampa@gmail.com (B.R.Z.); benedetto.farina@unier.it (B.F.)

2 Laboratorio de Neurociencias Moleculares e Integrativas, Escuela de Medicina, División Ciencias de la Salud, Universidad Anáhuac Mayab, Mérida 97302, Yucatán, Mexico; eric.murillo@anahuac.mx

3 Laboratory of Physical Activity Neuroscience, Physical Activity Sciences Postgraduate Program–Salgado de Oliveira University (UNIVERSO), Niterói 24030-060, RJ, Brazil; secm80@gmail.com

* Correspondence: claudio.imperatori@unier.it; Tel.: +06-6654-3873

Received: 21 January 2020; Accepted: 25 February 2020; Published: 28 February 2020

Abstract: An increasing body of experimental data have suggested that aberrant functional interactions between large-scale networks may be the most plausible explanation of psychopathology across multiple mental disorders, including substance-related and addictive disorders. In the current research, we have investigated the association between problematic cannabis use (PCU) and triple-network electroencephalographic (EEG) functional connectivity. Twelve participants with PCU and 24 non-PCU participants were included in the study. EEG recordings were performed during resting state (RS). The exact Low-Resolution Electromagnetic Tomography software (eLORETA) was used for all EEG analyses. Compared to non-PCU, PCU participants showed an increased delta connectivity between the salience network (SN) and central executive network (CEN), specifically, between the dorsal anterior cingulate cortex and right posterior parietal cortex. The strength of delta connectivity between the SN and CEN was positively and significantly correlated with higher problematic patterns of cannabis use after controlling for age, sex, educational level, tobacco use, problematic alcohol use, and general psychopathology ($r_p = 0.40$, $p = 0.030$). Taken together, our results show that individuals with PCU could be characterized by a specific dysfunctional interaction between the SN and CEN during RS, which might reflect the neurophysiological underpinnings of attentional and emotional processes of cannabis-related thoughts, memories, and craving.

Keywords: problematic cannabis use; triple network; EEG functional connectivity; eLORETA; resting state

1. Introduction

Cannabis is the most widely used illicit drug in Europe, with 18% and 9.3% of young people (i.e., the 15–24 age group) reporting having used cannabis in the last year and in the last month, respectively [1]. The lifetime prevalence of cannabis use disorder is about 6% [2], and the frequency of patients being treated for the first time for cannabis problems has dramatically increased over the last decade [1]. Therefore, cannabis use is considered a relevant topic that is gaining greater attention not only from a political point of view [1] but also from a scientific point of view, with a specific focus on the cognitive, behavioral, and neurobiological consequences associated to its use and abuse [3].

For example, research on animal models documented that while high concentrations of Δ9-tetrahydrocannabinol (THC), the main psychoactive constituent of cannabis [4], are necessary to impair memory and cognition in old rats, even low concentrations are deleterious in young animals [5]. Animal studies also showed that chronic THC exposure is associated with widespread neurochemical and neuroanatomical alterations in several brain areas, such as the limbic system and prefrontal cortex [6].

Similarly, human neuroimaging studies have shown that problematic cannabis use is related to different structural, functional, and neurophysiological brain alterations [7]. For instance, structural neuroimaging studies showed abnormalities in hippocampus volume and gray matter density associated with cannabis use [8]. Furthermore, Moreno-Alcázar et al. [9] recently reported that, compared to healthy controls, long-term heavy cannabis users showed increased gray matter volume in the basal ganglia and nucleus accumbens. A recent meta-analysis [3] on 35 task-related functional imaging studies also showed that cannabis use is associated with a decreased activity in brain areas involved in cognitive control process (e.g., the anterior cingulate cortex and dorsolateral prefrontal cortex (dlPFC)) and increased activity in brain structures involved in reward processing (e.g., the striatum). Lastly, electroencephalographic (EEG) studies showed that cannabis use is related to several neurophysiological abnormalities, such as increased cortical activation and connectivity, not only during drug cue exposure [10–12] but also during resting state (RS) condition [13–15].

Taken together, all these data are in line with the perspective that reward-related behaviors and addictive disorders are associated with dysfunctional dynamic interactions between large neural networks rather than alterations in single brain areas [16–19]. Within this modern view of the brain as a highly integrated and dynamic system, in the last years, a theoretical model has gained particular attention in the neuroscientific literature, the so-called triple network model [20]. This conceptualization underlines the crucial role of the synergistic interaction between large-scale networks in regulating the general access to cognitive functions [21] and conversely, it suggests that the dysfunctional communication within these neural systems is the most plausible explanation of psychopathology across multiple mental disorders [20,22].

In particular, the triple network model [20] focuses on the dynamic interaction among the default mode network (DMN), salience network (SN), and central executive network (CEN). While the DMN, centered on nodes in the medial prefrontal cortex (mPFC) and posterior cingulate cortex (PCC), is typically active during RS and involved in several higher-order integrative mental functions such as self-referential processing and mentalization [23,24], the CEN, anchored bilaterally in the dlPFC and posterior parietal cortex (PPC), is typically active during a wide range of cognitive tasks and involved in several mental functions such as working memory and problem solving [20,21]. The functional and dynamic switch between the DMN and CEN (i.e., between task-based and task-free states) is assured by the regular activity of the SN [25,26], which includes the dorsal anterior cingulate cortex (dACC) and bilateral anterior insula [20]. Indeed, this network plays a crucial role in filtering, detecting, and integrating relevant internal (e.g., autonomic input) and external (e.g., emotional information) salient stimuli in order to guide behavior [27,28].

In the last decade, an increasing body of experimental data has suggested that different aberrant functional interactions among the SN, CEN, and DMN may be considered potential neurophysiological biomarkers of different psychopathological phenomena emerging across several neuropsychiatric disorders, including substance-related and addictive disorders [20,22,29].

For example, it has been reported that, compared to the smoking state, nicotine abstinence is associated with lower SN–DMN connectivity, suggesting that a weaker network interaction contributes to smoke craving [30]. Decreased connectivity between the SN and DMN was also reported in cocaine-dependent individuals [31,32]. Furthermore, Li et al. showed that greater connectivity between the SN and DMN, as well as lower connectivity between the CEN and DMN, is associated with relapse behavior in heroin-dependent patients [33].

To the best of our knowledge, only one report has explored the association between cannabis use and triple network connectivity. In a functional magnetic resonance imaging (fMRI) study, Wall et al. [34] showed that in recreational cannabis users (i.e., not regular users) THC administration disrupts the DMN, where the PCC was the key brain region involved in the subjective experience of THC intoxication. Thus, the primary purpose of the current research was to extend these previous results examining the association between problematic cannabis use (PCU) and triple network EEG functional connectivity. Indeed, although fMRI is widely used to investigate brain functional connectivity, EEG is considered a suitable tool to investigate network properties [35,36], providing relevant data on functional interactions between dynamic neural systems in each frequency band [37,38].

2. Materials and Methods

2.1. Participants

Study participants were enrolled using advertising material posted around the university campus (i.e., a brief explanation of the study procedure including EEG procedure and questionnaire administration). The enrolment lasted from September to December 2019. Ninety-five undergraduate students who agreed to participate were screened for eligibility. All the individuals provided informed consent and contributed voluntarily to the study (i.e., they did not receive payment or academic credits). This research was approved by the ethics committee of the European University of Rome (Prot. N.008/19) in line with the Helsinki declaration standards.

Twelve participants (7 males and 5 females) with problematic cannabis use (PCU group) and twenty-four (9 males and 15 females) non-cannabis-using participants (non-PCU group) were finally enrolled. PCU individuals were enrolled if they met the following inclusion criteria: (i) Cannabis Abuse Screening Test (CAST) [39] total score ≥7, as recommended by Bastiani et al. [40] (see "self-report measures" section for details); (ii) frequency use of cannabis during the last 12 months ≥20 times [40]; (iii) age range 18–30 years old; (iv) negative past or current diagnosis of any psychiatric and/or neurological diseases (including head trauma); (v) right-handedness; (vi) negative psychoactive medications use and other illegal drugs consumption in the past two weeks prior to the EEG recordings.

Non-PCU group were included if they met the following inclusion criteria: (i) CAST total score = 0; (ii) frequency use of cannabis during the last 12 months = 0 times; (iii) age range 18–30 years old; (iv) negative past or current diagnosis of any psychiatric and/or neurological diseases (including head trauma); (v) right-handedness; (vi) negative psychoactive medications use and other illegal drugs consumption in the past two weeks prior to the EEG recordings.

2.2. Self-Report Measures

After the enrolment, all subjects were administered the CAST [39], a self-report measure of alcohol use problems (CAGE) [41], and the Symptom-Checklist-K-9 (SCL–K–9) [42], and they were asked screening questions according to a checklist developed for previous studies [43–46].

The CAST [39] is a 6-item self-report questionnaire widely used to assess problematic patterns of cannabis use within the past 12 months [47]. Items are scored on a 5-point Likert scale (from 0 = "never" to 4 = "very often"). The CAST includes two scoring options [48,49]: a binary version (i.e., computing the positive response thresholds that vary across items) and a full version (i.e., calculating the score using the full range of item responses). Good psychometric properties (e.g., high internal consistency) of both versions have been reported [48,49]. Satisfactory cross-cultural adaptation has been also documented [50,51]. In a sample of Italian young adults, using the Multiple Correspondence Analysis (MCA), Bastiani et al. [40] maximized item homogeneity of the CAST and obtained the best score in relation to the importance of the response categories for each item. Using this procedure, the authors showed that, compared to both the binary and the full version, the CAST MCA form had better psychometric properties and that the optimal cut-off score was 7 [40]. Therefore, in the current study, the CAST MCA version was used, and the Cronbach's α in our sample was 0.91.

The CAGE [41] is a 4-item self-report widely used to assess problematic alcohol use [41,52]. The acronym refers to the 4 dichotomous (yes = 1; no = 0) questions investigated by the questionnaire: (i) Cut down, (ii) Annoyed, (iii) Guilty, and (iv) Eye. The total score ranges from 0 to 4, and the recommended cut-off to screen problematic alcohol use is ≥2 [53]. Previous researches [53] reported that the CAGE has satisfactory psychometric properties (e.g., suitable correlations with other screening instruments). In the current research, we used the Italian adaptation of the CAGE [54], and the Cronbach's α in our sample was 0.68.

The Symptom-Checklist-K-9 (SCL–K–9) [42] is the short unidimensional version of the original Symptom Checklist-90-Revised (SCL–90–R) [55]. It is composed of the nine items of the SCL-90-R (rated on a 5-point Likert scale ranging from 0 = "not at all" to 4 = "extremely"), showing the highest discriminant power with the general level of psychopathology (i.e., the global severity index). Good psychometric properties (e.g., good reliability and good model fit), as well as significant correlations with other questionnaires assessing psychological distress, have been reported [56]. In the present study, we used the Italian adaptation of the SCL-K-9 [57], and the Cronbach's α in our sample was 0.86.

2.3. EEG Data Acquisition and Functional Connectivity Analysis

All EEG recordings were performed in the Cognitive and Clinical Psychology Laboratory of the European University of Rome. Eyes-closed RS EEG was recorded for at least 5 minutes. Study participants were invited to sit comfortably with their eyes closed in a quiet, semidarkened silent room; subjects were also instructed to avoid alcohol, caffeine, and cigarettes immediately before their experimental session (i.e., at least 4 h).

EEG data acquisition was performed using Micromed System Plus digital EEGraph (Micromed© S.p.A., Mogliano Veneto, TV, Italy) and 31 standard scalp leads, placed according to the 10-20 system. In this setting, Electro-oculogram and the Electrocardiogram were also acquired, and the reference electrodes were placed on the linked mastoids. As regards the EEG signal, it has been used a sampling frequency of 256 Hz and impedances were kept below 5KΩ before starting the recording and further controlled at the end of each experimental session. Other details about EEG recordings (e.g., A/D conversion and preamplifiers amplitude) can be found elsewhere [58,59]. Signal processing (i.e., filtering and artifact rejection procedure) was performed using EEGlab toolbox for MATLAB (The MathWorks, Inc). For filtering procedure, the "basic FIR filter" option was selected, and 0.2 Hz and 100 Hz were respectively the high-frequency filter and the low-frequency filter. Artifact rejection (i.e., removal of eye movements, blinks, cardiac pulses, muscular or movement activities) was performed visually on the raw EEG (for details, see [59–61]). At least 3 minutes of clean EEG data (not necessarily consecutive) were selected and analyzed for each subject. According to previous exact Low-Resolution Electromagnetic Tomography software (eLORETA) studies [43,45,62–66], artifact-free data were fragmented into epochs of 2 seconds for the EEG coherence analysis.

The exact Low-Resolution Electromagnetic Tomography software (eLORETA), a well-corroborated computer program able to detect electrocortical activity [67], was used for all EEG analyses. The eLORETA provides a "discrete, three-dimensional (3D) distributed, linear, weighted minimum norm inverse solution" [62]. Assuming that adjacent neuronal sources will be highly synchronized, the exact weights used in this software "endow the tomography with the property of exact localization to test point sources, yielding images of current density with exact localization albeit with low spatial resolution" [62]. The head model for the inverse solution uses the electric potential lead field computed with the boundary element method [68] averaged of a magnetic resonance image (MRI) data set. This forward equation "corresponds to an instantaneous discrete sampling of the measurement space (scalp electrodes) and the solution space (cortical voxels)" [67]. In other words, computations were performed using a realistic head model [68] determined according to the digitized MNI152 template provided by the Brain Imaging Center of the Montreal Neurological Institute (MNI) [69].

The standard electrode locations on the MNI152 scalp have been determined according to previous studies [70,71]. The three-dimensional spatial solution is limited to cortical gray matter, as determined by the probabilistic Talairach atlas [72], comprising 6239 voxels of 5 cubic mm spatial resolution (for details, see [62–64,73,74]). Only voxels that were unambiguously identified as cortical grey matter and those unequivocally felled within the brain compartment were considered by the software. Therefore, eLORETA images reflect the exact electrocortical activity at each voxel in neuroanatomic MNI space as the exact magnitude of the estimated current density [75]. Although the computations should be ideally performed on the exact head model, determined from each individual subject's MRI, the boundary element method is considered a suitable technique and it is one of the often-used realistic models in EEG source analysis [76]. Furthermore, compared to the previous version (e.g., sLORETA), the eLORETA is characterized by a correct localization even in the presence of structured noise [67,74]. Previous reports showed that the eLORETA provides a suitable localization agreement (the average depth localization error was 7 mm) with other neuroimaging methods [77–83], and also when a low number of electrodes were used (i.e., <30). The eLORETA is also characterized by no localization bias even in the presence of structured noise [62,67,84]. This software is also considered a suitable tool to investigate large brain network dynamics [35,36,38] by evaluating the modifications in the neuronal synchronization at varying time delays and frequencies [36]. As a matter of fact, compared to other brain-imaging methods, EEG time-series data provide a direct measure of postsynaptic potentials with millisecond temporal resolution [38,84], providing a relevant and precious complementary source of data for scholars and practitioners in a relatively ecological and economical way [85,86].

In the present study, the lagged phase synchronization (LPS) method [67,87] was used in order to investigate functional connectivity. The LPS evaluates "the similarity of two time series by means of the phases of the analyzed signal" [88] based on normalized Fourier transforms [63] with values ranging from 0 (i.e., no synchronization) to 1 (i.e., the maximum synchronization). Therefore, this approach is related to nonlinear functional connectivity, and it is considered to be accurately corrected, representing the synchrony of two signals after the removal of the instantaneous zero-lag component, which is characterized by several artifacts, such as volume conduction [63]. Although removing zero-lag phase synchronization could not completely remove volume conduction [89], the LPS is considered to include only physiological connectivity information and, compared to other connectivity indexes, it is also minimally affected by low spatial resolution [63,67]. For these reasons, the LPS is broadly used in clinical neurophysiology studies [62–65,88,90–92].

According to Li and coworkers [33], the triple network functional connectivity was investigated defining

9 Regions of Interest (ROIs; Table 1 and Figure 1). The LPS was calculated between all the ROIs (i.e., 81 connections) by the eLORETA, which also performed the source reconstruction [93,94]. According to previous reports [67,84], the "single nearest voxel" option (i.e., each ROI consisted of a single voxel, the closest to each seed) was chosen. In the current research, the following frequency bands were analyzed: delta (0.5–4 Hz); theta (4.5–7.5 Hz); alpha (8–13 Hz); beta (13.5–30 Hz); and gamma (30.5–60 Hz).

Table 1. eLORETA coordinates of the triple network.

Brain Network	Anatomical Structure	eLORETA MNI Coordinates [1] eLORETA Talairach Coordinates [1]		
		x	y	z
DMN	mPFC	0	55	25
		0	54	20
	PCC	0	−55	20
		0	−52	21
SN	dACC	0	20	35
		0	21	31
	Left AI	−45	15	−5
		−45	14	−5
	Right AI	50	15	−5
		50	14	−5
CEN	Left dlPFC	−45	20	35
		−45	21	31
	Right dlPFC	40	25	50
		40	27	45
	Left PPC	−40	−70	45
		−40	−66	45
	Right PPC	50	−60	40
		50	−56	40

Note: [1] coordinates referred to the ROI centroid; coordinates should be considered approximate due to the uncertain boundaries of the anatomical structures and brain activation patterns.

2.4. Statistical Analysis

EEG connectivity analyses were compared between PCU group and non-PCU group, for each frequency band, using the statistical nonparametric mapping (SnPM) methodology available in the eLORETA package. This procedure is based on the Fisher's permutation [95]. Correction of significance for multiple comparisons (i.e., between all ROIs for each frequency band) was performed using the nonparametric randomization procedure, included in the eLORETA software (for more details, see [64,73]). Briefly, this procedure computes 5000 data randomizations to determine the critical probability threshold of T-values [95,96] corresponding to a statistically corrected (i.e., after the multiple ROIs comparisons in each frequency) p-values ($p < 0.05$ and $p < 0.01$). Furthermore, the eLORETA software provides effect size thresholds for t-statistics corresponding to Cohen's d values [97]: small = 0.2, medium = 0.5, large = 0.8. Kolmogorov–Smirnov Z test and chi-squared test were performed to analyze differences between groups for continuous and dichotomous variables, respectively. The association between CAST total score and only statistically significant EEG connectivity data observed in the between-group comparison was evaluated using partial correlation (r_p) analyses, with age, sex, educational level, tobacco use, problematic alcohol use (i.e., CAGE ≥ 2), and SCL-K-9 total score as covariates. IBM SPSS Statistics for Windows, version 18.0 (Chicago, USA), has been used for the statistical analyses.

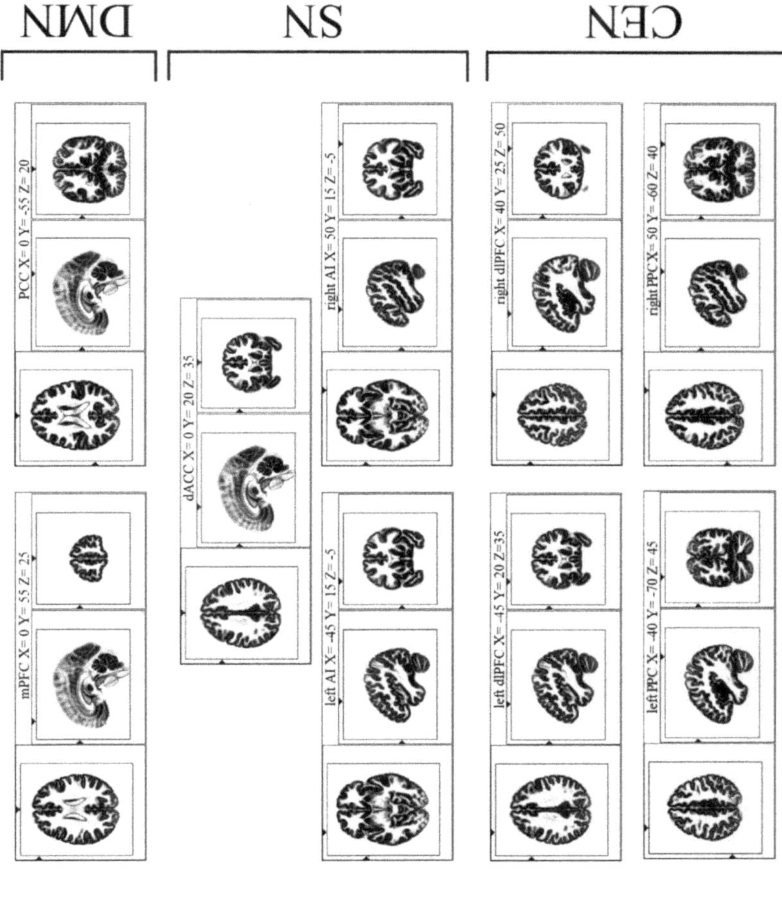

Figure 1. eLORETA ROIs of the triple network and Montreal Neurological Institute coordinates (Axial, Sagittal, and Coronal view). Abbreviations: eLORETA = exact Low Resolution Electromagnetic Tomography software; ROIs = Regions of Interests; mPFC = medial prefrontal cortex; PCC = posterior cingulate cortex; DMN = default mode network; dACC = dorsal anterior cingulate cortex; AI = anterior insula; SN = salience network; dlPFC = dorsolateral prefrontal cortex; PPC = posterior parietal cortex; CEN = central executive network.

3. Results

For all participants, suitable EEG recordings have been gained. In these recordings, no relevant modifications of the background rhythm frequency (e.g., focal abnormalities or evidence of drowsiness) were detected through a visual assessment of the EEG recordings. The average time analyzed was 248.83 ± 43.58 seconds (Min./Max.: 180/306) and 268.17 ± 38.72 seconds (Min./Max.: 180/318), respectively, for PCU and non-PCU participants (Z-test = 1.02, $p = 0.252$).

Differences between groups are reported in Table 2. No significant differences were observed for socio-demographic data or for general psychopathology, even though, compared to non-PCU, PCU participants reported more frequent tobacco use in the last 6 months, as well as more problematic alcohol use.

Table 2. Demographic and clinical data of participants (N = 36).

	PCU (N = 12)	Non-PCU (N = 24)	test	*p*
Variables				
Age–*M (SD)*	23.33 ± 3.47	21.21 ± 2.70	Z-test = 1.06	0.211
Educational level (years)–M ± SD	16.42 ± 1.51	15.54 ± 1.50	Z-test = 0.83	0.504
Men–N (%)	7 (58.3%)	9 (37.5%)	$\chi^2_1 = 1.41$	0.236
Tobacco use in the last 6 months–N (%)	8 (66.7%)	7 (29.2%)	$\chi^2_1 = 4.63$	0.031
CAST–*M (SD)*	10.25 ± 4.31	0.00 ± 0.00	-	-
CAGE–*M (SD)*	0.67 ± 1.07	0.04 ± 0.20	Z-test = 0.82	0.504
CAGE ≥ 2–N (%)	3 (25%)	0 (0%)	$\chi^2_1 = 6.55$	0.011
SCL-K-9–*M (SD)*	1.22 ± 0.97	0.73 ± 0.44	Z-test = 0.94	0.336

Note: PCU = problematic cannabis users; CAST = Cannabis Abuse Screening Test; CAGE = self-report measure of alcohol use problems; SCL-K-9 = Symptom-Checklist-K-9.

Functional Connectivity Results

In the comparison between PCU and non-PCU participants, the thresholds for significance, corrected for multiple comparisons, were T = ± 3.72 corresponding to $p < 0.05$, and T = ± 4.41, corresponding to $p < 0.01$. The effect sizes for T-threshold were 1.17, 2.92, and 4.67, corresponding, respectively, to small, medium, and large effect sizes.

Significant differences between groups were observed in delta band. PCU participants showed an increase of delta connectivity between the dACC and right PPC than non-PCU (T = 4.37, $p = 0.010$; Figure 2A). The strength of delta connectivity between the dACC and right PPC was positively and significantly correlated with the CAST total score after controlling for age, sex, educational level, tobacco use, problematic alcohol use, and general psychopathology ($r_p = 0.40$, $p = 0.030$; Figure 2B). The correlation between EEG connectivity data and CAST total score remains significant also when the seconds of analyzed EEG were added and considered ($r_p = 0.39$, $p = 0.038$).

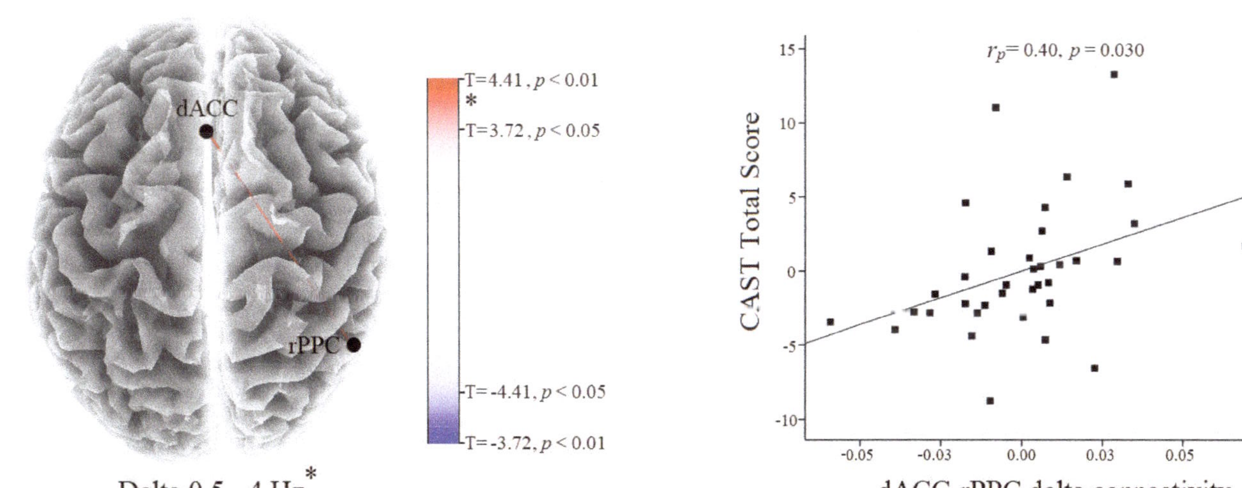

Figure 2. **Panel** (**A**) Results of the eLORETA between comparisons in delta frequency band. PCU individuals showed an increase of delta connectivity (red line) between the dACC and right PPC than non-PCU ($T = 4.37$, $p = 0.010$). **Panel** (**B**) Scatterplot of the correlation between CAST total score and delta connectivity between dACC and right PPC values adjusted for the effect of potentially competing factors (i.e., sex, age, education level, tobacco use, problematic alcohol use, and SCL-K-9 total score). Abbreviations: dACC = dorsal anterior cingulate cortex; rPPC = right posterior parietal cortex; CAST = Cannabis Abuse Screening Test; SCL-K-9 = Symptom-Checklist-K-9; PCU = problematic cannabis use.

No significant differences were detected in the other frequency bands. The most evident modifications of EEG connectivity observed in the theta band was noticed between the left anterior insula and the PCC (T = 2.45, p = 0.61). The most prominent modifications of EEG connectivity observed in the alpha band were reported between the mPFC and the PCC (T = −1.70, p = 0.99). The most relevant modifications of EEG connectivity observed in the beta band were detected between the dACC and the right PPC (T = 3.04, p = 0.23). Lastly, the most evident modifications of EEG connectivity observed in the gamma band were noticed between the left anterior insula and the PCC (T = 2.67, p = 0.45).

4. Discussion

The main aim of the current study was to investigate the association between PCU and triple network EEG functional connectivity. Compared to non-PCU, PCU participants showed an increase of delta connectivity between the SN and CEN, specifically, between the dACC and right PPC. Furthermore, SN–CN functional connectivity strength was positively correlated with CAST total score (i.e., higher connectivity was associated with higher problematic patterns of cannabis use), even when controlling for the presence of other variables (i.e., sex, age, educational level, general psychopathology, tobacco use, and problematic alcohol use). No significant association was observed among DMN hubs, suggesting that individuals with PCU could be characterized by a specific dysfunctional communication between the SN and CEN during RS.

In order to support a wide range of cognitive functions, both SN and CEN are conceptualized as task-positive networks interacting with each other [98,99]. Specifically, the SN detects and provides a selective amplification of relevant stimuli generating a top-down control input that activates the CEN in order to respond to salient information [28]. The dACC is considered a key region involved in reward-based decision making, which integrates various task-relevant stimuli and supports goal-directed behavior [100]. Furthermore, it is known that this brain area is crucial during craving-related experiences, not only in response to drug-cues [101] but also during RS condition [78]. On the other hand, the involvement of PPC in a wide range of cognitive tasks, such as attention, decision making, and episodic memory, is well documented [102].

Therefore, the increase of RS functional connectivity between the SN and CEN, detected in the present study, might reflect the tendency of PCU individuals to focus on reward-based decision making, triggered by attentional and emotional processes of cannabis-related thoughts, memories, and craving. Accordingly, this study pointed out an increase in SN–CEN connectivity in the delta band. This result is in accordance with previous neurophysiological studies reporting the involvement of delta frequency band in the brain reward system [103,104] and consequently in substance-related disorders, especially during withdrawal and craving states. For instance, the increase of frontal delta and theta power has been reported in crack-cocaine-dependent subjects during guided cocaine imagery [105] as well as in response to acute smoked cocaine self-administration [106]. Similarly, Li et al. [107] reported that delta-increased coherence between frontal and posterior regions was associated with cigarette cravings. Recently, Prashad et al. [13] also showed that cannabis users exhibited a greater cortico-cortical connectivity in both frontal and central regions in delta and theta frequencies band than noncannabis users.

Our results are not consistent with previous studies reporting functional connectivity alterations between DMN hubs and both SN and CEN nodes [30–33]. These differences could be related to several discrepancies in study designs (e.g., EEG vs. fMRI) and procedures (e.g., ROIs selection). However, it is also possible that specific substances are characterized not only by atypical neurophysiological signatures [13] but also by specific dysfunctional dynamic interactions between neural networks, which might also change according to the different behavioral states (i.e., intoxication, craving, bingeing, withdrawal, and relapse) associated with addiction [108]. This interpretation is purely hypothetical, but it could be investigated in future studies.

Although potentially interesting, the present findings should be evaluated taking into account some limits. The first limitation is the small sample size that reduces the generalizability of our findings and leads us to consider our study only as preliminary. Second, this is a cross-sectional report; thus, causal relationships between investigated variables cannot be established and should be examined through longitudinal and experimental studies. Third, our sample is composed of undergraduate students with no formal diagnosis of cannabis use disorder, which may be characterized by different EEG connectivity patterns within the triple network. Fourth, we did not assess triple network connectivity during drug cues exposure, making our interpretation specific to the RS condition (i.e., eyes closed). Furthermore, although we have excluded participants reporting psychoactive medication use and other illegal drug consumption, a formal urine toxicology screen was not performed. Lastly, it should be noted that abnormalities in grey matter have been reported in PCU, especially in long-term heavy cannabis users [9]. Therefore, although we have investigated young adults with PCU, it cannot completely be excluded that structural alterations might affect the forward modeling by means of different conduction delays and cortical thickness. Notwithstanding these limits, to the best of our knowledge, this is the first study that has examined the association between triple network EEG functional connectivity and PCU using a validated tool (i.e., eLORETA) to localize electrocortical activity and controlling for potential confounding variables.

Based on the results of the current research, future studies should design experimental paradigms using drug, compared to neutral, stimuli to broaden such findings concerning triple network connectivity during direct exposure to drug cues. Moreover, due to the association between PCU and other mental disorders [109], future research considering comorbidity with such disorders is needed to understand relationships among these variables and the neurophysiological mechanisms pointed out through this study. Lastly, future studies with larger samples, longitudinal, and/or experimental designs, and combing multimodal neuroimaging techniques, should be implemented in order to clarify long-term effects of PCU on both neurophysiological and neurocognitive point of view.

5. Conclusions

Taken together, our data would seem to suggest that individuals with PCU could be characterized by a trait-specific dysfunctional interaction between the SN and CEN (specifically between the dACC and right PPC) during RS. This result might reflect certain aspects of PCU such as attentional and emotional processes of cannabis-related thoughts, memories, and craving. Therefore, future investigations relating to the triple network model could provide novel insights into human behavior associated with addiction and substance-related disorder.

Author Contributions: Project administration: C.I. and B.F.; conceptualization: C.I., C.M., G.A.C., E.M.-R., S.M., and B.F.; methodology: C.I., C.M., G.A.C., A.P., E.M.-R., and S.M.; supervision: C.I., A.P., and B.F.; data curation, software and formal analysis: C.I., C.M., G.A.C., M.G., C.C., E.L., and B.R.Z.; writing—original draft preparation: C.I., C.M., G.A.C., and A.P.; writing—review and editing: E.M.-R., S.M., and B.F. All authors have read and agreed to the published version of the manuscript

Conflicts of Interest: The authors declare no conflicts of interest.

Abbreviations

eLORETA	exact Low Resolution Electromagnetic Tomography software
MNI	Montreal Neurological Institute
DMN	default mode network
mPFC	medial prefrontal cortex
PCC	posterior cingulate cortex
SN	salience network
dACC	dorsal anterior cingulate cortex
AI	anterior insula
CEN	central executive network
dlPFC	dorsolateral prefrontal cortex
PPC	posterior parietal cortex

References

1. European Monitoring Centre for Drugs and Drug Addiction. *European Drug Report 2019: Trends and Developments*; Publications Office of the European Union: Luxembourg, Belgium, 2019.
2. Hasin, D.S.; Kerridge, B.T.; Saha, T.D.; Huang, B.; Pickering, R.; Smith, S.M.; Jung, J.; Zhang, H.; Grant, B.F. Prevalence and correlates of dsm-5 cannabis use disorder, 2012–2013: Findings from the national epidemiologic survey on alcohol and related conditions-iii. *Am. J. Psychiatry* **2016**, *173*, 588–599. [CrossRef] [PubMed]
3. Yanes, J.A.; Riedel, M.C.; Ray, K.L.; Kirkland, A.E.; Bird, R.T.; Boeving, E.R.; Reid, M.A.; Gonzalez, R.; Robinson, J.L.; Laird, A.R.; et al. Neuroimaging meta-analysis of cannabis use studies reveals convergent functional alterations in brain regions supporting cognitive control and reward processing. *J. Psychopharmacol.* **2018**, *32*, 283–295. [CrossRef] [PubMed]
4. Hirst, R.A.; Lambert, D.G.; Notcutt, W.G. Pharmacology and potential therapeutic uses of cannabis. *Br. J. Anaesth.* **1998**, *81*, 77–84. [CrossRef] [PubMed]
5. Calabrese, E.J.; Rubio-Casillas, A. Biphasic effects of thc in memory and cognition. *Eur. J. Clin. Invest.* **2018**, *48*, e12920. [CrossRef] [PubMed]
6. Zehra, A.; Burns, J.; Liu, C.K.; Manza, P.; Wiers, C.E.; Volkow, N.D.; Wang, G.J. Cannabis addiction and the brain: A review. *J. Neuroimmune Pharmacol.* **2018**, *13*, 438–452. [CrossRef] [PubMed]
7. Bloomfield, M.A.P.; Hindocha, C.; Green, S.F.; Wall, M.B.; Lees, R.; Petrilli, K.; Costello, H.; Ogunbiyi, M.O.; Bossong, M.G.; Freeman, T.P. The neuropsychopharmacology of cannabis: A review of human imaging studies. *Pharmacol. Ther.* **2019**, *195*, 132–161. [CrossRef]
8. Nader, D.A.; Sanchez, Z.M. Effects of regular cannabis use on neurocognition, brain structure, and function: A systematic review of findings in adults. *Am. J. Drug Alcohol Abus.* **2018**, *44*, 4–18. [CrossRef]
9. Moreno-Alcazar, A.; Gonzalvo, B.; Canales-Rodriguez, E.J.; Blanco, L.; Bachiller, D.; Romaguera, A.; Monte-Rubio, G.C.; Roncero, C.; McKenna, P.J.; Pomarol-Clotet, E. Larger gray matter volume in the basal ganglia of heavy cannabis users detected by voxel-based morphometry and subcortical volumetric analysis. *Front. Psychiatry* **2018**, *9*, 175. [CrossRef]
10. Shevorykin, A.; Ruglass, L.M.; Melara, R.D. Frontal alpha asymmetry and inhibitory control among individuals with cannabis use disorders. *Brain Sci.* **2019**, *9*, 219. [CrossRef]
11. Asmaro, D.; Carolan, P.L.; Liotti, M. Electrophysiological evidence of early attentional bias to drug-related pictures in chronic cannabis users. *Addict. Behav.* **2014**, *39*, 114–121. [CrossRef]
12. Ruglass, L.M.; Shevorykin, A.; Dambreville, N.; Melara, R.D. Neural and behavioral correlates of attentional bias to cannabis cues among adults with cannabis use disorders. *Psychol. Addict. Behav.* **2019**, *33*, 69–80. [CrossRef] [PubMed]
13. Prashad, S.; Dedrick, E.S.; Filbey, F.M. Cannabis users exhibit increased cortical activation during resting state compared to non-users. *Neuroimage* **2018**, *179*, 176–186. [CrossRef] [PubMed]
14. Struve, F.A.; Patrick, G.; Straumanis, J.J.; Fitz-Gerald, M.J.; Manno, J. Possible eeg sequelae of very long duration marihuana use: Pilot findings from topographic quantitative eeg analyses of subjects with 15 to 24 years of cumulative daily exposure to thc. *Clin. Electroencephalogr.* **1998**, *29*, 31–36. [CrossRef] [PubMed]

15. Struve, F.A.; Straumanis, J.J.; Patrick, G.; Leavitt, J.; Manno, J.E.; Manno, B.R. Topographic quantitative eeg sequelae of chronic marihuana use: A replication using medically and psychiatrically screened normal subjects. *Drug Alcohol Depend.* **1999**, *56*, 167–179. [CrossRef]
16. Fingelkurts, A.A.; Kivisaari, R.; Autti, T.; Borisov, S.; Puuskari, V.; Jokela, O.; Kahkonen, S. Increased local and decreased remote functional connectivity at eeg alpha and beta frequency bands in opioid-dependent patients. *Psychopharmacology* **2006**, *188*, 42–52. [CrossRef] [PubMed]
17. Fingelkurts, A.A.; Kivisaari, R.; Autti, T.; Borisov, S.; Puuskari, V.; Jokela, O.; Kahkonen, S. Opioid withdrawal results in an increased local and remote functional connectivity at eeg alpha and beta frequency bands. *Neurosci. Res.* **2007**, *58*, 40–49. [CrossRef]
18. Zilverstand, A.; Huang, A.S.; Alia-Klein, N.; Goldstein, R.Z. Neuroimaging impaired response inhibition and salience attribution in human drug addiction: A systematic review. *Neuron* **2018**, *98*, 886–903. [CrossRef]
19. Bechara, A. Decision making, impulse control and loss of willpower to resist drugs: A neurocognitive perspective. *Nat. Neurosci.* **2005**, *8*, 1458–1463. [CrossRef]
20. Menon, V. Large-Scale brain networks and psychopathology: A unifying triple network model. *Trends Cogn. Sci.* **2011**, *15*, 483–506. [CrossRef]
21. Bressler, S.L.; Menon, V. Large-Scale brain networks in cognition: Emerging methods and principles. *Trends Cogn. Sci.* **2010**, *14*, 277–290. [CrossRef]
22. Menon, B. Towards a new model of understanding—The triple network, psychopathology and the structure of the mind. *Med. Hypotheses* **2019**, *133*, 109385. [CrossRef] [PubMed]
23. Andrews-Hanna, J.R.; Smallwood, J.; Spreng, R.N. The default network and self-generated thought: Component processes, dynamic control, and clinical relevance. *Ann. N. Y. Acad. Sci.* **2014**, *1316*, 29–52. [CrossRef] [PubMed]
24. Andrews-Hanna, J.R. The brain's default network and its adaptive role in internal mentation. *Neuroscientist* **2012**, *18*, 251–270. [CrossRef] [PubMed]
25. Goulden, N.; Khusnulina, A.; Davis, N.J.; Bracewell, R.M.; Bokde, A.L.; McNulty, J.P.; Mullins, P.G. The salience network is responsible for switching between the default mode network and the central executive network: Replication from dcm. *Neuroimage* **2014**, *99*, 180–190. [CrossRef] [PubMed]
26. Sridharan, D.; Levitin, D.J.; Menon, V. A critical role for the right fronto-insular cortex in switching between central-executive and default-mode networks. *Proc. Natl. Acad. Sci. USA* **2008**, *105*, 12569–12574. [CrossRef]
27. Seeley, W.W.; Menon, V.; Schatzberg, A.F.; Keller, J.; Glover, G.H.; Kenna, H.; Reiss, A.L.; Greicius, M.D. Dissociable intrinsic connectivity networks for salience processing and executive control. *J. Neurosci.* **2007**, *27*, 2349–2356. [CrossRef]
28. Menon, V.; Uddin, L.Q. Saliency, switching, attention and control: A network model of insula function. *Brain Struct. Funct.* **2010**, *214*, 655–667. [CrossRef]
29. Sutherland, M.T.; McHugh, M.J.; Pariyadath, V.; Stein, E.A. Resting state functional connectivity in addiction: Lessons learned and a road ahead. *Neuroimage* **2012**, *62*, 2281–2295. [CrossRef]
30. Lerman, C.; Gu, H.; Loughead, J.; Ruparel, K.; Yang, Y.; Stein, E.A. Large-Scale brain network coupling predicts acute nicotine abstinence effects on craving and cognitive function. *JAMA Psychiatry* **2014**, *71*, 523–530. [CrossRef]
31. Liang, X.; He, Y.; Salmeron, B.J.; Gu, H.; Stein, E.A.; Yang, Y. Interactions between the salience and default-mode networks are disrupted in cocaine addiction. *J. Neurosci.* **2015**, *35*, 8081–8090. [CrossRef]
32. Geng, X.; Hu, Y.; Gu, H.; Salmeron, B.J.; Adinoff, B.; Stein, E.A.; Yang, Y. Salience and default mode network dysregulation in chronic cocaine users predict treatment outcome. *Brain* **2017**, *140*, 1513–1524. [CrossRef] [PubMed]
33. Li, Q.; Liu, J.; Wang, W.; Wang, Y.; Li, W.; Chen, J.; Zhu, J.; Yan, X.; Li, Y.; Li, Z.; et al. Disrupted coupling of large-scale networks is associated with relapse behaviour in heroin-dependent men. *J. Psychiatry Neurosci.* **2018**, *43*, 48–57. [CrossRef] [PubMed]
34. Wall, M.B.; Pope, R.; Freeman, T.P.; Kowalczyk, O.S.; Demetriou, L.; Mokrysz, C.; Hindocha, C.; Lawn, W.; Bloomfield, M.A.; Freeman, A.M.; et al. Dissociable effects of cannabis with and without cannabidiol on the human brain's resting-state functional connectivity. *J. Psychopharmacol.* **2019**, *33*, 822–830. [CrossRef] [PubMed]
35. Neuner, I.; Arrubla, J.; Werner, C.J.; Hitz, K.; Boers, F.; Kawohl, W.; Shah, N.J. The default mode network and eeg regional spectral power: A simultaneous fmri-eeg study. *PLoS ONE* **2014**, *9*, e88214. [CrossRef]

36. Thatcher, R.W.; North, D.M.; Biver, C.J. Loreta eeg phase reset of the default mode network. *Front. Hum. Neurosci.* **2014**, *8*, 529. [CrossRef]
37. Srinivasan, R.; Winter, W.R.; Ding, J.; Nunez, P.L. Eeg and meg coherence: Measures of functional connectivity at distinct spatial scales of neocortical dynamics. *J. Neurosci. Methods* **2007**, *166*, 41–52. [CrossRef]
38. Whitton, A.E.; Deccy, S.; Ironside, M.L.; Kumar, P.; Beltzer, M.; Pizzagalli, D.A. Electroencephalography source functional connectivity reveals abnormal high-frequency communication among large-scale functional networks in depression. *Biol. Psychiatry Cogn. Neurosci. Neuroimaging* **2018**, *3*, 50–58. [CrossRef]
39. Legleye, S.; Karila, L.; Beck, F.; Reynaud, M. Validation of the cast, a general population cannabis abuse screening test. *J. Subst. Use* **2007**, *17*, 233–242. [CrossRef]
40. Bastiani, L.; Siciliano, V.; Curzio, O.; Luppi, C.; Gori, M.; Grassi, M.; Molinaro, S. Optimal scaling of the cast and of sds scale in a national sample of adolescents. *Addict. Behav.* **2013**, *38*, 2060–2067. [CrossRef]
41. Ewing, J.A. Detecting alcoholism. The cage questionnaire. *JAMA* **1984**, *252*, 1905–1907. [CrossRef]
42. Klaghofer, R.; Brähler, E. Konstruktion und teststatistische prüfung einer kurzform der scl-90-r (construction and test statistical evaluation of a short version of the scl-90–r). *Z. Klin. Psychol. Psychiatr. Psychother.* **2001**, *49*, 115–124.
43. Imperatori, C.; Farina, B.; Valenti, E.M.; Di Poce, A.; D'Ari, S.; De Rossi, E.; Murgia, C.; Carbone, G.A.; Massullo, C.; Della Marca, G. Is resting state frontal alpha connectivity asymmetry a useful index to assess depressive symptoms? A preliminary investigation in a sample of university students. *J. Affect. Disord.* **2019**, *257*, 152–159. [CrossRef] [PubMed]
44. Adenzato, M.; Imperatori, C.; Ardito, R.B.; Valenti, E.M.; Marca, G.D.; D'Ari, S.; Palmiero, L.; Penso, J.S.; Farina, B. Activating attachment memories affects default mode network in a non-clinical sample with perceived dysfunctional parenting: An eeg functional connectivity study. *Behav. Brain Res.* **2019**, *372*, 112059. [CrossRef] [PubMed]
45. Imperatori, C.; Farina, B.; Adenzato, M.; Valenti, E.M.; Murgia, C.; Marca, G.D.; Brunetti, R.; Fontana, E.; Ardito, R.B. Default mode network alterations in individuals with high-trait-anxiety: An eeg functional connectivity study. *J. Affect. Disord.* **2019**, *246*, 611–618. [CrossRef]
46. Imperatori, C.; Massullo, C.; Carbone, G.A.; Farina, B.; Colmegna, F.; Riboldi, I.; Giacomo, E.D.; Clerici, M.; Dakanalis, A. Electroencephalographic (eeg) alterations in young women with high subclinical eating pathology levels: A quantitative eeg study. *Eat. Weight Disord.* **2019**. [CrossRef]
47. Casajuana, C.; Lopez-Pelayo, H.; Balcells, M.M.; Miquel, L.; Colom, J.; Gual, A. Definitions of risky and problematic cannabis use: A systematic review. *Subst. Use Misuse* **2016**, *51*, 1760–1770. [CrossRef]
48. Legleye, S.; Kraus, L.; Piontek, D.; Phan, O.; Jouanne, C. Validation of the cannabis abuse screening test in a sample of cannabis inpatients. *Eur. Addict. Res.* **2012**, *18*, 193–200. [CrossRef]
49. Legleye, S.; Piontek, D.; Kraus, L. Psychometric properties of the cannabis abuse screening test (cast) in a french sample of adolescents. *Drug Alcohol Depend.* **2011**, *113*, 229–235. [CrossRef]
50. Legleye, S.; Eslami, A.; Bougeard, S. Assessing the structure of the cast (cannabis abuse screening test) in 13 European countries using multigroup analyses. *Int. J. Methods Psychiatr. Res.* **2017**, *26*. [CrossRef]
51. Cantillano, V.; Del Villar, P.; Contreras, L.; Martinez, D.; Zuzulich, M.S.; Ramirez, C.; Pons, C.; Bashford, J. Psychometric properties of the spanish version of the cannabis use problems identification test among Chilean University students: A validation study. *Drug Alcohol Depend.* **2017**, *170*, 32–36. [CrossRef]
52. Mayfield, D.; McLeod, G.; Hall, P. The cage questionnaire: Validation of a new alcoholism screening instrument. *Am. J. Psychiatry* **1974**, *131*, 1121–1123. [PubMed]
53. Dhalla, S.; Kopec, J.A. The cage questionnaire for alcohol misuse: A review of reliability and validity studies. *Clin. Invest. Med.* **2007**, *30*, 33–41. [CrossRef] [PubMed]
54. Agabio, R.; Marras, P.; Gessa, G.L.; Carpiniello, B. Alcohol use disorders, and at-risk drinking in patients affected by a mood disorder, in Cagliari, Italy: Sensitivity and specificity of different questionnaires. *Alcohol Alcohol.* **2007**, *42*, 575–581. [CrossRef] [PubMed]
55. Derogatis, L. *The Scl-90-R Manual*; Clinical Psychometric Research Unit, Johns Hopkins University School of Medicine: Baltimore, MD, USA, 1977.
56. Petrowski, K.; Schmalbach, B.; Kliem, S.; Hinz, A.; Brahler, E. Symptom-checklist-k-9: Norm values and factorial structure in a representative German sample. *PLoS ONE* **2019**, *14*, e0213490. [CrossRef]

57. Imperatori, C.; Bianciardi, E.; Niolu, C.; Fabbricatore, M.; Gentileschi, P.; Di Lorenzo, G.; Siracusano, A.; Innamorati, M. The Symptom-Checklist-K-9 (SCL-K-9) discriminates between overweight/obese patients with and without significant binge eating pathology: Psychometric Properties of an Italian version. *Nutrients* **2020**, *12*, 647. [CrossRef]
58. Imperatori, C.; Brunetti, R.; Farina, B.; Speranza, A.M.; Losurdo, A.; Testani, E.; Contardi, A.; Della Marca, G. Modification of eeg power spectra and eeg connectivity in autobiographical memory: A sloreta study. *Cogn. Process.* **2014**, *15*, 351–361. [CrossRef]
59. Imperatori, C.; Farina, B.; Brunetti, R.; Gnoni, V.; Testani, E.; Quintiliani, M.I.; Del Gatto, C.; Indraccolo, A.; Contardi, A.; Speranza, A.M.; et al. Modifications of eeg power spectra in mesial temporal lobe during n-back tasks of increasing difficulty. A sloreta study. *Front. Hum. Neurosci.* **2013**, *7*, 109. [CrossRef]
60. Imperatori, C.; Farina, B.; Quintiliani, M.I.; Onofri, A.; Castelli Gattinara, P.; Lepore, M.; Gnoni, V.; Mazzucchi, E.; Contardi, A.; Della Marca, G. Aberrant eeg functional connectivity and eeg power spectra in resting state post-traumatic stress disorder: A sloreta study. *Biol. Psychol.* **2014**, *102*, 10–17. [CrossRef]
61. Imperatori, C.; Della Marca, G.; Amoroso, N.; Maestoso, G.; Valenti, E.M.; Massullo, C.; Carbone, G.A.; Contardi, A.; Farina, B. Alpha/Theta neurofeedback increases mentalization and default mode network connectivity in a non-clinical sample. *Brain Topogr.* **2017**, *30*, 822–831. [CrossRef]
62. Canuet, L.; Tellado, I.; Couceiro, V.; Fraile, C.; Fernandez-Novoa, L.; Ishii, R.; Takeda, M.; Cacabelos, R. Resting-State network disruption and apoe genotype in Alzheimer's disease: A lagged functional connectivity study. *PLoS ONE* **2012**, *7*, e46289. [CrossRef]
63. Hata, M.; Kazui, H.; Tanaka, T.; Ishii, R.; Canuet, L.; Pascual-Marqui, R.D.; Aoki, Y.; Ikeda, S.; Kanemoto, H.; Yoshiyama, K.; et al. Functional connectivity assessed by resting state eeg correlates with cognitive decline of Alzheimer's disease—An eloreta study. *Clin. Neurophysiol.* **2016**, *127*, 1269–1278. [CrossRef] [PubMed]
64. Kitaura, Y.; Nishida, K.; Yoshimura, M.; Mii, H.; Katsura, K.; Ueda, S.; Ikeda, S.; Pascual-Marqui, R.D.; Ishii, R.; Kinoshita, T. Functional localization and effective connectivity of cortical theta and alpha oscillatory activity during an attention task. *Clin. Neurophysiol. Pract.* **2017**, *2*, 193–200. [CrossRef] [PubMed]
65. Di Lorenzo, G.; Daverio, A.; Ferrentino, F.; Santarnecchi, E.; Ciabattini, F.; Monaco, L.; Lisi, G.; Barone, Y.; Di Lorenzo, C.; Niolu, C.; et al. Altered resting-state eeg source functional connectivity in schizophrenia: The effect of illness duration. *Front. Hum. Neurosci.* **2015**, *9*, 234. [CrossRef] [PubMed]
66. Imperatori, C.; Fabbricatore, M.; Innamorati, M.; Farina, B.; Quintiliani, M.I.; Lamis, D.A.; Mazzucchi, E.; Contardi, A.; Vollono, C.; Della Marca, G. Modification of eeg functional connectivity and eeg power spectra in overweight and obese patients with food addiction: An eloreta study. *Brain Imaging Behav.* **2015**, *9*, 703–716. [CrossRef] [PubMed]
67. Pascual-Marqui, R.D.; Lehmann, D.; Koukkou, M.; Kochi, K.; Anderer, P.; Saletu, B.; Tanaka, H.; Hirata, K.; John, E.R.; Prichep, L.; et al. Assessing interactions in the brain with exact low-resolution electromagnetic tomography. *Philos. Trans. A Math. Phys. Eng. Sci.* **2011**, *369*, 3768–3784. [CrossRef]
68. Fuchs, M.; Kastner, J.; Wagner, M.; Hawes, S.; Ebersole, J.S. A standardized boundary element method volume conductor model. *Clin. Neurophysiol.* **2002**, *113*, 702–712. [CrossRef]
69. Mazziotta, J.; Toga, A.; Evans, A.; Fox, P.; Lancaster, J.; Zilles, K.; Woods, R.; Paus, T.; Simpson, G.; Pike, B.; et al. A probabilistic atlas and reference system for the human brain: International consortium for brain mapping (ICBM). *Philos. Trans. R. Soc. Lond. B Biol. Sci.* **2001**, *356*, 1293–1322. [CrossRef]
70. Jurcak, V.; Tsuzuki, D.; Dan, I. 10/20, 10/10, and 10/5 systems revisited: Their validity as relative head-surface-based positioning systems. *Neuroimage* **2007**, *34*, 1600–1611. [CrossRef]
71. Oostenveld, R.; Praamstra, P. The five percent electrode system for high-resolution eeg and erp measurements. *Clin. Neurophysiol.* **2001**, *112*, 713–719. [CrossRef]
72. Lancaster, J.L.; Woldorff, M.G.; Parsons, L.M.; Liotti, M.; Freitas, C.S.; Rainey, L.; Kochunov, P.V.; Nickerson, D.; Mikiten, S.A.; Fox, P.T. Automated talairach atlas labels for functional brain mapping. *Hum. Brain Mapp.* **2000**, *10*, 120–131. [CrossRef]
73. Hata, M.; Hayashi, N.; Ishii, R.; Canuet, L.; Pascual-Marqui, R.D.; Aoki, Y.; Ikeda, S.; Sakamoto, T.; Iwata, M.; Kimura, K.; et al. Short-Term meditation modulates eeg activity in subjects with post-traumatic residual disabilities. *Clin. Neurophysiol. Pract.* **2019**, *4*, 30–36. [CrossRef] [PubMed]
74. Jatoi, M.A.; Kamel, N.; Malik, A.S.; Faye, I. Eeg based brain source localization comparison of sloreta and eloreta. *Australas. Phys. Eng. Sci. Med.* **2014**, *37*, 713–721. [CrossRef] [PubMed]

75. Takahashi, H.; Rissling, A.J.; Pascual-Marqui, R.; Kirihara, K.; Pela, M.; Sprock, J.; Braff, D.L.; Light, G.A. Neural substrates of normal and impaired preattentive sensory discrimination in large cohorts of nonpsychiatric subjects and schizophrenia patients as indexed by mmn and p3a change detection responses. *Neuroimage* **2013**, *66*, 594–603. [CrossRef] [PubMed]
76. Michel, C.M.; Brunet, D. Eeg source imaging: A practical review of the analysis steps. *Front. Neurol.* **2019**, *10*, 325. [CrossRef]
77. De Ridder, D.; Vanneste, S.; Kovacs, S.; Sunaert, S.; Dom, G. Transient alcohol craving suppression by rtms of dorsal anterior cingulate: An fmri and loreta eeg study. *Neurosci. Lett.* **2011**, *496*, 5–10. [CrossRef]
78. Huang, Y.; Mohan, A.; De Ridder, D.; Sunaert, S.; Vanneste, S. The neural correlates of the unified percept of alcohol-related craving: A fmri and eeg study. *Sci. Rep.* **2018**, *8*, 923. [CrossRef]
79. Kirino, E. Three-Dimensional stereotactic surface projection in the statistical analysis of single photon emission computed tomography data for distinguishing between alzheimer's disease and depression. *World J. Psychiatry* **2017**, *7*, 121–127. [CrossRef]
80. Muller, T.J.; Federspiel, A.; Horn, H.; Lovblad, K.; Lehmann, C.; Dierks, T.; Strik, W.K. The neurophysiological time pattern of illusionary visual perceptual transitions: A simultaneous eeg and fmri study. *Int. J. Psychophysiol.* **2005**, *55*, 299–312. [CrossRef]
81. Horacek, J.; Brunovsky, M.; Novak, T.; Skrdlantova, L.; Klirova, M.; Bubenikova-Valesova, V.; Krajca, V.; Tislerova, B.; Kopecek, M.; Spaniel, F.; et al. Effect of low-frequency rtms on electromagnetic tomography (loreta) and regional brain metabolism (pet) in schizophrenia patients with auditory hallucinations. *Neuropsychobiology* **2007**, *55*, 132–142. [CrossRef]
82. Pizzagalli, D.A.; Oakes, T.R.; Fox, A.S.; Chung, M.K.; Larson, C.L.; Abercrombie, H.C.; Schaefer, S.M.; Benca, R.M.; Davidson, R.J. Functional but not structural subgenual prefrontal cortex abnormalities in melancholia. *Mol. Psychiatry* **2004**, *9*, 393–405. [CrossRef]
83. Zumsteg, D.; Wennberg, R.A.; Treyer, V.; Buck, A.; Wieser, H.G. H2(15)o or 13nh3 pet and electromagnetic tomography (loreta) during partial status epilepticus. *Neurology* **2005**, *65*, 1657–1660. [CrossRef] [PubMed]
84. Canuet, L.; Ishii, R.; Pascual-Marqui, R.D.; Iwase, M.; Kurimoto, R.; Aoki, Y.; Ikeda, S.; Takahashi, H.; Nakahachi, T.; Takeda, M. Resting-State eeg source localization and functional connectivity in schizophrenia-like psychosis of epilepsy. *PLoS ONE* **2011**, *6*, e27863. [CrossRef]
85. Todder, D.; Levine, J.; Abujumah, A.; Mater, M.; Cohen, H.; Kaplan, Z. The quantitative electroencephalogram and the low-resolution electrical tomographic analysis in posttraumatic stress disorder. *Clin. EEG Neurosci.* **2012**, *43*, 48–53. [CrossRef] [PubMed]
86. Toppi, J.; Borghini, G.; Petti, M.; He, E.J.; De Giusti, V.; He, B.; Astolfi, L.; Babiloni, F. Investigating cooperative behavior in ecological settings: An eeg hyperscanning study. *PLoS ONE* **2016**, *11*, e0154236. [CrossRef] [PubMed]
87. Pascual-Marqui, R.D. Coherence and phase synchronization: Generalization to pairs of multivariate time series, and removal of zero-lag contributions. *arXiv* **2007**, arXiv:0706.1776v3. Available online: http://arxiv.org/pdf/0706.1776 (accessed on 12 July 2007).
88. Olbrich, S.; Trankner, A.; Chittka, T.; Hegerl, U.; Schonknecht, P. Functional connectivity in major depression: Increased phase synchronization between frontal cortical eeg-source estimates. *Psychiatry Res.* **2014**, *222*, 91–99. [CrossRef]
89. Palva, J.M.; Wang, S.H.; Palva, S.; Zhigalov, A.; Monto, S.; Brookes, M.J.; Schoffelen, J.M.; Jerbi, K. Ghost interactions in meg/eeg source space: A note of caution on inter-areal coupling measures. *Neuroimage* **2018**, *173*, 632–643. [CrossRef]
90. Pagani, M.; Di Lorenzo, G.; Verardo, A.R.; Nicolais, G.; Monaco, L.; Lauretti, G.; Russo, R.; Niolu, C.; Ammaniti, M.; Fernandez, I.; et al. Neurobiological correlates of emdr monitoring—An eeg study. *PLoS ONE* **2012**, *7*, e45753. [CrossRef]
91. Takamiya, A.; Hirano, J.; Yamagata, B.; Takei, S.; Kishimoto, T.; Mimura, M. Electroconvulsive therapy modulates resting-state eeg oscillatory pattern and phase synchronization in nodes of the default mode network in patients with depressive disorder. *Front. Hum. Neurosci.* **2019**, *13*, 1. [CrossRef]
92. Ramyead, A.; Kometer, M.; Studerus, E.; Koranyi, S.; Ittig, S.; Gschwandtner, U.; Fuhr, P.; Riecher-Rossler, A. Aberrant current source-density and lagged phase synchronization of neural oscillations as markers for emerging psychosis. *Schizophr. Bull.* **2015**, *41*, 919–929. [CrossRef]

93. Pascual-Marqui, R.D.; Michel, C.M.; Lehmann, D. Low resolution electromagnetic tomography: A new method for localizing electrical activity in the brain. *Int. J. Psychophysiol.* **1994**, *18*, 49–65. [CrossRef]
94. Pascual-Marqui, R.D.; Michel, C.M.; Lehmann, D. Segmentation of brain electrical activity into microstates: Model estimation and validation. *IEEE Trans. Biomed. Eng.* **1995**, *42*, 658–665. [CrossRef] [PubMed]
95. Nichols, T.E.; Holmes, A.P. Nonparametric permutation tests for functional neuroimaging: A primer with examples. *Hum. Brain Mapp.* **2002**, *15*, 1–25. [CrossRef] [PubMed]
96. Winkler, A.M.; Webster, M.A.; Brooks, J.C.; Tracey, I.; Smith, S.M.; Nichols, T.E. Non-Parametric combination and related permutation tests for neuroimaging. *Hum. Brain Mapp.* **2016**, *37*, 1486–1511. [CrossRef]
97. Cohen, J. *Statistical Power Analysis for the Behavioral Sciences*, 2nd ed.; Erlbaum: Hillsdale, NJ, USA, 1988.
98. Cocchi, L.; Zalesky, A.; Fornito, A.; Mattingley, J.B. Dynamic cooperation and competition between brain systems during cognitive control. *Trends Cogn. Sci.* **2013**, *17*, 493–501. [CrossRef]
99. Elton, A.; Gao, W. Divergent task-dependent functional connectivity of executive control and salience networks. *Cortex* **2014**, *51*, 56–66. [CrossRef]
100. Heilbronner, S.R.; Hayden, B.Y. Dorsal anterior cingulate cortex: A bottom-up view. *Annu. Rev. Neurosci.* **2016**, *39*, 149–170. [CrossRef]
101. Kuhn, S.; Gallinat, J. Common biology of craving across legal and illegal drugs—A quantitative meta-analysis of cue-reactivity brain response. *Eur. J. Neurosci.* **2011**, *33*, 1318–1326. [CrossRef]
102. Sestieri, C.; Shulman, G.L.; Corbetta, M. The contribution of the human posterior parietal cortex to episodic memory. *Nat. Rev. Neurosci.* **2017**, *18*, 183–192. [CrossRef]
103. Knyazev, G.G. Motivation, emotion, and their inhibitory control mirrored in brain oscillations. *Neurosci. Biobehav. Rev.* **2007**, *31*, 377–395. [CrossRef]
104. Knyazev, G.G. Eeg delta oscillations as a correlate of basic homeostatic and motivational processes. *Neurosci. Biobehav. Rev.* **2012**, *36*, 677–695. [CrossRef] [PubMed]
105. Reid, M.S.; Prichep, L.S.; Ciplet, D.; O'Leary, S.; Tom, M.; Howard, B.; Rotrosen, J.; John, E.R. Quantitative electroencephalographic studies of cue-induced cocaine craving. *Clin. Electroencephalogr.* **2003**, *34*, 110–123. [CrossRef]
106. Reid, M.S.; Flammino, F.; Howard, B.; Nilsen, D.; Prichep, L.S. Topographic imaging of quantitative eeg in response to smoked cocaine self-administration in humans. *Neuropsychopharmacology* **2006**, *31*, 872–884. [CrossRef] [PubMed]
107. Li, X.; Ma, R.; Pang, L.; Lv, W.; Xie, Y.; Chen, Y.; Zhang, P.; Chen, J.; Wu, Q.; Cui, G.; et al. Delta coherence in resting-state eeg predicts the reduction in cigarette craving after hypnotic aversion suggestions. *Sci. Rep.* **2017**, *7*, 2430. [CrossRef] [PubMed]
108. Parvaz, M.A.; Alia-Klein, N.; Woicik, P.A.; Volkow, N.D.; Goldstein, R.Z. Neuroimaging for drug addiction and related behaviors. *Rev. Neurosci.* **2011**, *22*, 609–624. [CrossRef]
109. Satre, D.D.; Bahorik, A.; Zaman, T.; Ramo, D. Psychiatric disorders and comorbid cannabis use: How common is it and what is the clinical impact? *J. Clin. Psychiatry* **2018**, *79*. [CrossRef]

Article

Frontal Alpha Asymmetry and Inhibitory Control among Individuals with Cannabis Use Disorders

Alina Shevorykin [1,*], Lesia M. Ruglass [2] and Robert D. Melara [3]

1. Pace University, Department of Mental Health Counseling and Psychology, 861 Bedford Road, Marks 22, Pleasantville, NY 10570, USA
2. Rutgers University, Center of Alcohol and Substance Use Studies, Graduate School of Applied and Professional Psychology, 607 Allison Road, Smithers Hall, 222, Piscataway, NJ 08854, USA
3. The City College of New York, CUNY, Department of Psychology, 160 Convent Avenue, NAC 7120, New York, NY 10031, USA

* Correspondence: alina113@gmail.com; Tel.: +1-347-522-0852

Received: 29 June 2019; Accepted: 28 August 2019; Published: 29 August 2019

Abstract: To better understand the biopsychosocial mechanisms associated with development and maintenance of cannabis use disorder (CUD), we examined frontal alpha asymmetry (FAA) as a measure of approach bias and inhibitory control in cannabis users versus healthy nonusers. We investigated: (1) whether FAA could distinguish cannabis users from healthy controls; (2) whether there are cue-specific FAA effects in cannabis users versus controls; and (3) the time course of cue-specific approach motivation and inhibitory control processes. EEG data were analyzed from forty participants (CUD (n = 20) and controls (n = 20)) who completed a modified visual attention task. Results showed controls exhibited greater relative right hemisphere activation (indicating avoidance/withdrawal motivation) when exposed to cannabis cues during the filtering task. By contrast, cannabis users exhibited greater relative left activation (approach) to all cues (cannabis, positive, negative, and neutral), reflecting a generalized approach motivational tendency, particularly during later stages of inhibitory control processes. The difference between cannabis users and controls in FAA was largest during mid- to late processing stages of all cues, indicating greater approach motivation during later stages of information processing among cannabis users. Findings suggest FAA may distinguish cannabis users from healthy controls and shows promise as a measure of inhibitory control processes in cannabis users.

Keywords: cannabis use disorder; cue reactivity; craving; inhibitory control; frontal alpha asymmetry; EEG; cannabinoids

1. Introduction

Among individuals age 12 and older, cannabis is the most widely used illicit drug, with an estimated 24 million people reporting past month use in 2016 [1]. The largest increases from 2002 to 2016 were among adults age 26 and older [1]. Studies suggest an increase in permissive attitudes towards cannabis use and reduced perceptions of cannabis-related harm may underlie increases in cannabis use among adults [2,3]. However, studies also show that both short-term and long-term/chronic cannabis use are associated with detrimental psychological and physical effects. For example, acute negative effects may include impairments in attention, short-term memory, and motor coordination, increasing risk for accidental injuries [4,5]. Long-term or heavy use of cannabis may increase the risk for developing a cannabis use disorder (CUD) [6], which may contribute to lasting structural and functional brain changes. The potency of cannabis has also been increasing significantly over the past decade and serves as a significant risk factor for the onset of CUD symptoms [7,8]. Data from two waves of the National Epidemiological Survey on Alcohol and Related Conditions [9] revealed that, among cannabis users, three out of 10 evidenced a Diagnostic and Statistical Manual of Mental Disorders, 4th Edition (DSM-IV) past-year CUD, with rates of CUD doubling from 2001–2002 to 2012–2013. Given these increases and associated consequences, there is continued need to better understand the underlying processes associated with the development and maintenance of CUD, which is critical to the development of novel interventions [2].

Evidence suggests there are several key brain-based mechanisms involved in the maintenance of CUD, including drug cue reactivity, attentional bias, craving, and inhibitory control deficits—all of which underlie drug seeking, consumption, and relapse among those trying to abstain [10,11]. The dual process theory of addictive behaviors integrates these mechanisms by arguing that the approach (appetitive) system and self-regulatory (executive function/control) systems become imbalanced throughout the addiction process [12–14]. The approach-oriented (appetitive) system underlies the automatic behavioral tendency to approach one's drug of choice (i.e., approach bias). This automatic approach tendency is presumed to be a function of sensitization that occurs as a result of repeated and persistent drug use contributing to a heightened response to drugs and conditioned drug cues [15,16]. In tandem, a deficit in the executive control system, and reduced inhibitory control in particular, makes it difficult to resist or inhibit the impulse to approach and use drugs [13,14,17].

Over the last several decades, an emerging body of research has positioned frontal alpha asymmetry (FAA), measured via electroencephalogram (EEG), as a promising neural index of the approach motivational system [18–20]. FAA is the difference between left and right alpha activity/activation in the frontal cortical areas of the brain. FAA has been widely studied as a psychophysiological index in research on motivation, cognition, and psychopathology [21–24]. Alpha activity is cortical EEG activity in the alpha frequency band (8–13 Hz) recorded over a period of time and is believed to reflect an individual's tendency or predisposition to engage in certain motivational or emotional responses [25]. Alpha activation, on the other hand, is a task-related change in alpha activity [25], which has been investigated in relationship to current emotions and behaviors [26,27].

The approach/withdrawal motivation model of EEG asymmetry suggests that greater relative left frontal activity is associated with approach-related tendencies, and greater relative right activity is associated with withdrawal-related tendencies [28]. A majority of the studies to date have examined the link between FAA and various psychopathological conditions, including depression [21,29–31] and ADHD [32,33]. Overall, these studies lend support to the motivational theory that both a behavioral inhibition system (BIS) and a behavioral activation system (BAS) drive our emotions and behaviors. Evidence suggests that a novel/aversive stimulus results in the organization of cognitive resources for removal or rejection of the stimulus (i.e., behavioral inhibition), while an incentive/appetitive stimulus results in the organization of cognitive resources to attain the desired stimulus (i.e., behavioral approach) [34].

Despite its relevance for understanding addictive behaviors, there has been a dearth of studies examining FAA among those who use alcohol/drugs or have a substance use disorder. In one of the few studies examining FAA among substance using individuals, Gable and colleagues [35], in a nonclinical sample of 42 college students who reported alcohol use in the past month, found that, among those with high impulsivity, there was relatively greater left frontal alpha asymmetry in response to alcohol cues. The authors speculated that the inhibitory control system may serve a regulatory function in the neural response to alcohol cues. Likewise, Bowley and colleagues [36] found a trend among college students for left frontal activity enhancement after exposure to beer stimuli, suggesting enhanced approach motivation [36]. Others have also investigated the role of FAA in areas such as attentional narrowing and inhibitory control in alcohol-related cognitions [35], as well as craving and cue reactivity in nicotine dependence [37]. A majority of these studies, however, were conducted with college students with varying levels of alcohol use or with individuals with nicotine dependence. Thus, the findings may not be generalizable to those with other substance use or use disorders such as CUD. To our knowledge, there have been no studies to date that have examined FAA among those with CUD. Moreover, the studies reviewed have all examined FAA averaged over time by stimulus; it thus remains unclear whether approach motivation and inhibitory control processes are static or shifting over time during the course of information processing [32].

The aims of our study were thus threefold: (1) to examine whether FAA could distinguish individuals with CUD from healthy controls; (2) to examine if there are cue-specific FAA effects in cannabis users versus healthy controls; and (3) to determine the time course of approach motivation and inhibitory control processes during processing of cannabis cues. Given the dearth of studies examining the processes underlying CUD, this study represents a novel and important area of research, with potential implications for our understanding of the neurobiological mechanisms of CUD with implications for development of treatment interventions.

2. Materials and Methods

Data for this analysis came from a recently completed lab-based experimental study that utilized EEG and ERP (event-related potential) to examine the time course of attentional bias and cue reactivity among individuals with CUD compared to healthy controls. See Ruglass et al. [38] for details. The parent study examined ERPs as indices of attentional bias to cannabis cues in cannabis users. By contrast, the current study leveraged the EEG data collected to examine FAA as a measure of approach motivation in cannabis users.

2.1. Participants

Forty participants were recruited from printed flyers, online advertisements, and word of mouth (see Table 1 for demographic characteristics). For full details on inclusion and exclusion criteria, see Ruglass et al. [38]. The current study included cannabis smokers (n = 20, M_{age} = 26.2, SD = 8.53) who were physically healthy English-speaking adults and were diagnosed with current CUD (abuse or dependence). Similarly, the study included healthy controls (n = 20, M_{age} = 28, SD = 10.87) who were physically healthy English-speaking adults and did not meet criteria for any current or past psychiatric or substance use disorders according to the DSM-IV [39]. Participants in the CUD group were excluded if they had any other current or past psychiatric disorder, or a positive drug test for any substance other than cannabis. Participants were excluded from the healthy control group if they had a positive drug test for any drug. Participants were also excluded from the study if they reported suicidality or homicidality, history of seizures, organic mental syndrome, or they refused to be audio recorded. All participants self-reported normal or corrected normal visual acuity. Informed consent was obtained from each participant. The Institutional Review Board of the City University of New York reviewed and approved all materials and procedures. Participants were compensated $100 in cash for their time, effort, and transportation for both sessions.

Table 1. Demographic characteristics.

Variable	CUD (n = 20) M (SD) or %	Controls (n = 20) M (SD) or %	t statistics/χ^2
Baseline Session			
Age	26.2 (8.53)	28 (10.87)	$t = 0.58, p = 0.56$
Education[1]	13.85 (1.63)	14.85 (1.35)	$t = 2.11, p = 0.041$
Sex (% males)	80%	75%	$\chi^2 = 0.143, p = 0.705$
Marital Status (% Single)	100% (n = 20)	95% (n = 19)	$\chi^2 = 2.105, p = 0.35$

Table 1. *Cont.*

Variable	CUD (n = 20) M (SD) or %	Controls (n = 20) M (SD) or %	t statistics/χ^2
	Race/Ethnicity		
Black/African American	50% (n = 10)	45% (n = 9)	$\chi^2 = 1.26, p = 0.74$
Hispanic/Latino	30% (n = 6)	25% (n = 5)	
White	20% (n = 4)	25% (n = 5)	
Other	0% (n = 0)	5% (n = 1)	
	Employment		
Full-time	45% (n = 9)	25% (n = 5)	$\chi^2 = 6.56, p = 0.161$
Part-time	30% (n = 6)	50% (n = 10)	
Student	15% (n = 3)	25% (n = 5)	
Unemployed	10% (n = 2)	0%	
	Cannabis Use		
Past week use of cannabis (# of days)	6.4 (1.85)	N/A	N/A
Past 90 days use of cannabis (# of joints)	246.1 (183.35)	N/A	N/A

[1] $p < 0.05$.

2.2. Materials and Procedure

After completing a phone screen, participants completed a baseline interview and an experimental session on two separate days. Participants first completed a urine toxicology screen and were administered a series of diagnostic and clinical measures during both sessions (see Ruglass et al. [38] for details). During the experimental session, while inside an electrically and acoustically shielded Industrial Acoustics Company (New York, USA) chamber, each participant completed 24 blocks of experimental trials (and one or more blocks of practice trials) in a modified version of the visual flanker task [40] called the temporal flanker paradigm (see Figure 1) while their electroencephalographic (EEG) responses were recorded. The visual flanker task is a traditional cognitive method for measuring the extent to which distractor stimuli draw attention away from and affect the processing of target stimuli [41]. The current study investigated whether there is a difference in attentional control to various cues between cannabis users and nonusers as measured by FAA elicited in a version of the visual flanker task called the temporal flanker paradigm. The experimental session lasted approximately 3 hours, including EEG preparation and short breaks.

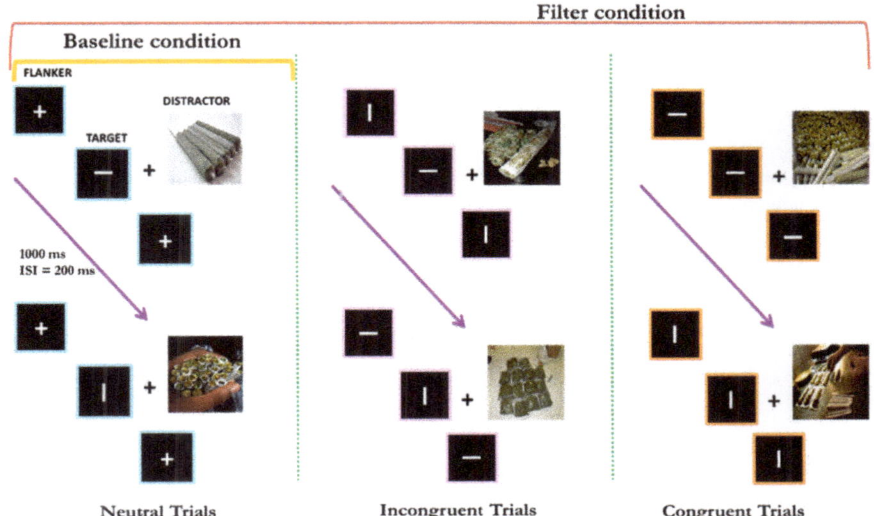

Figure 1. Modified flanker task, made up of a block of 80 trials, each consisting of a fixation square followed by the first flanker, target, and second flanker (stimulus displays), presented sequentially for 150 ms separated by a random interstimulus interval (153–390 ms). The target was represented by a vertical or horizontal line, or a cross superimposed on a cannabis-related picture or a neutral, positive, or negative image from the International Affective Picture System (IAPS) [42].

Stimuli were created in Presentation®(Neurobehavioral Systems). Each task in the visual flanker paradigm was made up of a block of 80 trials, each consisting of a fixation square (0.67°) followed by a first flanker, target, and second flanker (stimulus displays), presented sequentially for 150 ms separated by a random interstimulus interval (153–390 ms), see Figure 1. The target was represented by a vertical or horizontal line, or a cross superimposed on one of the following images: a cannabis-related picture, or a colored picture from the International Affective Picture System (IAPS) [42] that was pre-judged to be neutral (e.g., door scene), positive (e.g., beach scene), or negative (e.g., natural disaster scene). IAPS pictures' valence and arousal ratings range from 1 (low pleasure or low arousal, respectively) to 9 (high pleasure or high arousal, respectively). Neutral images were considered those with a mean rating of five. Average valence/arousal of IAPS stimuli included in this study were 5.20/3.07 for neutral cues, 7.23/5.24 for positive cues, and 2.81/5.50 for negative cues. Cannabis-related pictures were collected from free online sources and included images of cannabis, joints, and cannabis paraphernalia (e.g., pipe, rolling paper). Cannabis pictures were matched with IAPS pictures in terms of size. Line stimuli, subtending a visual angle of 0.57°, appeared in gray on a black background; cues subtended 10.85° (V) × 9.74° (H) of the visual angle.

Overall, each participant completed 24 blocks of trials, consisting of 3 repeating sets of 8 blocks: 4 baseline and 4 filtering for each image type (neutral, positive, negative, and cannabis). The order of blocks was balanced across participants. Each block consisted of 80 trials, in which participants were asked to respond by clicking the mouse key as quickly and accurately as possible to the orientation of the target line, while ignoring the flanker lines and the surrounding image. Assignment of line orientation to response keys was counterbalanced across participants, and tasks were divided into baseline and filtering [43].

In the baseline task, distractors were held constant across trials, such that two crosses always flanking the target and a single image (cue), drawn from one of four cue sets (neutral, positive, negative, and cannabis), appearing on each trial in a block. Distractors were held constant in order to create a baseline measurement of attention with minimal distraction. Participants immediately repeated the task if they did not reach the required 80% level of accuracy for the baseline condition. For the filtering task, distractors appeared randomly, such that target and flanker lines matched in terms of direction (congruent trials) in 40% of trials (32 of 80) and did not match in terms of direction (incongruent trials) in 40% of trials (32 of 80), while in 20% of trials (16 of 80), the flankers were crosses (neutral trials). The use of congruent and incongruent distractors in the filtering task introduces stimulus conflict as an attentional requirement of the task, thereby creating a measurement of inhibitory control when compared to participants' baseline performance. Participants immediately repeated the task if they did not reach the required 60% level of accuracy in the filtering condition. All eight images from one of the four cue sets (neutral, positive, negative, cannabis) appeared randomly an equal number of times during each block of filtering trials, creating four distinct filtering tasks. See Ruglass et al. [38] for further details.

2.3. Data Recording and Analysis

EEG recordings were collected, using an ANT neuro system (ANT, Philadelphia, PA, USA) in a high-density (128 electrodes) montage arranged in a cap, continuously at a sampling rate of 512 Hz. An electrooculogram (EOG) was used to monitor blinks and other eye movements from two electrode montages, one electrode was placed on the infra- and supra-orbital ridges of the right eye (VEOG), and the other was placed on the outer canthi of each eye (HEOG). Trials in which mastoid activity was greater than 100 μV were excluded. Trials with excessive blinks, eye movements, or other movement artifacts were defined as z-values on the VEOG, HEOG, and lowermost scalp channels exceeding 4.5 in a frequency band between 1 and 140 Hz; a MATLAB routine [44] was used to remove artifact trials.

Stimulus-locked waveforms (sweep time = 2000 ms) were referenced to linked mastoids band-pass filtered between 0.1 and 30 Hz. Induced alpha power (8–13 Hz) was extracted using a Morlet wavelet transform (spectral bandwidth = 6–8 Hz; wavelet duration = 80–106 ms) individually on each trial of each task (baseline, filtering) to each cue (neutral, cannabis, positive, negative) in each of three time epochs (0–200, 201–400, and 401–800 ms), separately for each of three pairs of lateral electrode locations: frontal (F7, F8), midfrontal (F3, F4), and midline frontocentral (FC3, FC4). Alpha power was log transformed [45] to derive a composite measure of FAA [25,46,47], as follows:

$$\text{FAA} = (\ln[\alpha F8] + \ln[\alpha F4] + \ln[\alpha FC4])/3 - (\ln[\alpha F7] + \ln[\alpha F3] + \ln[\alpha FC3])/3 \quad (1)$$

where α is induced alpha power at the corresponding frontal electrode locations. Higher FAA scores indicate relatively higher left cortical activity [48].

We performed mixed model analyses of variance (ANOVAs) on FAA scores using Statistica®software, with Group (2 levels: cannabis smokers, healthy controls) as the between-subjects factor and Task (2 levels: baseline, filtering), Cue (4 levels: neutral, cannabis, positive, negative), Frequency (6 levels: 8–13 Hz), and Epoch (3 levels: early, middle, and late) as within-subject factors. To guard against violations of the sphericity assumption with repeated-measures data, all main effects and interactions reported as significant were reliable after Greenhouse–Geisser correction [49].

3. Results

ANOVA of FAA uncovered a significant main effect of Task, $F(1,38) = 19.00$, $p < 0.001$, MSe = 0.07, $\eta^2 = 0.22$. FAA was significantly more positive in the baseline task (0.06) than in the filtering task (0.03). There was also a main effect of Frequency, $F(5,190) = 9.07$, $p < 0.001$, MSe = 0.01, $\eta^2 = 0.05$. Post-hoc analysis (Newman–Keuls, 0.05 criterion level) revealed less left-sided activation at 8 Hz (low alpha) than at the other five alpha frequencies. Moreover, the difference in FAA between baseline and filtering tasks was larger at lower alpha (8–9 Hz) than higher alpha (12–13 Hz), $F(5,190) = 2.52$, $p = 0.03$, MSe = 0.001, $\eta^2 = 0.002$, particularly at the later time epochs, $F(10,380) = 7.75$, $p < 0.001$, MSe = 0.0002, $\eta^2 = 0.003$. As shown in Figure 2, the larger task difference at low alpha was especially prominent to cannabis cues relative to neutral, positive, and negative cues, leading to a significant Task x Cue x

Frequency interaction F(15,570) = 2.63, $p < 0.001$, MSe = 0.001, $\eta^2 = 0.002$. Furthermore, the effect of cannabis cues on low alpha during the filtering task was evident only in nonusers, creating a Group × Task × Cue × Frequency interaction, F(15,570) = 1.92, $p = 0.02$, MSe = 0.001, $\eta^2 = 0.003$.

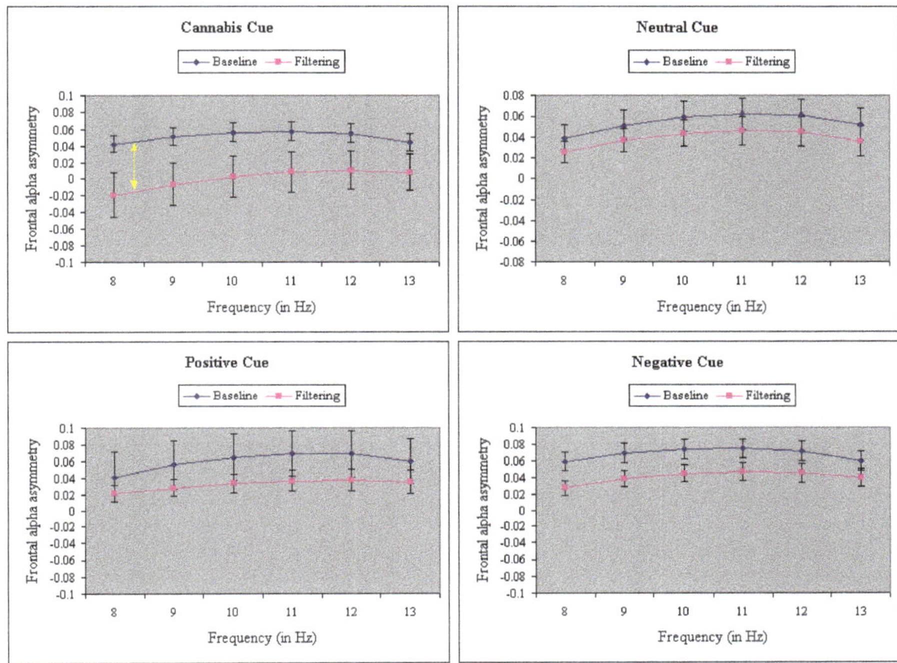

Figure 2. A significant three-way interaction of task (baseline vs. filtering), cue (cannabis, neutral, positive, and negative) and frequency (8–13 Hz) on frontal alpha asymmetry. Larger task difference at low alpha was especially prominent to cannabis cues relative to neutral, positive, and negative cues. Error bars represent the standard error of the mean.

Figure 3 depicts the four-way interaction in a pairing of the neutral and cannabis cue conditions only. As one can see, only to cannabis cues in control participants during filtering does FAA show right-sided hemisphere (avoidance) activation. Control participants show left-sided (approach) activation in the neutral, positive, and negative conditions, while cannabis users show left-sided activation (approach) in all conditions (neutral, positive, negative, and cannabis). As shown in Figure 4, the group difference in FAA was largest during the middle (201–400 ms) and late time epochs (401–800 ms), F(2,76) = 3.23, $p < 0.05$, MSe = 0.07, $\eta^2 = 0.07$.

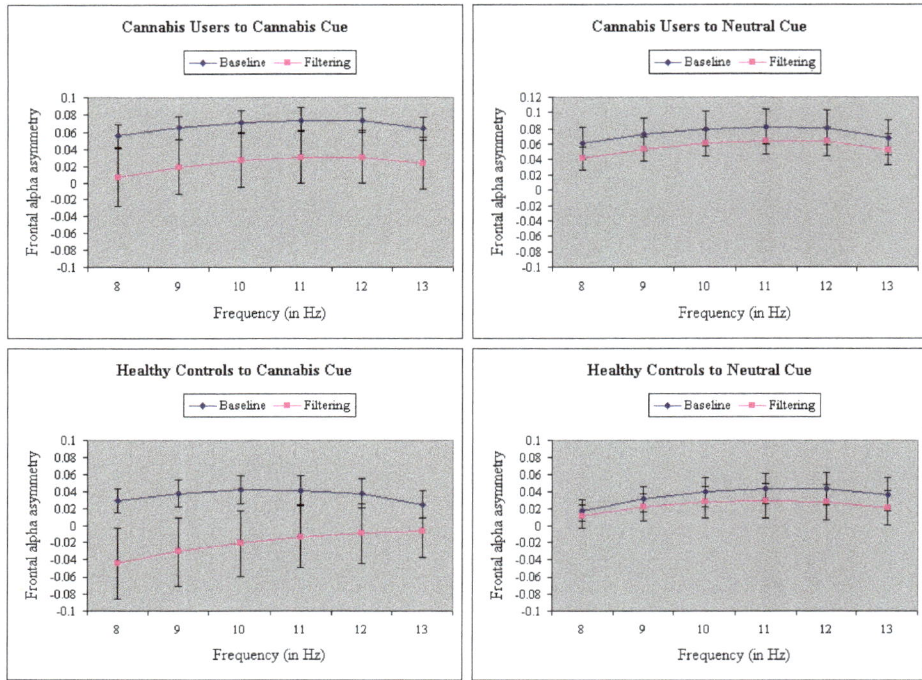

Figure 3. A significant four-way interaction in group (cannabis users vs. controls), task (baseline vs. filtering), cues (cannabis vs. neutral) and frequency (8–13 Hz) on frontal alpha asymmetry. Control participants show right-sided hemisphere (avoidance) frontal alpha asymmetry (FAA) activation only to cannabis cues during the filtering task, and they show left-sided (positive) activation in the other conditions, while cannabis users show left-sided activation (approach) in all conditions. Error bars represent the standard error of the mean.

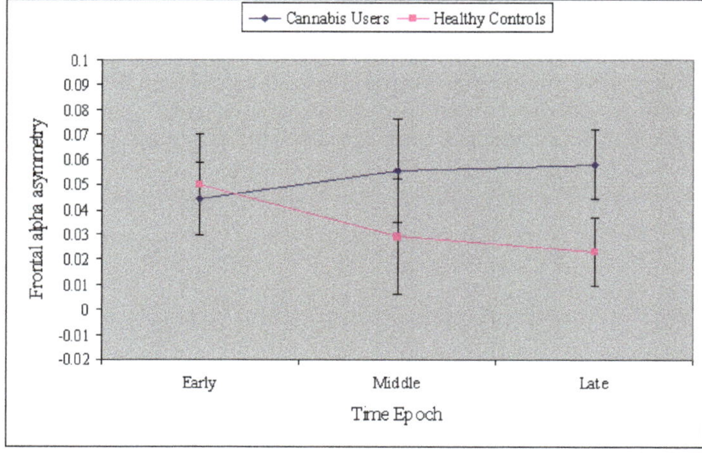

Figure 4. Group difference (cannabis users vs. controls) in FAA. The largest group difference was during the middle and late time epochs. Error bars represent the standard error of the mean. See the online article for the color version of this figure.

4. Discussion

The current study examined three related questions: (1) whether FAA could distinguish individuals with cannabis use disorders from healthy controls; (2) whether there were cue-specific FAA effects; and (3) when in the time course after cue presentation (early, middle, or late) the differences in FAA (an index of approach motivation and inhibitory control processes) are greatest between individuals with cannabis users and healthy controls.

Results revealed that the healthy controls/nonusers exhibited greater relative right hemisphere activation (typically indicative of avoidance/withdrawal motivation; Davidson, 1993; Davidson et al., 1990) when exposed to cannabis cues during the filtering task, especially during early frequencies. By contrast, cannabis users showed greater relative left activation (indicative of approach motivation) during all conditions (neutral, positive, negative, and cannabis). It is possible that healthy controls withdraw their attention to potent (cannabis) distractors as a way to enhance inhibitory control and improve performance. Healthy controls may also perceive cannabis cues as unpleasant or aversive (particularly when there are multiple cannabis cues being presented randomly), triggering the withdrawal/avoidance system. These findings are consistent with the theory of behavioral inhibition system (BIS) [34] and may reflect participants' organization of cognitive resources for avoidance of aversive stimuli.

Counter to expectations, the cannabis cues did not elicit greater approach motivation among cannabis users. Instead, cannabis users exhibited greater relative left hemisphere activation (approach) across all conditions (neutral, positive, negative, and cannabis and neutral cues), suggesting positive feelings and/or high engagement with all stimuli. Cannabis users also evidenced higher FAA (greater relative left activation) compared to the healthy controls during the middle and late stages of processing of all cues, reflecting a generalized approach motivational tendency, particularly during the later stages of inhibitory control processes. It is possible that, due to an altered reward processing system secondary to structural (particularly in the prefrontal cortex) and functional changes in the brain as a result of long-standing and continual cannabis use [50], the cues elicited a generalized approach tendency among cannabis users.

Conversely, it is possible that the cannabis cues were not salient enough for the cannabis users—particularly during the latter phases of cue processing where more consciously controlled inhibitory processes are at play—to activate greater approach-related tendencies above and beyond those activated for the neutral, positive, or negative cues. It is also possible that the neutral cues were not perceived as "neutral" by our CUD participants, contributing to similar levels of left activation. Indeed, FAA may index both motivational (approach versus avoid) and affective processes (positive or negative) [18]. Thus, cannabis users' greater relative left activation to all cues suggests more positive feelings and greater engagement with all stimuli. Studies also suggest that individual differences in substance use duration and severity and trait impulsivity may play important roles in the degree to which there is greater left activation to appetitive cues (e.g., substance use cues) compared to neutral cues [51–53]. For example, Mechin and colleagues (2016) found a positive correlation between trait impulsivity and left activation to alcohol cues, even after controlling for recent drinking behaviors, highlighting the importance of examining individual difference variables. Future studies that examine personality differences such as impulsivity, among cannabis users and level of severity of cannabis use disorder and their influence on FAA are critical to further understand FAA processes among this population.

Limitations

A few limitations should be mentioned. The generalizability of our findings is limited to nontreatment-seeking individuals with CUD. Moreover, our small sample size may have limited our ability to detect significant effects among cannabis users. Future research is needed with a larger sample size and follow-up timepoints for replication of findings, examination of individual differences in the associations between cue exposure and FAA, as well as determination of whether greater relative left FAA predicts increased craving and future cannabis use among those with cannabis use disorders.

5. Conclusions

Despite limitations, this study is one of the first to investigate FAA among those with cannabis use disorder and contributes to the growing literature on the relationship between FAA and motivational or inhibitory control processes, especially in substance users, by highlighting the usefulness of FAA as a measure of motivational processes. Methodological strengths include the careful matching of controls to cannabis users, the lack of psychiatric and other drug use comorbidities, biomedical verification of cannabis and other substance use, as well as measurement of neural activity via EEG. Results suggest that FAA holds significant promise as a measure of attentional and motivational processes in cannabis users, with promising areas of future research, including utilizing FAA as a transdiagnostic marker that distinguishes cannabis users from healthy controls; FAA as a measure or mechanism of cue reactivity and specifically the impact of cues on attention; and the potential for attentional or approach-bias modification training to influence motivation and inhibitory control processes in substance users.

Author Contributions: Conceptualization, A.S., L.M.R., and R.D.M.; Methodology, A.S., L.M.R., and R.D.M.; Software, A.S. and R.D.M.; Validation, A.S., L.M.R., and R.D.M.; Formal analysis, A.S., L.M.R., and R.D.M.; Investigation, A.S., L.M.R., and R.D.M.; Writing—original draft preparation, A.S., L.M.R., and R.D.M.; writing—review and editing, A.S., L.M.R., and R.D.M.; Visualization, A.S. and R.D.M.; Project administration, A.S.; Funding acquisition, L.M.R.

Funding: This research was funded by grants from the City College of New York City Seeds Fund and the American Psychological Association's Office of Ethnic Minority Affairs (PI: Lesia M. Ruglass).

Acknowledgments: The content is solely the responsibility of the authors and does not necessarily represent the official views of the City College of New York and the American Psychological Association.

Conflicts of Interest: The authors declare no conflict of interest.

References

1. Azofeifa, A.; Schauer, G.; McAfee, T.; Grant, A.; Lyerla, R.; Mattson, M.E. National Estimates of Marijuana Use and Related Indicators—National Survey on Drug Use and Health, United States, 2002–2014. *MMWR Surveill. Summ.* **2016**, *65*, 1–28. [CrossRef] [PubMed]
2. Carliner, H.; Brown, Q.L.; Sarvet, A.L.; Hasin, D.S. Cannabis use, attitudes, and legal status in the U.S.: A review. *Prev. Med.* **2017**, *104*, 13–23. [CrossRef] [PubMed]
3. Okaneku, J.; Vearrier, D.; McKeever, R.G.; LaSala, G.S.; Greenberg, M.I. Change in perceived risk associated with marijuana use in the United States from 2002 to 2012. *Clin. Toxicol.* **2015**, *53*, 151–155. [CrossRef] [PubMed]
4. Volkow, N.D.; Baler, R.D.; Compton, W.M.; Susan, R.B. Adverse Health Effects of Marijuana Use. *N. Engl. J. Med.* **2016**, *370*, 2219–2227. [CrossRef] [PubMed]
5. Volkow, N.D.; Swanson, J.M.; Evins, A.E.; DeLisi, L.E.; Meier, M.H.; González, R.; Bloomfield, M.A.P.; Curran, H.V.; Baler, R. Effects of Cannabis Use on Human Behavior, Including Cognition, Motivation, and Psychosis: A Review. *JAMA Psychiatry* **2016**, *73*, 292–297. [CrossRef] [PubMed]
6. Hall, W.; Degenhardt, L. The adverse health effects of chronic cannabis use. *Drug Test. Anal.* **2014**, *6*, 39–45. [CrossRef]
7. Arterberry, B.J.; Padovano, H.T.; Foster, K.T.; Zucker, R.A.; Hicks, B.M. Higher average potency across the United States is associated with progression to first cannabis use disorder symptom. *Drug Alcohol. Depend.* **2019**, *195*, 186–192. [CrossRef]
8. ElSohly, M.A.; Mehmedic, Z.; Foster, S.; Gon, C.; Chandra, S.; Church, J.C. Changes in cannabis potency over the last 2 decades (1995–2014): Analysis of current data in the United States. *Biol. Psychiatry* **2016**, *79*, 613–619. [CrossRef]
9. Hasin, D.S.; Saha, T.D.; Kerridge, B.T.; Goldstein, R.B.; Chou, S.P.; Zhang, H.; Jung, J.; Pickering, R.P.; Ruan, W.J.; Smith, S.M.; et al. Prevalence of Marijuana Use Disorders in the United States Between 2001–2002 and 2012–2013. *JAMA Psychiatry* **2015**, *72*, 1235–1242. [CrossRef]

10. Koob, G.F.; Volkow, N.D. Neurobiology of addiction: A neurocircuitry analysis. *Lancet Psychiatry* **2016**, *3*, 760–773. [CrossRef]
11. Field, M.; Cox, W. Attentional bias in addictive behaviors: A review of its development, causes, and consequences. *Drug Alcohol. Depend.* **2008**, *97*, 1–20. [CrossRef] [PubMed]
12. Evans, J.S.B.T.; Coventry, K. A Dual-Process Approach to Behavioral Addiction: The Case of Gambling. In *Handbook of Implicit Cognition and Addiction*; SAGE Publications, Inc.: Thousand Oaks, CA, USA, 2006; pp. 29–44.
13. Watson, P.; De Wit, S.; Hommel, B.; Wiers, R.W. Motivational Mechanisms and Outcome Expectancies Underlying the Approach Bias toward Addictive Substances. *Front. Psychol.* **2012**, *3*, 440. [CrossRef]
14. Wills, T.A.; Bantum, E.O.; Pokhrel, P.; Maddock, J.E.; Ainette, M.G.; Morehouse, E.; Fenster, B. A dual-process model of early substance use: Tests in two diverse populations of adolescents. *Health Psychol.* **2013**, *32*, 533–542. [CrossRef] [PubMed]
15. Robinson, T. The neural basis of drug craving: An incentive-sensitization theory of addiction. *Brain Res. Rev.* **1993**, *18*, 247–291. [CrossRef]
16. Robinson, T.E.; Berridge, K.C. The incentive sensitization theory of addiction: Some current issues. *Philos. Trans. R. Soc. B Boil. Sci.* **2008**, *363*, 3137–3146. [CrossRef]
17. Wiers, R.W.; Bartholow, B.D.; Wildenberg, E.V.D.; Thush, C.; Engels, R.C.; Sher, K.J.; Grenard, J.; Ames, S.L.; Stacy, A.W. Automatic and controlled processes and the development of addictive behaviors in adolescents: A review and a model. *Pharmacol. Biochem. Behav.* **2007**, *86*, 263–283. [CrossRef] [PubMed]
18. Harmon-Jones, E.; Gable, P.A. On the role of asymmetric frontal cortical activity in approach and withdrawal motivation: An updated review of the evidence. *Psychophysiology* **2018**, *55*, e12879. [CrossRef]
19. Harmon-Jones, E.; Gable, P.A.; Peterson, C.K. The role of asymmetric frontal cortical activity in emotion-related phenomena: A review and update. *Biol. Psychol.* **2010**, *84*, 451–462. [CrossRef]
20. Davidson, R.J.; Ekman, P.; Saron, C.D.; Senulis, J.A.; Friesen, W.V. Approach-withdrawal and cerebral asymmetry: Emotional expression and brain physiology: I. *J. Pers. Soc. Psychol.* **1990**, *58*, 330–341. [CrossRef]
21. Blackhart, G.C.; Minnix, J.A.; Kline, J.P. Can EEG asymmetry patterns predict future development of anxiety and depression? A preliminary study. *Biol. Psychol.* **2006**, *72*, 46–50. [CrossRef]
22. Papousek, I.; Reiser, E.M.; Weber, B.; Freudenthaler, H.H.; Schulter, G. Frontal brain asymmetry and affective flexibility in an emotional contagion paradigm. *Psychophysiology* **2012**, *49*, 489–498. [CrossRef] [PubMed]
23. Sutton, S.K.; Davidson, R.J. Prefrontal Brain Asymmetry: A Biological Substrate of the Behavioral Approach and Inhibition Systems. *Psychol. Sci.* **1997**, *8*, 204–210. [CrossRef]
24. Wheeler, R.E.; Davidson, R.J.; Tomarken, A.J. Frontal brain asymmetry and emotional reactivity: A biological substrate of affective style. *Psychophysiology* **2007**, *30*, 82–89. [CrossRef]
25. Smith, E.E.; Reznik, S.J.; Stewart, J.L.; Allen, J.J.B. Assessing and conceptualizing frontal EEG asymmetry: An updated primer on recording, processing, analyzing, and interpreting frontal alpha asymmetry. *Int. J. Psychophysiol.* **2017**, *111*, 98–114. [CrossRef] [PubMed]
26. Coan, J.A.; Allen, J.J.; Harmon-Jones, E. Voluntary facial expression and hemispheric asymmetry over the frontal cortex. *Psychophysiology* **2001**, *38*, 912–925. [CrossRef]
27. Killeen, L.A.; Teti, D.M. Mothers' frontal EEG asymmetry in response to infant emotion states and mother–infant emotional availability, emotional experience, and internalizing symptoms. *Dev. Psychopathol.* **2012**, *24*, 9–21. [CrossRef] [PubMed]
28. Davidson, R.J. Cerebral asymmetry and emotion: Conceptual and methodological conundrums. *Cogn. Emot.* **1993**, *7*, 115–138. [CrossRef]
29. Allen, J.J.; Reznik, S.J. Frontal EEG Asymmetry as a Promising Marker of Depression Vulnerability: Summary and Methodological Considerations. *Curr. Opin. Psychol.* **2015**, *4*, 93–97. [CrossRef]
30. Davidson, R.J. Anterior electrophysiological asymmetries, emotion, and depression: Conceptual and methodological conundrums. *Psychophysiology* **1998**, *35*, 607–614. [CrossRef]
31. Mathersul, D.; Williams, L.M.; Hopkinson, P.J.; Kemp, A.H. Investigating models of affect: Relationships among EEG alpha asymmetry, depression, and anxiety. *Emotion* **2008**, *8*, 560–572. [CrossRef]
32. Ellis, A.J.; Kinzel, C.; Salgari, G.C.; Loo, S.K. Frontal Alpha Asymmetry Predicts Inhibitory Processing in Youth with Attention Deficit/Hyperactivity Disorder. *Neuropsychologia* **2017**, *102*, 45–51. [CrossRef] [PubMed]

33. Keune, P.M.; Wiedemann, E.; Schneidt, A.; Schönenberg, M. Frontal brain asymmetry in adult attention-deficit/hyperactivity disorder (ADHD): Extending the motivational dysfunction hypothesis. *Clin. Neurophysiol.* **2015**, *126*, 711–720. [CrossRef] [PubMed]
34. Carver, C.S.; White, T.L. Behavioral inhibition, behavioral activation, and affective responses to impending reward and punishment: The BIS/BAS Scales. *J. Pers. Soc. Psychol.* **1994**, *67*, 319–333. [CrossRef]
35. Gable, P.A.; Mechin, N.C.; Neal, L.B. Booze cues and attentional narrowing: Neural correlates of virtual alcohol myopia. *Psychol. Addict. Behav.* **2016**, *30*, 377–382. [CrossRef] [PubMed]
36. Bowley, C.; Faricy, C.; Hegarty, B.; Johnstone, S.J.; Smith, J.L.; Kelly, P.J.; Rushby, J.A. The effects of inhibitory control training on alcohol consumption, implicit alcohol-related cognitions and brain electrical activity. *Int. J. Psychophysiol.* **2013**, *89*, 342–348. [CrossRef]
37. Knott, V.J.; Naccache, L.; Cyr, E.; Fisher, D.J.; McIntosh, J.F.; Millar, A.M.; Villeneuve, C.M. Craving-Induced EEG Reactivity in Smokers: Effects of Mood Induction, Nicotine Dependence and Gender. *Neuropsychobiology* **2008**, *58*, 187–199. [CrossRef]
38. Ruglass, L.M.; Shevorykin, A.; Dambreville, N.; Melara, R.D. Neural and behavioral correlates of attentional bias to cannabis cues among adults with cannabis use disorders. *Psychol. Addict. Behav.* **2019**, *33*, 69–80. [CrossRef]
39. American Psychiatric Association. *Diagnostic and Statistical Manual of Mental Disorders*, 4th ed.; American Psychiatric Association: Philadelphia, PA, USA, 2000.
40. Eriksen, B.A.; Eriksen, C.W. Effects of noise letters upon the identification of a target letter in a nonsearch task. *Percept. Psychophys.* **1974**, *16*, 143–149. [CrossRef]
41. Melara, R.D.; Singh, S.; Hien, D.A. Neural and Behavioral Correlates of Attentional Inhibition Training and Perceptual Discrimination Training in a Visual Flanker Task. *Front. Hum. Neurosci.* **2018**, *12*, 191. [CrossRef]
42. Lang, P.J.; Bradley, M.M.; Cuthbert, B.N. *International Affective Picture System (IAPS): Instruction Manual and Affective Ratings*; The Center for Research in Psychophysiology, University of Florida: Gainesville, FL, USA, 1999.
43. Garner, W.R. The stimulus in information processing. In *Sensation and Measurement: Papers in Honor of S.S. Stevens*; Moskowitz, H.R., Scharf, B., Stevens, J.C., Eds.; Springer: New York, NY, USA, 1974; pp. 77–90.
44. Oostenveld, R.; Fries, P.; Maris, E.; Schoffelen, J.-M. FieldTrip: Open source software for advanced analysis of MEG, EEG, and invasive electrophysiological data. *Comput. Intell. Neurosci.* **2011**, *2011*, 1–9. [CrossRef]
45. Gasser, T.; Bächer, P.; Möcks, J. Transformations towards the normal distribution of broad band spectral parameters of the EEG. *Electroencephalogr. Clin. Neurophysiol.* **1982**, *53*, 119–124. [CrossRef]
46. Quaedflieg, C.W.; Smulders, F.T.; Meyer, T.; Peeters, F.; Merckelbach, H.; Smeets, T. The validity of individual frontal alpha asymmetry EEG neurofeedback. *Soc. Cogn. Affect. Neurosci.* **2016**, *11*, 33–43. [CrossRef] [PubMed]
47. Meyer, T.; Quaedflieg, C.W.E.M.; Giesbrecht, T.; Meijer, E.H.; Abiad, S.; Smeets, T. Frontal EEG asymmetry as predictor of physiological responses to aversive memories. *Psychophysiology* **2014**, *51*, 853–865. [CrossRef] [PubMed]
48. Davidson, R.J.; Chapman, J.P.; Chapman, L.J.; Henriques, J.B. Asymmetrical Brain Electrical Activity Discriminates Between Psychometrically-Matched Verbal and Spatial Cognitive Tasks. *Psychophysiology* **1990**, *27*, 528–543. [CrossRef] [PubMed]
49. Greenhouse, S.W.; Geisser, S. On methods in the analysis of profile data. *Psychometrika* **1959**, *24*, 95–112. [CrossRef]
50. Pattij, T.; Wiskerke, J.; Schoffelmeer, A.N. Cannabinoid modulation of executive functions. *Eur. J. Pharmacol.* **2008**, *585*, 458–463. [CrossRef] [PubMed]
51. Cousijn, J.; Goudriaan, A.E.; Ridderinkhof, K.R.; van den Brink, W.; Veltman, D.J.; Wiers, R.W. Neural responses associated with cue-reactivity in frequent cannabis users. *Addict. Biol.* **2013**, *18*, 570–580. [CrossRef] [PubMed]
52. Mechin, N.; Gable, P.A.; Hicks, J.A. Frontal asymmetry and alcohol cue reactivity: Influence of core personality systems. *Psychophysiology* **2016**, *53*, 1224–1231. [CrossRef]
53. Sjoerds, Z.; Brink, W.V.D.; Beekman, A.T.F.; Penninx, B.W.J.H.; Veltman, D.J. Cue Reactivity Is Associated with Duration and Severity of Alcohol Dependence: An fMRI Study. *PLoS ONE* **2014**, *9*, e84560. [CrossRef]

Article

Assessing the Role of Cannabis Use on Cortical Surface Structure in Adolescents and Young Adults: Exploring Gender and Aerobic Fitness as Potential Moderators

Ryan M. Sullivan [1], Alexander L. Wallace [1], Natasha E. Wade [2], Ann M. Swartz [3] and Krista M. Lisdahl [1,*]

1 Department of Psychology, University of Wisconsin-Milwaukee, Milwaukee, WI 53211, USA; rmsul@uwm.edu (R.M.S.); walla228@uwm.edu (A.L.W.)
2 Department of Psychiatry, University of California, San Diego, La Jolla, CA 92093, USA
3 Department of Kinesiology, University of Wisconsin-Milwaukee, Milwaukee, WI 53211, USA; aswartz@uwm.edu
* Correspondence: krista.medina@gmail.com; Tel.: +1-414-229-7159

Received: 17 January 2020; Accepted: 20 February 2020; Published: 22 February 2020

Abstract: Cannabis use in adolescents and young adults is linked with aberrant brain structure, although findings to date are inconsistent. We examined whether aerobic fitness moderated the effects of cannabis on cortical surface structure and whether gender may play a moderating role. Seventy-four adolescents and young adults completed three-weeks of monitored abstinence, aerobic fitness testing, and structural magnetic resonance imaging (sMRI). Whole-sample linear regressions examined the effects of gender, VO_2 max, cannabis use, and their interactions on the surface area (SA) and local gyrification index (LGI). Cannabis use was associated with greater cuneus SA. Gender-by-cannabis predicted precuneus and frontal SA, and precentral, supramarginal, and frontal LGI; female cannabis users demonstrated greater LGI, whereas male cannabis users demonstrated decreased LGI compared to non-users. Aerobic fitness was positively associated with various SA and LGI regions. Cannabis-by-aerobic fitness predicted cuneus SA and occipital LGI. These findings demonstrate that aerobic fitness moderates the impact of cannabis on cortical surface structure, and gender differences are evident. These moderating factors may help explain inconsistencies in the literature and warrant further investigation. Present findings and aerobic fitness literature jointly suggest aerobic intervention may be a low-cost avenue for improving cortical surface structure, although the impact may be gender-specific.

Keywords: cannabis; gyrification; surface area; cortical surface structure; aerobic fitness; gender

1. Introduction

Cannabis is the most used "illicit" substance worldwide with estimated lifetime prevalence rate of 16.9%, with the highest rates in the United States and New Zealand [1]. Specifically within the United States, cannabis is the second most commonly used substance within adolescents and young adults [2]. Approximately 29.7% of adolescents (Grades 8, 10, and 12) [3] and 52% of young adults (aged 18–25) [4] have used cannabis within their lifetime; and, these prevalence rates can vary by state policies [5]. Heavy and chronic cannabis use is associated with adverse psychopathological [6], neurocognitive, and aberrant brain morphology outcomes [7]. Yet, distinct structural changes are less understood [8] and results are not always consistent [7]. Therefore, further research is necessary to elucidate potential moderators of cannabis effects that may explain individual differences.

Cannabis contains cannabinoids, including delta-9-tetrahydrocannabinol (THC) and cannabidiol (CBD). THC is the main psychoactive ingredient and interacts with the endocannabinoid system, affecting the brain through binding to its cannabinoid receptor type 1 (CB1), which is widely distributed throughout the cerebral cortex and principally involved in neuromodulation [9]. Repeated and regular exogenous cannabis exposure can affect CB1 binding [10] and downregulation [11], as well as effects on the brain structure and function (e.g., white matter integrity, functional connectivity, and cerebral blood flow) [12]. Frequent cannabis use, especially high-potency THC products, is associated with structural alterations in medial-temporal, frontal, limbic, and cerebellar regions [7,8,13], however, the extant literature regarding definitive effects on brain structure have garnered mixed findings [14]. Specific investigations into volumetric indices (i.e., cortical thickness and volume) has shown differences between cannabis users and non-users in several areas, including, the hippocampus [15,16], prefrontal cortex [17], right amygdala [18,19], right fusiform [20], orbitofrontal cortex [21,22], inferior parietal cortex [21], anterior cingulate [23,24], precentral gyrus [23,25], superior frontal gyrus [23], right thalamus [25], and cerebellum [26]; see review [27]. Lorenzetti, Chye, Silva, Solowij and Roberts [27] noted further investigation is needed to elucidate potentially mitigating variables in this relationship. Nonetheless, reviews largely center volumetric indices due to a large proportion of literature examining these outcomes. However, less is known on the relationship between cortical surface structure (i.e., cortical gyrification and surface area) and cannabis use and thus is the focus of the present study.

Cortical folding or local gyrification index (LGI) optimizes cortico-connectivity [28]. LGI has been shown to continue development well into adulthood [29] and it is hypothesized to be more susceptible to environmental factors, such as exogenous drug exposure [30–33]. Cortical surface area (SA) reaches peak levels within adolescence [34] and decreases with age [29], with development largely attributable to heritability [35]. As stated previously, repeated exogenous cannabis exposure can cause alterations in the endocannabinoid system, which plays a role in neuromodulation, pruning, and white matter development. Thus, it is postulated this exposure could disrupt developmental trajectories of LGI and SA, however, this association may be more complex than previously conceptualized. Despite this, only a few studies have examined LGI and SA in cannabis users. Mata, et al. [36] found decreased LGI (i.e., flattening) in the bilateral temporal lobes and left prefrontal cortex within cannabis users; however, no global hemispheric differences in SA were observed. Previous ROI analyses from our lab found decreased LGI in frontal, medial, and ventral medial poles in regular young adult cannabis users, and marginal differences in SA—indicating that cannabis may impact LGI more diffusely compared to SA [37]. Filbey, et al. [38] found no differences in LGI between early and late onset of cannabis users (i.e., age of onset 16.5 and 19 years old, respectively); however, users with earlier onset showed a significant relationship between heavier and longer duration of cannabis use and decreased LGI in prefrontal regions. In a six-year prospective longitudinal study in younger adolescents (aged 12–14 at baseline), decreases in bilateral medial orbitofrontal cortex and right insula SA were found after alcohol and cannabis initiation, though the alcohol-only group demonstrated the most robust findings [39]. Lastly, a recent multi-site analysis in 261 cannabis users and controls found no differences within SA and LGI in regard to cannabis use, cannabis dependence, and age of onset [40]; however, this sample consisted of adult participants (rather than adolescents and young adults) and fine-tuned patterns of use were not ascertained. Furthermore, a majority of these studies were methodologically observational did not examine abstained users to elucidate chronic rather than acute effects. Consequently, the relationship between cannabis and cortical morphometry remains unclear; one potential reason for these inconsistent findings may be moderating factors that put some individuals at a higher risk for negative structural consequences compared to others.

Aerobic exercise (AE) has been linked with positive impacts on the brain. Increasing overall AE in animal models has been linked with increased c-Fos expression [41], brain-derived neurotrophic factors (BDNF) [42], cell proliferation in the hippocampus [43], decreased inflammatory response [44] and oxidative stress [45], and blocking deleterious alcohol-related effects on the hippocampus and dentate gyrus [46,47]. In humans, the interplay between overall aerobic fitness and brain health has been well-established in older adults [43–50], but less is known in younger adults or adolescents. Previous research has linked levels of aerobic fitness with better neuropsychological performance in adolescents [51,52] and young adults [53,54], brain structure in children [55], and volumetric differences in young adults [56,57]. Interestingly, depending on levels of aerobic fitness (i.e., high versus low), adolescents display differing functional activation despite similar behavioral performance [58]. Therefore, AE can serve as a potential moderator for brain morphometry in youth. Further, acute AE releases naturally-occurring endocannabinoids [59]; therefore, it is theorized that AE may boost endocannabinoid signaling, potentially counteracting the downregulation effects of exogenous cannabis use [60]. Supporting this hypothesis, our lab recently reported that highly aerobically-fit cannabis users demonstrated better performance in neurocognitive measures compared to users with low aerobic fitness [53]. Further, an AE intervention with sedentary cannabis users found a decrease in craving and use at completion of the intervention [61]. Despite this, few studies explore the relationship between AE and cannabis use on structural brain outcomes.

Another frequent moderating factor in cannabis use research more broadly is gender or sex differences [62–66]. Neuroimaging evidence has pointed towards gender differences in the relationship between chronic cannabis use and structural outcomes [15,17,19,26]; see review [13]. However, none of the aforementioned studies examining LGI or SA outcomes considered gender as a moderator [36–40]. Gender differences are observed within typical assessments of cortical surface structure indices. For example, females exhibit greater cortical complexities (i.e., LGI) overall and in more frontal regions compared to males [67,68], whereas males generally have exhibited larger SA compared to females [69–71]. Thus, an investigation into gender differences in cannabis effects is warranted. Finally, gender differences are seen in the effects of AE on cognitive outcomes [72,73]. Consequently, the possibility of gender differences in the impact of AE and cannabis on cortical surface structure is plausible.

The aim of the present study is to look at both aerobic fitness and gender as potentially moderating the relationship between regular cannabis use and cortical surface structure indices, LGI and SA, in adolescents and young adults following three weeks of monitored cannabis abstinence. We expect to see associations between greater aerobic fitness level and increased SA and LGI, and for cannabis users who are aerobically fit to demonstrate fewer cortical abnormalities. We will also examine whether gender moderates the relationship between cannabis and brain morphometry.

2. Materials and Methods

2.1. Participants

Participants were recruited for a larger parent study through advertisements and flyers in the local community and college. Seventy-four participants in the present study (cannabis users = 36, non-using controls = 38) were between the ages of 16 and 26 years (M = 21.1, SD = 2.6), were generally split for gender (44.6% female), and racial identities consisted of predominantly: Caucasian (64.9%), Asian (10.8%), Multi-racial (10.8%), and African-American (8.1%). (See Table 1).

Participants were included in the parent study if they were right-handed, spoke English, and were willing to abstain from substance use over a three-week period. Exclusion criteria for the parent study included having an independent DSM-IV Axis I (mood, anxiety, psychotic, or attention) disorder, current use of psychoactive medications, major medical or neurological disorders (including metabolic disorders), loss of consciousness >2 min, history of learning disability or intellectual disability, prenatal medical issues or premature birth (gestation <35 weeks), MRI contraindications (pregnancy, claustrophobia, metal in body), reported significant prenatal alcohol exposure (≥4 drinks in a day or ≥6 drinks in a week), prenatal illicit drug exposure, or prenatal nicotine exposure (average > 5 cigarettes per day longer than 1 month), elevated Physical Activity Readiness Questionnaire (PARQ) [74] scores indicating difficulty engaging in VO_2 max testing, or excessive other illicit drug use (>20 times of lifetime use for each drug category, including cannabis use for non-using control participants).

Participants were also balanced at screening for active vs. sedentary physical activity, based on International Physical Activity Questionnaire (IPAQ) [75] score.

The present sample consisted of cannabis users who are categorized as current users who used cannabis at least 44 times in the last year (i.e., nearly weekly) and at least 100 lifetime uses (i.e., nearly weekly for two years). Non-using controls used cannabis no more than 5 times in the past year and less than 20 times in their lifetime [53,66,76,77]. If participants from the parent study (parent study total $N = 100$) did not meet these defined group thresholds, or if participants did not complete VO_2 max or MRI protocol, they were excluded from the present study (excluded from present study $N = 26$).

Table 1. Demographics.

	Cannabis Users			Non-Using Controls		
	All	Male	Female	All	Male	Female
N	36	23	13	38	18	20
M (SD) or %						
Age (years)	21.4 (2.3)	21.4 (2.4)	21.4 (2.0)	20.8 (2.8)	20.5 (3.1)	21.0 (2.6)
Race (% Caucasian)	58.3%	65.2%	46.2%	71.1%	72.2%	70.0%
Ethnicity (% Non-Hisp)	77.8%	78.3%	76.9%	86.8%	94.4%	80.0%
Educational Attainment (years)	14.0 (1.6)	13.9 (1.8)	14.1 (1.3)	14.1 (2.4)	14.0 (2.9)	14.2 (1.9)
Past yr Alcohol Use [a,*]	338.7 (300.8)	376.6 (306.2)	271.6 (290.5)	100.6 (173.6)	141.8 (225.1)	63.5 (101.5)
Past yr Tobacco Use [a,*]	214.6 (483.7)	311.8 (585.1)	42.8 (68.1)	0.5 (1.97)	0.2 (0.43)	0.7 (2.7)
Cotinine Level [b,*]	2.0 (1.8)	2.3 (2.1)	1.5 (1.0)	1.1 (0.6)	1.1 (0.6)	1.1 (0.6)
Past yr Cannabis Use [a,*]	428.2 (440.4)	499.9 (510.7)	301.5 (245.4)	0.36 (1.2)	0.7 (1.6)	0.1 (0.22)
Lifetime Cannabis Use [a,*]	1189.6 (1372.3)	1419.7 (1621.6)	782.5 (625.0)	1.5 (2.9)	1.2 (2.3)	1.8 (3.5)
Age at Regular Cannabis Use Onset (years)	17.5 (1.7)	17.4 (1.9)	17.8 (1.3)	– [c]	– [c]	– [c]
Cannabis Abstinence Length (days) [d]	31.1 (22.9)	34.3 (27.9)	25.5 (6.5)	– [c]	– [c]	– [c]
VO_2 maximum [e,^]	43.7 (9.0)	47.9 (6.6)	36.1 (7.7)	41.4 (9.8)	47.9 (8.8)	35.5 (6.3)
VO_2 maximum (%ile) [f]	– [c]	69.4%	37.2%	– [c]	68.9%	33.9%
Body Fat (%) [g,^]	19.1% (8.5)	15.6% (6.9)	25.3% (7.7)	21.6% (10.0)	13.6% (6.1)	28.7% (7.0)

* $p < 0.001$ between cannabis users and non-using controls. ^ $p < 0.001$ between males and females. [a] Measured in standard uses on TLFB [78]. [b] Measured at VO_2 maximum testing session. [c] Not applicable. [d] Calculated from TLFB last cannabis use date and date of sMRI. [e] Raw values. [f] Gender-specific percentiles were calculated with mean age, using ASCM norms [79,80]. [g] Body Fat was ascertained in the same session at which VO_2 maximum testing was conducted.

2.2. Procedures

Data was ascertained from a larger parent study examining the independent and interactive effects of cannabis use and aerobic fitness on neurocognitive outcomes in adolescents and young adults (R01 DA030354; PI: Lisdahl); all aspects of the protocol were approved by the University of Wisconsin-Milwaukee IRB (Study #: PRO00016025). Potential participants who expressed interest in the parent study were asked, over the phone, for demographic information (including age, gender, race/ethnicity, and educational attainment) and screened through an initial semi-structured interview for independent lifetime and past-year Axis I Disorders other than substance use disorder. If determined eligible, study staff obtained a written consent from participants who were aged 18 or older at the start of participation or obtained written assert after parent consent was obtained for minors under 18 years old. Additionally, parents of participants were consented for a parent-administered phone interview that screened for medical, psychiatric and prenatal history before an in-person session was conducted.

Participants who were eligible for the study came in for five study sessions over the course of three weeks. The first three sessions occurred one week apart and consisted of a brief neuropsychological battery (explained in more detail in Wallace, et al. [81]) and urinary drug analysis. Sessions four and five occurred at least one week after session three and consisted of ascertaining body composition, aerobic fitness VO_2 maximum (VO_2 max) testing, and then a sMRI that occurred within 24 to 48 h of each other. During the entire study period, participants were asked to abstain from cannabis, alcohol, and other drug use (other than tobacco), which was confirmed through urine, breath, and sweat toxicology screening, which was administered to all participants across all study sessions. If they tested positive for illicit drug use, showed an increase in THCCOOH levels, or had a breath alcohol concentration greater than 0.000 at the start of session two or three (i.e., before VO_2 max and sMRI procedures), participants were asked to conduct the session after a week of abstinence. Participants were not allowed to complete session four (VO_2 max) or session five (MRI scan) if they tested positive for any illicit drug use, a rise in THCCOOH levels, or had a breath alcohol concentration greater than 0.000, and instead were allowed to continue their involvement in the study from session one. Participants who used tobacco were asked to abstain from use an hour before the MRI scan.

2.3. Measures

2.3.1. Detailed Phone Screen

Physical Activity—Extent of physical activity was assessed with the IPAQ [75] and physical ability to engage in VO_2 max testing was assessed with the PAR-Q [74].

Lifetime Substance Use Patterns—Overall patterns of drug and alcohol use were determined by administering the Customary Drinking and Drug Use Record (CDDR) [82] at baseline to measure frequency of cannabis, alcohol, nicotine, and other drug use, SUD symptoms, and the age of onset for first time and regular (i.e., weekly) use.

Mini Psychiatric Interview—The Mini International Psychiatric Interview (MINI) [83] or MINI-Kid [84] was administered to participants and parents of minors to screen out for psychiatric comorbidities.

2.3.2. Session Measures

Past Year Substance Use—A modified version of the Timeline Follow-Back (TLFB) was conducted by trained research assistants to assess substance use patterns on a week-by-week basis capturing use within the past year, while providing memory cues such as personal events and holidays [66,78]. Substances were measured by standard units [alcohol (standard drinks), nicotine (number of cigarettes and hits of chew/snuff/pipe/cigar/hookah), cannabis (all methods converted to joints or mg in concentrates), ecstasy (number of tablets), sedatives (number of pills or hits of GHB), stimulants (cocaine and methamphetamine use converted to mg and number of amphetamine pills), hallucinogens (number of hits or occasions of ketamine/salvia/shrooms/other hallucinogens), opioids (number of hits of heroin/opium), and inhalants (number of hits)]. Days of cannabis abstinence at scan were calculated from date of last cannabis use based on the TLFB and date of scan.

Verifying Drug Abstinence—Participants were expected to remain abstinent from alcohol and other drugs (except tobacco) throughout the course of the study, thus abstinence was evaluated at each session through urine toxicology. The ACCUTEST SplitCup 10 Panel drug test measures amphetamines, barbiturates, benzodiazepines, cocaine, ecstasy, methadone, methamphetamines, opiates, PCP, and THC. Urine samples were also tested using NicAlert to test cotinine level, a metabolite of nicotine. Participants also wore PharmChek Drugs of Abuse Patches, which continuously monitor sweat toxicology for the presence of cocaine, benzoylecgonine, heroin, 6MAM, morphine, codeine, amphetamines, methamphetamine, delta-9-tetrahydrocannabinol (THC), and phencyclidine. Participants additionally underwent breathalyzer screens to test for alcohol use at the start of each session.

Anthropometric Measures—Height and weight were measured in light clothes and without shoes. Body Mass Index was calculated as weight divided by height squared (kg/m^2).

Body Fat Percentage—an electrical bioimpedance analysis system was utilized to measure body fat percentage [The Tanita Body Composition Analyzer, TBF-300 (Tanita Corporation, Tokyo, Japan)].

VO_2 Maximum—Participants were instructed to refrain from food and caffeine for 4 hours prior to the exercise test. Prior to each exercise test, the metabolic measurement system, ParvoMedics TrueOne 2400 (ParvoMedics, Salt Lake City, UT, USA) was calibrated using a two-point calibration for the gas analyzers (room air and certified gas: 4.008% CO_2, 15.98% O_2, balance N_2) and a 3 Liter syringe for the pneumotachometer. Participants were fitted with the rubber mouthpiece connected to a Hans Rudolf 2700 series two-way nonrebreathing valve (Kansas City, MO, USA), nose clip, and heart rate strap (Polar Wearlink 31, Kempele, Finland) for the measurement of heart rate and collection of expired gases. Participants completed a maximal incremental exercise test on a treadmill (Full Vision Inc., TMX425C Trackmaster, Newton, KS, USA) following the Bruce Protocol until volitional fatigue. Expired gases were measured continuously using a ParvoMedics TrueOne 2400 metabolic measurement system (ParvoMedics, Salt Lake City, UT, USA). Criteria for determination of attainment of VO_2 max were based on those recommended [85]. Metabolic data were averaged over 1 min and exported for analysis.

MRI Acquisition—Structural MRI scans were acquired on a 3T Signa LX MRI scanner (GE Healthcare, Waukesha, WI, USA) using a 32-channel quadrature transmit/receive head coil. High-resolution anatomical images were acquired using a T1-weighted spoiled gradient-recalled at steady-state (SPGR) pulse sequence (TR = 8.2 ms, TE = 3.4 s, TI = 450 and flip angle of 12°). The in-plane resolution of the anatomical images was 256 × 256 with a square field of view (FOV) of 240 mm. One hundred fifty slices were acquired at 1 mm thickness.

Processing Pipeline—Participant structural scans were processed in a standard processing pipeline within FreeSurfer version 5.3 (http://surfer.nmr.mgh.harvard.edu/fswiki/recon-all). T1-weighted 3D anatomical datasets underwent an automated pipeline for motion correction, nonparametric non-uniform intensity normalization, Montreal Neurologic Institute transformation, removal of non-brain materials, skull-stripping, and topology correction. Preprocessed scans were visually examined, and manual edits were made when appropriate. Surface based data created from FreeSurfer's automated pipeline was utilized for both surface area and LGI analyses. LGI data were computed from pial surface files in accordance with Schaer, et al. [86].

2.4. Statistical Analysis

Differences in demographic variables were examined using ANOVAs and Chi-square tests in *R* [87]. A series of multivariate regressions were run with cannabis group, gender, VO_2 max levels, and their interactions (cannabis group*gender, cannabis group*VO_2 max, and cannabis group*VO_2 max*gender) as the independent variables of interest; covariates included past year alcohol use (i.e., total standard drinks) and cotinine levels at the time of VO_2 max testing. Separate regressions were run measuring for SA [88] and LGI [89]. Analyses were done separately between each hemisphere (right and left) and smoothed with a global Gaussian blur at FWHM of 10 for all analyses. Corrections for multiple comparisons were made using Monte Carlo simulations at a cluster wise probability (*cwp*) of $p = 0.05$, correcting across both hemispheric spaces, and smoothed at FWHM of 20 for SA and 25 for LGI corrections. Decisions about statistical significance were made at $p < 0.01$ for all analyses. Regions that meet statistical significance were annotated using the Desikan-Killiany Atlas [90].

3. Results

3.1. Demographic Data

There were no significant differences between cannabis user and non-user groups in age [$F(1,72) = 1.22$, $p = 0.27$], gender [$\chi^2(1) = 1.43$, $p = 0.23$], race [$\chi^2(6) = 5.87$, $p = 0.44$], ethnicity [$\chi^2(2) = 2.69$, $p = 0.26$], educational attainment [$F(1,72) = 0.08$, $p = 0.78$], VO_2 max [$F(1,72) = 1.10$, $p = 0.30$], and body fat percentage [$F(1,72) = 1.27$, $p = 0.27$]. There was a significant difference in amount of lifetime [$F(1,72) = 28.5$, $p < 0.001$] and past year [$F(1,72) = 35.9$, $p < 0.001$] cannabis use, alcohol consumed within the past year [$F(1,72) = 17.6$, $p < 0.001$], past year tobacco use [$F(1,72) = 7.46$, $p = 0.008$], and cotinine levels at VO_2 max testing [$F(1,72) = 9.24$, $p = 0.003$]; past year alcohol use and cotinine levels were included as covariates in all analyses.

There were no significant differences between males and females for age [$F(1,72) = 0.05$, $p = 0.83$], race [$\chi^2(6) = 8.23$, $p = 0.22$], ethnicity [$\chi^2(2) = 1.48$, $p = 0.48$], educational attainment [$F(1,72) = 0.18$, $p = 0.67$], past year tobacco use [$F(1,72) = 3.82$, $p = 0.06$], cotinine levels at VO_2 max testing [$F(1,72) = 2.54$, $p = 0.12$], or past year [$F(1,72) = 3.56$, $p = 0.06$] and lifetime [$F(1,72) = 3.57$, $p = 0.06$] cannabis use. Significant differences between genders were observed for past year alcohol use [$F(1,72) = 4.29$, $p = 0.04$], VO_2 max [$F(1,72) = 52.62$, $p < 0.001$], and body fat percentage [$F(1,72) = 60.9$, $p < 0.001$].

Within cannabis users, there was no significant differences between genders in past year alcohol use [$F(1,34) = 1.01$, $p = 0.32$], past year tobacco use [$F(1,34) = 2.69$, $p = 0.11$], cotinine levels at VO_2 max testing [$F(1,34) = 1.59$, $p = 0.22$], past year [$F(1,34) = 1.72$, $p = 0.20$] or lifetime [$F(1,34) = 1.83$, $p = 0.19$] cannabis use, age of first regular cannabis use onset [$F(1,34) = 0.37$, $p = 0.55$], or days of cannabis abstinence prior to sMRI [$F(1,34) = 1.26$, $p = 0.27$].

3.2. Primary Analyses

Whole-sample analyses were conducted examining the effects of cannabis, VO_2 max, gender and their interactions on SA and LGI, while covarying for past year alcohol use and cotinine level at time of VO_2 max testing. (See Tables 2 and 3).

Table 2. Surface Area findings.

	t	Size (mm^2)	x	y	z	*cwp*
Cannabis						
Left Cuneus	2.639	1706.64	−4.1	−78.6	19.1	0.006
Cannabis*Gender						
Left Precuneus	−3.306	1718.12	−10.3	−54.6	46.7	0.006
Left Rostral Middle Frontal	−2.299	2348.85	−44.7	27	31.4	0.0006
Right Superior Frontal	−3.491	1819.72	11.5	9	36.9	0.003
Right Superior Frontal	−2.248	2007.88	23.1	0.4	61.6	0.002
VO_2						
Left Superior Parietal	4.654	1673.98	−28.4	−64.4	39.9	0.007
Left Inferior Parietal	4.236	2535.22	−45.1	−63.9	10	0.0001
Right Inferior Parietal	3.894	3235.39	47.3	−59.7	29.4	0.0001
Right Inferior Temporal	3.268	2877.28	51.4	−55.2	−15.1	0.0001
Cannabis*VO_2						
Left Cuneus	−3.724	2736.75	−4.4	−77.1	21.6	0.0001

Table 3. Gyrification findings.

	t	Size (mm^2)	x	y	z	cwp
Cannabis*Gender						
Left Precentral	−2.894	4993.5	−37.6	−12.2	62.1	0.0001
Left Lateral Orbitofrontal	−2.533	3240.22	−18.8	51.8	−13.8	0.0004
Right Supramarginal	−3.784	4763.28	49.6	−41.1	40.9	0.0001
VO_2						
Left Superior Temporal	5.174	10682.85	−64.5	−25.2	4.2	0.0001
Right Lateral Orbitofrontal	3.272	13062.8	39	27.3	−9	0.0001
Right Inferior Parietal	2.78	2718.55	34.2	−73.4	37.6	0.0015
Cannabis*VO_2						
Left Lateral Occipital	−3.712	3297.29	−28.2	−95.1	−12.7	0.0004
Cannabis*VO_2*Gender						
Right Supramarginal	2.572	2207.65	48.8	−40.5	40.3	0.009

3.2.1. Cannabis Results

Surface area. Cannabis users demonstrated significantly larger SA in the left cuneus compared to controls [$t(58) = 2.64$, $cwp = 0.006$] (See Figure 1). *Gyrification.* There was no main effect of cannabis group observed for LGI.

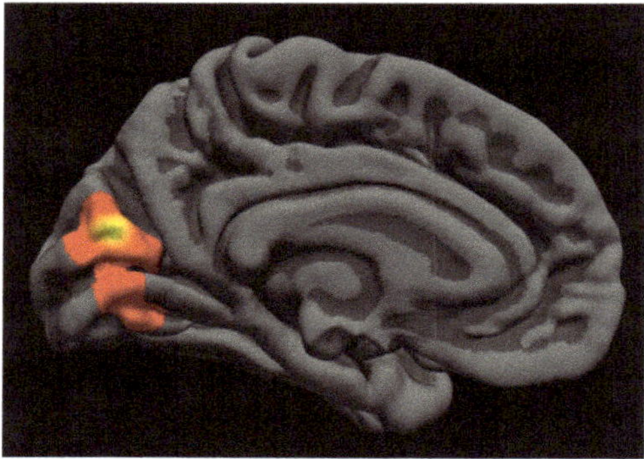

Figure 1. Cannabis Findings. Medial view of cannabis finding within left cuneus SA, with larger area in cannabis users.

3.2.2. Cannabis*Gender

Surface area. There was a significant interaction between cannabis group and gender in the left precuneus [$t(58) = -3.31$, $cwp = 0.006$], left rostral middle frontal [$t(58) = -2.30$, $cwp = 0.0006$], and two right superior frontal [$t(58) = -3.49$, $cwp = 0.003$; $t(58) = -2.25$, $cwp = 0.002$] regions for SA. Female cannabis users had increased SA in these regions compared to non-using females, whereas male cannabis users demonstrated decreased SA compared to non-using males, except in the second right rostral middle region, where cannabis using males and females exhibited less SA compared to non-using males and females. *Gyrification.* There was a significant interaction between cannabis group and gender in the left precentral [$t(58) = -2.89$, $cwp = 0.0001$], left lateral orbitofrontal [$t(58) = -2.53$, $cwp = 0.0004$], and right supramarginal [$t(58) = -3.78$, $cwp = 0.0001$] regions for LGI. Female cannabis users had increased LGI in these regions compared to non-using females and male cannabis users had decreased LGI compared to non-using males; except in left lateral orbitofrontal region, where female cannabis users displayed slightly less gyrification compared to non-using females and male cannabis users had decreased LGI compared to non-using males. (See Figure 2).

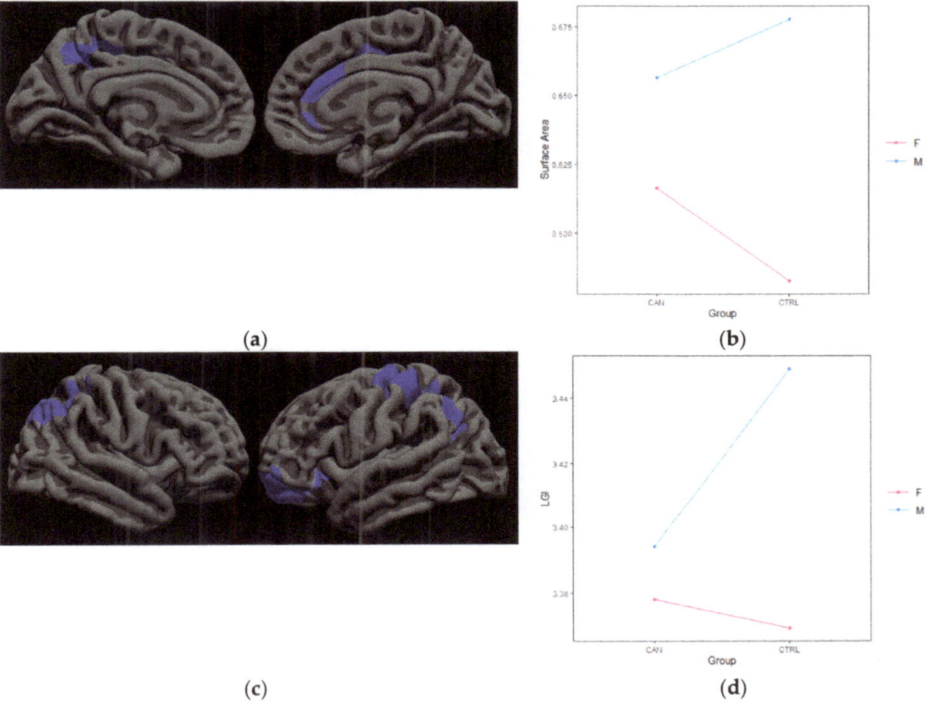

Figure 2. Cannabis*Gender Findings. (**a**) Medial view of significant interaction between cannabis group and gender in left cuneus, left rostral middle frontal (not pictured), and right superior frontal SA. (**b**) Generally, female cannabis users demonstrated more SA in these regions compared to non-using females, whereas male cannabis users demonstrated less SA compared to non-using males (left rostral middle frontal SA finding depicted). (**c**) Lateral view of significant interaction between cannabis group and gender in right supramarginal, left precentral, and left lateral orbitofrontal LGI. (**d**) Generally, female cannabis users had more LGI in this region compared to non-using females, whereas male cannabis users had less LGI compared to non-using males (left precentral LGI finding depicted).

3.2.3. VO_2 Results

Surface area. In both the cannabis users and non-users, significant relationships between greater VO_2 max and larger SA were found in the left superior parietal [$t(58) = 4.65$, $cwp = 0.007$], left inferior parietal [$t(58) = 4.24$, $cwp = 0.0001$], right inferior parietal [$t(58) = 3.89$, $cwp = 0.0001$], and right inferior temporal [$t(58) = 3.27$, $cwp = 0.0001$] regions. *Gyrification.* Participants displayed a significant relationship between greater VO_2 max and greater LGI in the left superior temporal [$t(58) = 5.17$, $cwp = 0.0001$], right lateral orbitofrontal [$t(58) = 3.27$, $cwp = 0.0001$], and right inferior parietal [$t(58) = 2.78$, $cwp = 0.0015$] regions. (See Figure 3).

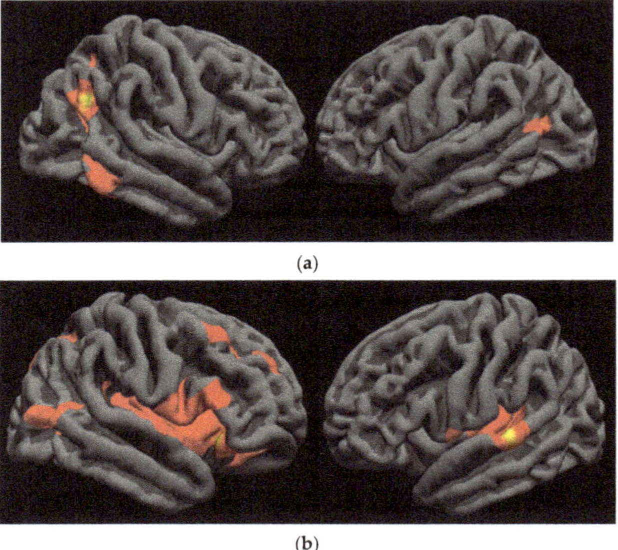

(**a**)

(**b**)

Figure 3. VO_2 Findings. (**a**) Lateral view of VO_2 finding observed in right inferior parietal, right inferior temporal, left superior parietal (not pictured), and left inferior parietal SA. Greater VO_2 was associated with more area in these regions. (**b**) Lateral view of VO_2 finding observed in right lateral orbitofrontal, right inferior parietal, and left superior temporal LGI. Greater VO_2 was associated with more gyrification in these regions.

3.2.4. Cannabis*VO_2

Surface area. There was a significant interaction between VO_2 max and cannabis group in the left cuneus [$t(58) = -3.72$, $cwp = 0.0001$] region; non-using controls demonstrated a strong positive relationship between VO_2 max and increased SA, whereas cannabis users demonstrated a negative relationship. *Gyrification.* There was a significant interaction between VO_2 max and cannabis group in the left lateral occipital region [$t(58) = -3.71$, $cwp = 0.0004$]; non-using controls demonstrated a positive relationship between VO_2 max and increased LGI, whereas no trend was observed for cannabis users. (See Figure 4).

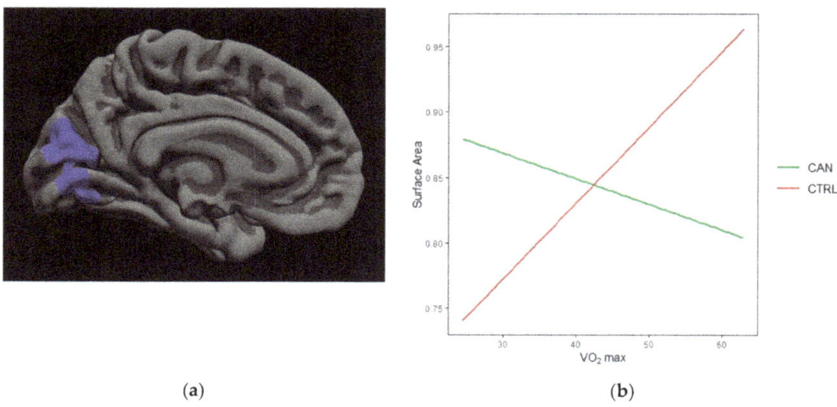

(**a**) (**b**)

Figure 4. *Cont.*

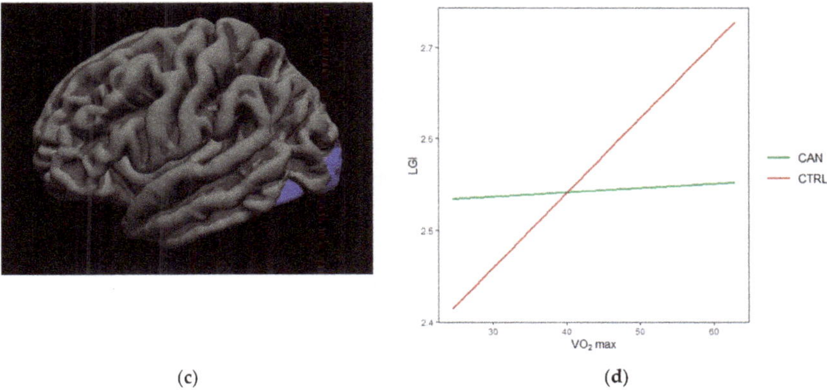

(c) (d)

Figure 4. Cannabis*VO_2 Findings. (**a**) Medial view of significant interaction between VO_2 and cannabis group in left cuneus SA; (**b**) non-using controls demonstrated a positive relationship between VO_2 and increased area, whereas cannabis users demonstrated a negative relationship. (**c**) Lateral view of significant interaction between VO_2 and cannabis group in left lateral occipital LGI; (**d**) non-using controls demonstrated a positive relationship between VO_2 and increased LGI, whereas no trend was observed for cannabis users.

3.2.5. Cannabis*VO_2*Gender

In order to further characterize potential gender differences, exploratory three-way interaction analyses were investigated. Results revealed an interaction between group, gender, and VO_2 was shown for LGI in the right supramarginal region [$t(58) = 2.57$, $cwp = 0.009$]. Representing a potential area of further investigation with larger sample sizes.

4. Discussion

Worldwide, cannabis use rates are known to vary across countries [1] and age of onset is typically between 18 to 19 years old [91]. Additionally, given the ongoing policy debate surrounding cannabis in the United States, its long-term effects in youth are of increasing scientific and clinical interest. To date, findings regarding the impact of regular use on brain morphometric outcomes have not been entirely consistent [12]. Reasons underlying these differential findings include demographic differences in samples, especially regarding gender, or other potential resilience or risk factors including extent of aerobic fitness [53,60,92]. Here, we attempt to clarify the confusion regarding cannabis effects by examining two potential moderators: aerobic fitness and gender. We found that after three weeks of monitored drug abstinence, cannabis users demonstrated greater SA in the cuneus. Notably, we also found significant interactions between gender, aerobic fitness, and cannabis use on SA and LGI outcomes in prefrontal and parietal cortical regions. Generally speaking, males appeared more sensitive to cannabis impacts following abstinence, and further, participants demonstrated a positive relationship between aerobic fitness and cortical surface structure more broadly.

After accounting for gender, alcohol, cotinine, and aerobic fitness level, the whole-group findings demonstrated increased SA in the cannabis users compared to controls in a region denoted as the cuneus in the left hemisphere. The cuneus is functionally connected with parietal and other occipital regions for the purposes of integrating visual information [93] and contains CB1 receptors [94]. This region has a non-linear trajectory of neuronal maturation [95] and thus larger SA in cannabis users may be evident of delayed development in this region compared to non-using counterparts. Further, cannabis users have previously demonstrated abnormal dose–response blood-oxygenated-level-dependent signaling in the cuneus [96] along with aberrant functional activation in the occipital region more broadly [13,97]. The present analysis builds on this literature by demonstrating SA in this region is structurally different between adolescent and young adult cannabis users and non-users. Yet, this is inconsistent with prior findings demonstrating either no differences between cannabis users and controls in cuneus SA [36,38,40], or research reporting decreased LGI in prefrontal and temporal lobes in cannabis users [36,37] and early onset cannabis use [38], and reduced SA in comorbid cannabis and alcohol-using adolescents [39]. These inconsistencies may be due to differences in gender distribution across the studies, all of which skewed male [55% (current study), 56% [39], 64% [37], 67% [40], 73% [38], and 77% [36] male], varying levels of aerobic fitness in the sample (current sample was balanced for recent physical activity levels), or accounting for aerobic fitness levels within the statistical design. In addition, these inconsistencies may be due to differing age of samples [36,38], with younger samples undergoing greater neurodevelopment; or, due to analysis design—e.g., whole-brain analyses compared to ROI analyses [37]. None of the prior studies examining these outcomes tested whether gender or aerobic fitness moderated these effects or controlled for aerobic fitness level. Although we found the relationship between cannabis use and cortical surface structure was moderated by two-way interactions which were observed between gender, cannabis and aerobic fitness in frontal, cingulate, and parietal regions; regions that have been found to be abnormal in previous studies [21,23,24,36,37,39]. Thus, we will focus on these novel interactions.

Male cannabis users had lower SA in left precuneus, rostral middle frontal, and right superior frontal and lower LGI in left precentral, lateral orbitofrontal, and right supramarginal regions compared to non-using males—after accounting for alcohol use, cotinine level, and aerobic fitness level. Of note, our sample underwent three-weeks of monitored abstinence from cannabis and other drugs of abuse before structural scans were conducted; thus, THC and other exogenous cannabinoids were metabolized out of the system [98], representing chronic rather than residual associations. These fronto-parietal cortical findings are consistent with several studies reporting abnormal brain morphometry in cannabis users, including reduced volume [21,23,24,99] and lesser cortical gyrification [36,37] in samples that were primarily male. These overall interactions did uncover SA abnormalities in the male cannabis users, which is inconsistent with prior null or marginal SA findings [37,40,100], however, future gender-stratified analyses are needed to ascertain gender-specific mechanisms in the relationship between cortical surface structure and cannabis. In females, cannabis use was linked with greater SA in left precuneus, rostral middle frontal, and right superior frontal and greater LGI in left precentral and supramarginal regions compared to non-using females. Interestingly, female cannabis users also exhibited slightly reduced SA in another right superior frontal region and within left lateral orbitofrontal LGI. The increased SA and LGI findings could suggest that cannabis use in adolescent and young adult females is either advantageous, or potentially represents delayed pruning, as the female trajectory of pruning has an earlier rate compared to males [101]. Of note, our groups did not differ in age and we are capturing a cross-sectional snapshot potentially depicting a delay due to cannabis use specifically in females, whereas, smaller volumes in male cannabis users could be interpreted as detrimental effects associated with use. Interpretation of the present female findings are consistent with prior studies reporting greater brain volumes in female adolescent cannabis users [17,19]. More pronounced differences in males is consistent with prior findings that male cannabis users were more vulnerable to neurocognitive deficits in sequencing ability compared to females [66]. Interestingly, across genders, LGI differences were equally apparent compared to SA differences; indicating that LGI, an understudied structural index in addiction literature, is potentially susceptible to environmental influences [30–33] and is a viable avenue for further investigation. It is notable that our prior work has found cognitive deficits in psychomotor speed, working memory, sustained attention, and inhibitory control in an overlapping sample of both male and female young adult cannabis users [53,66,76], findings that are consistent with recent meta-analysis [102] and longitudinal [103] studies. Thus, structural deviancies may represent a mechanism for downstream cognitive functioning, although future studies are needed to assess whether structural changes directly impact function.

These differential gender patterns in cannabis findings may be due to multiple reasons. One potential underlying cause is differential substance use patterns. While male or female cannabis-using groups did not statistically differ from one another in their use patterns, our male users had more cannabis use on average compared to female users. This is consistent with previous literature indicating males have more severe use patterns compared to females [63]. Moreover, male users may also be more prone to using greater individual doses or more potent THC products [104], although the current study cannot address that possibility. Male sensitivity may also be due to differences in CB1 receptor density, as greater density in males is observed in animal models [105,106]. These gender differences may also be influenced in part by inherent sexual dimorphism in neurodevelopment [107,108] and the introduction of cannabis into these staggered developmental trajectories. Another possibility is that there was less power to detect differences in females, as their sample size was smaller overall (female: $n = 33$ vs. male: $n = 41$) and within cannabis users (female: $n = 13$ vs. male: $n = 23$); thus, additional research utilizing larger samples specifically examining the relationship between gender and cannabis use on these morphometric outcomes is needed.

Regarding aerobic fitness, results demonstrated several main effects of VO_2 max across both cortical surface structure indices in both the cannabis users and non-users. Greater VO_2 max was associated with increased SA in left superior and inferior parietal, and right inferior parietal and inferior temporal regions; and, increased LGI in left superior temporal, and right lateral orbitofrontal and inferior parietal regions. These findings indicate that in both cannabis users and non-users, VO_2 max has a strong positive relationship with cortical surface structural indices. This is consistent with previous analyses in our lab showcasing a positive link between AE and neurocognitive outcomes [53], and recent reviews of brain morphological outcomes in aerobic fitness literature [109]. Further, these findings suggest that, similar to non-users, cannabis users appear to have a positive link between aerobic fitness and brain structure in several regions of interest. When examining the interaction between cannabis and aerobic fitness, an interaction was observed in left cuneus SA and left lateral occipital LGI, with non-using controls exhibiting positive associations between increased indices with greater VO_2 max compared to cannabis users who demonstrated either a flat or negative relationship. This finding suggests that the impact of aerobic fitness may be less pronounced in regular cannabis users in these particular regions. Albeit, main effects of VO_2 max suggest that both groups generally had a positive relationship between aerobic fitness and brain morphometry; which is consistent with prior studies in adolescents, young adults, and older adults [51,110,111]. On balance, we previously reported that aerobic fitness was linked with superior visual memory, verbal fluency, and sequencing ability and highly fit cannabis users performed better on psychomotor speed, visual memory and sequencing ability compared to low-fit users [53]. It is important to note that our participants had no comorbid metabolic conditions (e.g., hyperlipidemia, hypertension, diabetes) suggesting benefits of aerobic fitness even in physically healthy youth. Possible mechanisms supporting positive effects of aerobic fitness and brain structure are likely multi-factorial. As aforementioned, engaging in AE releases BDNF [42], vascular growth factors [112], insulin-like growth factor-1 (IGF-1) [113], neurogenesis [114], improved catecholaminergic signaling [115], increased c-FOS expression [41], is linked with increased hippocampal volume [116], and reduces inflammation and oxidative stress [44,45]. Future studies are needed to tease apart these potential underlying mechanisms.

As we have postulated previously [53,60], AE intervention may be a plausible avenue to explore in further studies aiming to reduce or ameliorate neurocognitive deviances associated with repeated and chronic cannabis use [92]. Indeed, other groups have reported that aerobically-fit cannabis users reported reduced craving and fewer symptoms of cannabis use disorder compared to unfit users [61,117]. One potential explanation of the interactive association between cannabis use and AE could be more aerobically-fit users are metabolizing exogenous cannabinoids out of the body faster, thus, mechanistically dampening the impact on cortical surface structure integrity and neurocognition more broadly. Previous literature has shown mixed findings in acute AE increasing cannabinoid metabolites (i.e., THCCOOH) [118,119]. Further, AE releases endocannabinoids [120–122], which may help mitigate the negative impact of repeated exogenous cannabis exposure. Notable for future studies, it is hypothesized this relationship may be gender-specific; thus, research examining the potential use of aerobic interventions in cannabis users should prioritize investigation of potential gender differences.

There are potential weaknesses to note. First, causality cannot be determined from the present sample, sMRI scans were obtained after regular cannabis use was established. Second, the sample size for female cannabis users was relatively small compared to male cannabis users; still, findings support gender-specific cannabis and aerobic fitness associations through the interactions uncovered. This lends additional evidence for the need to examine sex as a potential moderator of cannabis effects, which has largely been understudied in cortical surface structure indices. Findings may not generalize to other samples of cannabis users with substantially different use patterns, e.g., length of abstinence, age of regular use onset, or other substance use. Third, concerns have been raised within the literature regarding divergent findings based on different brain morphometry detection algorithms and software [123]. While several validation studies of Freesurfer's surface indices have been published [89,124,125], future work should examine how differing surface-based analyses may produce slightly different regions of interest from cannabis use in adolescents and young adults. Fourth, the relationship between cannabis use and cortical surface structure has other potential moderators of interest; two important ones are genetic factors [126–128] and psychiatric comorbidities [129–131]. The ability to simultaneously examine several potential moderators of cannabis effects on neurocognitive outcomes will soon be available with the large-scale, prospective, longitudinal Adolescent Brain Cognitive Development (ABCD) Study [132] (www.abcdstudy.org/), which has enrolled over 11,800 youth. Lastly, our average VO_2 max was lower than age-based norms [79], so assessing a more aerobically fit group may demonstrate even stronger associations with cortical surface structure.

The current study found that cannabis users had larger cuneus SA and male cannabis users exhibited smaller SA and less complex LGI in frontal, cingulate and parietal regions, even after three weeks of monitored abstinence, compared to male non-users. In contrast, female cannabis users generally demonstrated increased SA and LGI in the aforementioned regions. Prospective, longitudinal studies, such as the ABCD Study, are needed to address whether abnormalities in LGI were caused by cannabis or due to premorbid factors, and identifying the potentially gender-specific developmental trajectories in the impact of cannabis use on the brain. We also found that both cannabis users and non-using controls had a significant link between increased aerobic fitness and more complex LGI and larger SA in frontal, parietal, and temporal regions. These findings, combined with our prior report of superior cognitive functioning in aerobically-fit cannabis users [53], support the notion that it is viable to investigate whether enhancing aerobic fitness, through AE, may be a feasible prevention or ameliorative tool aimed at reducing the impact of chronic cannabis on neurocognitive outcomes in adolescents and young adults.

Author Contributions: Conceptualization, R.M.S. and K.M.L.; Data curation, N.E.W. and K.M.L.; Formal analysis, R.M.S. and A.L.W.; Funding acquisition, K.M.L.; Investigation, K.M.L.; Methodology, R.M.S., A.L.W., N.E.W. and K.M.L.; Supervision, K.M.L.; Writing—original draft, R.M.S.; Writing—review & editing, A.L.W., N.E.W., A.M.S. and K.M.L. All authors have read and agreed to the published version of the manuscript.

Funding: This work was supported by the National Institute on Drug Abuse to KML (R01DA030354; U01DA041025) and the National Institute of Health to NEW (T32AA013525; P.I.: Riley/Tapert).

Conflicts of Interest: The authors declare no conflict of interest.

References

1. Degenhardt, L.; Bharat, C.; Glantz, M.D.; Sampson, N.A.; Al-Hamzawi, A.; Alonso, J.; Andrade, L.H.; Bunting, B.; Cia, A.; de Girolamo, G.; et al. Association of Cohort and Individual Substance Use with Risk of Transitioning to Drug Use, Drug Use Disorder, and Remission from Disorder: Findings From the World Mental Health Surveys. *JAMA Psychiatry* **2019**, *76*, 708–720. [CrossRef]

2. Schulenberg, J.E.; Johnston, L.D.; O'Malley, P.M.; Bachman, J.G.; Miech, R.A.; Patrick, M.E. *Monitoring the Future National Survey Results on Drug Use, 1975–2018: Volume II, College Students and Adults Ages 19–60*; Institute for Social Research: Ann Arbor, MI, USA, 2019. [CrossRef]
3. Johnston, L.D.; Miech, R.A.; O'Malley, P.M.; Bachman, J.G.; Schulenberg, J.E.; Patrick, M.E. *Monitoring the Future National Survey Results on Drug Use, 1975–2018: Overview, Key Findings on Adolescent Drug Use*; Institute for Social Research: Ann Arbor, MI, USA, 2019. [CrossRef]
4. Han, B.; Compton, W.M.; Blanco, C.; Jones, C.M. Time since first cannabis use and 12-month prevalence of cannabis use disorder among youth and emerging adults in the United States. *Addiction* **2019**, *114*, 698–707. [CrossRef]
5. Carliner, H.; Brown, Q.L.; Sarvet, A.L. Hasin, D.S. Cannabis use, attitudes, and legal status in the U.S.: A review. *Prev. Med.* **2017**, *104*, 13–23. [CrossRef] [PubMed]
6. Hasin, D.S.; Kerridge, B.T.; Saha, T.D.; Huang, B.; Pickering, R.; Smith, S.M.; Jung, J.; Zhang, H.; Grant, B.F. Prevalence and correlates of DSM-5 cannabis use disorder, 2012–2013: Findings from the National Epidemiologic Survey on Alcohol and Related Conditions-III. *Am. J. Psychiatry* **2016**, *173*, 588–599. [CrossRef] [PubMed]
7. Lisdahl, K.M.; Wright, N.E.; Kirchner-Medina, C.; Maple, K.E.; Shollenbarger, S. Considering Cannabis: The Effects of Regular Cannabis Use on Neurocognition in Adolescents and Young Adults. *Curr. Addict. Rep.* **2014**, *1*, 144–156. [CrossRef] [PubMed]
8. Lorenzetti, V.; Solowij, N.; Fornito, A.; Lubman, D.I.; Yucel, M. The association between regular cannabis exposure and alterations of human brain morphology: An updated review of the literature. *Curr. Pharm. Des.* **2014**, *20*, 2138–2167. [CrossRef] [PubMed]
9. Mechoulam, R.; Parker, L.A. The endocannabinoid system and the brain. *Annu. Rev. Psychol.* **2013**, *64*, 21–47. [CrossRef]
10. Villares, J. Chronic use of marijuana decreases cannabinoid receptor binding and mRNA expression in the human brain. *Neuroscience* **2007**, *145*, 323–334. [CrossRef]
11. Hirvonen, J.; Goodwin, R.S.; Li, C.T.; Terry, G.E.; Zoghbi, S.S.; Morse, C.; Pike, V.W.; Volkow, N.D.; Huestis, M.A.; Innis, R.B. Reversible and regionally selective downregulation of brain cannabinoid CB1 receptors in chronic daily cannabis smokers. *Mol. Psychiatry* **2012**, *17*, 642–649. [CrossRef]
12. Bloomfield, M.A.P.; Hindocha, C.; Green, S.F.; Wall, M.B.; Lees, R.; Petrilli, K.; Costello, H.; Ogunbiyi, M.O.; Bossong, M.G.; Freeman, T.P. The neuropsychopharmacology of cannabis: A review of human imaging studies. *Pharmacol. Ther.* **2019**, *195*, 132–161. [CrossRef]
13. Batalla, A.; Bhattacharyya, S.; Yucel, M.; Fusar-Poli, P.; Crippa, J.A.; Nogue, S.; Torrens, M.; Pujol, J.; Farre, M.; Martin-Santos, R. Structural and functional imaging studies in chronic cannabis users: A systematic review of adolescent and adult findings. *PLoS ONE* **2013**, *8*, e55821. [CrossRef] [PubMed]
14. Rocchetti, M.; Crescini, A.; Borgwardt, S.; Caverzasi, E.; Politi, P.; Atakan, Z.; Fusar-Poli, P. Is cannabis neurotoxic for the healthy brain? A meta-analytical review of structural brain alterations in non-psychotic users. *Psychiatry Clin. Neurosci.* **2013**, *67*, 483–492. [CrossRef]
15. Ashtari, M.; Avants, B.; Cyckowski, L.; Cervellione, K.L.; Roofeh, D.; Cook, P.; Gee, J.; Sevy, S.; Kumra, S. Medial temporal structures and memory functions in adolescents with heavy cannabis use. *J. Psychiatr. Res.* **2011**, *45*, 1055–1066. [CrossRef] [PubMed]
16. Medina, K.L.; Schweinsburg, A.D.; Cohen-Zion, M.; Nagel, B.J.; Tapert, S.F. Effects of alcohol and combined marijuana and alcohol use during adolescence on hippocampal volume and asymmetry. *Neurotoxicol. Teratol.* **2007**, *29*, 141–152. [CrossRef] [PubMed]
17. Medina, K.L.; McQueeny, T.; Nagel, B.J.; Hanson, K.L.; Yang, T.T.; Tapert, S.F. Prefrontal cortex morphometry in abstinent adolescent marijuana users: Subtle gender effects. *Addict. Biol.* **2009**, *14*, 457–468. [CrossRef] [PubMed]
18. Yucel, M.; Solowij, N.; Respondek, C.; Whittle, S.; Fornito, A.; Pantelis, C.; Lubman, D.I. Regional brain abnormalities associated with long-term heavy cannabis use. *Arch. Gen. Psychiatry* **2008**, *65*, 694–701. [CrossRef]
19. McQueeny, T.; Padula, C.B.; Price, J.; Medina, K.L.; Logan, P.; Tapert, S.F. Gender effects on amygdala morphometry in adolescent marijuana users. *Behav. Brain Res.* **2011**, *224*, 128–134. [CrossRef]
20. Mashhoon, Y.; Sava, S.; Sneider, J.T.; Nickerson, L.D.; Silveri, M.M. Cortical thinness and volume differences associated with marijuana abuse in emerging adults. *Drug Alcohol Depend.* **2015**, *155*, 275–283. [CrossRef]

21. Price, J.S.; McQueeny, T.; Shollenbarger, S.; Browning, E.L.; Wieser, J.; Lisdahl, K.M. Effects of marijuana use on prefrontal and parietal volumes and cognition in emerging adults. *Psychopharmacology* **2015**, *232*, 2939–2950. [CrossRef]
22. Filbey, F.M.; Aslan, S.; Calhoun, V.D.; Spence, J.S.; Damaraju, E.; Caprihan, A.; Segall, J. Long-term effects of marijuana use on the brain. *Proc. Natl. Acad. Sci. USA* **2014**, *111*, 16913–16918. [CrossRef]
23. Lisdahl, K.M.; Tamm, L.; Epstein, J.N.; Jernigan, T.; Molina, B.S.; Hinshaw, S.P.; Swanson, J.M.; Newman, E.; Kelly, C.; Bjork, J.M.; et al. The impact of ADHD persistence, recent cannabis use, and age of regular cannabis use onset on subcortical volume and cortical thickness in young adults. *Drug Alcohol Depend.* **2016**, *161*, 135–146. [CrossRef] [PubMed]
24. Maple, K.E.; Thomas, A.M.; Kangiser, M.M.; Lisdahl, K.M. Anterior cingulate volume reductions in abstinent adolescent and young adult cannabis users: Association with affective processing deficits. *Psychiatry Res. Neuroimaging* **2019**, *288*, 51–59. [CrossRef] [PubMed]
25. Matochik, J.A.; Eldreth, D.A.; Cadet, J.L.; Bolla, K.I. Altered brain tissue composition in heavy marijuana users. *Drug Alcohol Depend.* **2005**, *77*, 23–30. [CrossRef] [PubMed]
26. Medina, K.L.; Nagel, B.J.; Tapert, S.F. Abnormal cerebellar morphometry in abstinent adolescent marijuana users. *Psychiatry Res.* **2010**, *182*, 152–159. [CrossRef] [PubMed]
27. Lorenzetti, V.; Chye, Y.; Silva, P.; Solowij, N.; Roberts, C.A. Does regular cannabis use affect neuroanatomy? An updated systematic review and meta-analysis of structural neuroimaging studies. *Eur. Arch. Psychiatry Clin. Neurosci.* **2019**, *269*, 59–71. [CrossRef] [PubMed]
28. White, T.; Su, S.; Schmidt, M.; Kao, C.Y.; Sapiro, G. The development of gyrification in childhood and adolescence. *Brain Cognit.* **2010**, *72*, 36–45. [CrossRef] [PubMed]
29. Hogstrom, L.J.; Westlye, L.T.; Walhovd, K.B.; Fjell, A.M. The structure of the cerebral cortex across adult life: Age-related patterns of surface area, thickness, and gyrification. *Cereb. Cortex* **2013**, *23*, 2521–2530. [CrossRef]
30. Infante, M.A.; Moore, E.M.; Bischoff-Grethe, A.; Migliorini, R.; Mattson, S.N.; Riley, E.P. Atypical cortical gyrification in adolescents with histories of heavy prenatal alcohol exposure. *Brain Res.* **2015**, *1624*, 446–454. [CrossRef]
31. Jednorog, K.; Altarelli, I.; Monzalvo, K.; Fluss, J.; Dubois, J.; Billard, C.; Dehaene-Lambertz, G.; Ramus, F. The influence of socioeconomic status on children's brain structure. *PLoS ONE* **2012**, *7*, e42486. [CrossRef]
32. Lohmann, G.; von Cramon, D.Y.; Steinmetz, H. Sulcal variability of twins. *Cereb. Cortex* **1999**, *9*, 754–763. [CrossRef]
33. Kuhn, S.; Witt, C.; Banaschewski, T.; Barbot, A.; Barker, G.J.; Buchel, C.; Conrod, P.J.; Flor, H.; Garavan, H.; Ittermann, B.; et al. From mother to child: Orbitofrontal cortex gyrification and changes of drinking behaviour during adolescence. *Addict. Biol.* **2016**, *21*, 700–708. [CrossRef]
34. Schnack, H.G.; van Haren, N.E.; Brouwer, R.M.; Evans, A.; Durston, S.; Boomsma, D.I.; Kahn, R.S.; Hulshoff Pol, H.E. Changes in thickness and surface area of the human cortex and their relationship with intelligence. *Cereb. Cortex* **2015**, *25*, 1608–1617. [CrossRef] [PubMed]
35. Panizzon, M.S.; Fennema-Notestine, C.; Eyler, L.T.; Jernigan, T.L.; Prom-Wormley, E.; Neale, M.; Jacobson, K.; Lyons, M.J.; Grant, M.D.; Franz, C.E.; et al. Distinct genetic influences on cortical surface area and cortical thickness. *Cereb. Cortex* **2009**, *19*, 2728–2735. [CrossRef] [PubMed]
36. Mata, I.; Perez-Iglesias, R.; Roiz-Santianez, R.; Tordesillas-Gutierrez, D.; Pazos, A.; Gutierrez, A.; Vazquez-Barquero, J.L.; Crespo-Facorro, B. Gyrification brain abnormalities associated with adolescence and early-adulthood cannabis use. *Brain Res.* **2010**, *1317*, 297–304. [CrossRef]
37. Shollenbarger, S.G.; Price, J.; Wieser, J.; Lisdahl, K. Impact of cannabis use on prefrontal and parietal cortex gyrification and surface area in adolescents and emerging adults. *Dev. Cognit. Neurosci.* **2015**, *16*, 46–53. [CrossRef] [PubMed]
38. Filbey, F.M.; McQueeny, T.; DeWitt, S.J.; Mishra, V. Preliminary findings demonstrating latent effects of early adolescent marijuana use onset on cortical architecture. *Dev. Cognit. Neurosci.* **2015**, *16*, 16–22. [CrossRef]
39. Infante, M.A.; Courtney, K.E.; Castro, N.; Squeglia, L.M.; Jacobus, J. Adolescent Brain Surface Area Pre- and Post-Cannabis and Alcohol Initiation. *J. Stud. Alcohol Drugs* **2018**, *79*, 835–843. [CrossRef]
40. Chye, Y.; Suo, C.; Lorenzetti, V.; Batalla, A.; Cousijn, J.; Goudriaan, A.E.; Martin-Santos, R.; Whittle, S.; Solowij, N.; Yucel, M. Cortical surface morphology in long-term cannabis users: A multi-site MRI study. *Eur. Neuropsychopharmacol.* **2019**, *29*, 257–265. [CrossRef]

41. Sim, Y.J.; Kim, H.; Shin, M.S.; Chang, H.K.; Shin, M.C.; Ko, I.G.; Kim, K.J.; Kim, T.S.; Kim, B.K.; Rhim, Y.T.; et al. Effect of postnatal treadmill exercise on c-Fos expression in the hippocampus of rat pups born from the alcohol-intoxicated mothers. *Brain Dev.* **2008**, *30*, 118–125. [CrossRef]
42. Huang, T.; Larsen, K.T.; Ried-Larsen, M.; Moller, N.C.; Andersen, L.B. The effects of physical activity and exercise on brain-derived neurotrophic factor in healthy humans: A review. *Scand. J. Med. Sci. Sports* **2014**, *24*, 1–10. [CrossRef]
43. Hill, M.N.; Titterness, A.K.; Morrish, A.C.; Carrier, E.J.; Lee, T.T.; Gil-Mohapel, J.; Gorzalka, B.B.; Hillard, C.J.; Christie, B.R. Endogenous cannabinoid signaling is required for voluntary exercise-induced enhancement of progenitor cell proliferation in the hippocampus. *Hippocampus* **2010**, *20*, 513–523. [CrossRef] [PubMed]
44. Sakurai, T.; Izawa, T.; Kizaki, T.; Ogasawara, J.E.; Shirato, K.; Imaizumi, K.; Takahashi, K.; Ishida, H.; Ohno, H. Exercise training decreases expression of inflammation-related adipokines through reduction of oxidative stress in rat white adipose tissue. *Biochem. Biophys. Res. Commun.* **2009**, *379*, 605–609. [CrossRef] [PubMed]
45. Radak, Z.; Kumagai, S.; Taylor, A.W.; Naito, H.; Goto, S. Effects of exercise on brain function: Role of free radicals. *Appl. Physiol. Nutr. Metab.* **2007**, *32*, 942–946. [CrossRef] [PubMed]
46. Leasure, J.L.; Jones, M. Forced and voluntary exercise differentially affect brain and behavior. *Neuroscience* **2008**, *156*, 456–465. [CrossRef]
47. Helfer, J.L.; Goodlett, C.R.; Greenough, W.T.; Klintsova, A.Y. The effects of exercise on adolescent hippocampal neurogenesis in a rat model of binge alcohol exposure during the brain growth spurt. *Brain Res.* **2009**, *1294*, 1–11. [CrossRef]
48. Thomas, A.G.; Dennis, A.; Bandettini, P.A.; Johansen-Berg, H. The effects of aerobic activity on brain structure. *Front. Psychol.* **2012**, *3*, 86. [CrossRef]
49. Smith, P.J.; Blumenthal, J.A.; Hoffman, B.M.; Cooper, H.; Strauman, T.A.; Welsh-Bohmer, K.; Browndyke, J.N.; Sherwood, A. Aerobic exercise and neurocognitive performance: A meta-analytic review of randomized controlled trials. *Psychosom. Med.* **2010**, *72*, 239–252. [CrossRef]
50. Bherer, L.; Erickson, K.I.; Liu-Ambrose, T. A review of the effects of physical activity and exercise on cognitive and brain functions in older adults. *J. Aging Res.* **2013**, *2013*, 657508. [CrossRef]
51. Herting, M.M.; Chu, X. Exercise, cognition, and the adolescent brain. *Birth Defects Res.* **2017**, *109*, 1672–1679. [CrossRef]
52. Wengaard, E.; Kristoffersen, M.; Harris, A.; Gundersen, H. Cardiorespiratory Fitness Is Associated with Selective Attention in Healthy Male High-School Students. *Front. Hum. Neurosci.* **2017**, *11*, 330. [CrossRef]
53. Wade, N.E.; Wallace, A.L.; Swartz, A.M.; Lisdahl, K.M. Aerobic Fitness Level Moderates the Association Between Cannabis Use and Executive Functioning and Psychomotor Speed Following Abstinence in Adolescents and Young Adults. *J. Int. Neuropsychol. Soc.* **2019**, *25*, 134–145. [CrossRef] [PubMed]
54. Hwang, J.; Castelli, D.M.; Gonzalez-Lima, F. The positive cognitive impact of aerobic fitness is associated with peripheral inflammatory and brain-derived neurotrophic biomarkers in young adults. *Physiol. Behav.* **2017**, *179*, 75–89. [CrossRef] [PubMed]
55. Chaddock, L.; Pontifex, M.B.; Hillman, C.H.; Kramer, A.F. A review of the relation of aerobic fitness and physical activity to brain structure and function in children. *J. Int. Neuropsychol. Soc.* **2011**, *17*, 975–985. [CrossRef] [PubMed]
56. Schwarb, H.; Johnson, C.L.; Daugherty, A.M.; Hillman, C.H.; Kramer, A.F.; Cohen, N.J.; Barbey, A.K. Aerobic fitness, hippocampal viscoelasticity, and relational memory performance. *NeuroImage* **2017**, *153*, 179–188. [CrossRef]
57. Whiteman, A.S.; Young, D.E.; Budson, A.E.; Stern, C.E.; Schon, K. Entorhinal volume, aerobic fitness, and recognition memory in healthy young adults: A voxel-based morphometry study. *NeuroImage* **2016**, *126*, 229–238. [CrossRef]
58. Herting, M.M.; Nagel, B.J. Differences in brain activity during a verbal associative memory encoding task in high- and low-fit adolescents. *J. Cognit. Neurosci.* **2013**, *25*, 595–612. [CrossRef]
59. Koltyn, K.F.; Brellenthin, A.G.; Cook, D.B.; Sehgal, N.; Hillard, C. Mechanisms of exercise-induced hypoalgesia. *J. Pain* **2014**, *15*, 1294–1304. [CrossRef]
60. Lisdahl, K.M.; Gilbart, E.R.; Wright, N.E.; Shollenbarger, S. Dare to delay? The impacts of adolescent alcohol and marijuana use onset on cognition, brain structure, and function. *Front. Psychiatry* **2013**, *4*, 53. [CrossRef]

61. Buchowski, M.S.; Meade, N.N.; Charboneau, E.; Park, S.; Dietrich, M.S.; Cowan, R.L.; Martin, P.R. Aerobic exercise training reduces cannabis craving and use in non-treatment seeking cannabis-dependent adults. *PLoS ONE* **2011**, *6*, e17465. [CrossRef]
62. Schlienz, N.J.; Budney, A.J.; Lee, D.C.; Vandrey, R. Cannabis Withdrawal: A Review of Neurobiological Mechanisms and Sex Differences. *Curr. Addict. Rep.* **2017**, *4*, 75–81. [CrossRef]
63. Khan, S.S.; Secades-Villa, R.; Okuda, M.; Wang, S.; Perez-Fuentes, G.; Kerridge, B.T.; Blanco, C. Gender differences in cannabis use disorders: Results from the National Epidemiologic Survey of Alcohol and Related Conditions. *Drug Alcohol Depend.* **2013**, *130*, 101–108. [CrossRef] [PubMed]
64. Fogel, J.S.; Kelly, T.H.; Westgate, P.M.; Lile, J.A. Sex differences in the subjective effects of oral Delta(9)-THC in cannabis users. *Pharm. Biochem. Behav.* **2017**, *152*, 44–51. [CrossRef] [PubMed]
65. Cooper, Z.D.; Haney, M. Investigation of sex-dependent effects of cannabis in daily cannabis smokers. *Drug Alcohol Depend.* **2014**, *136*, 85–91. [CrossRef] [PubMed]
66. Lisdahl, K.M.; Price, J.S. Increased marijuana use and gender predict poorer cognitive functioning in adolescents and emerging adults. *J. Int. Neuropsychol. Soc.* **2012**, *18*, 678–688. [CrossRef]
67. Luders, E.; Narr, K.L.; Thompson, P.M.; Rex, D.E.; Jancke, L.; Steinmetz, H.; Toga, A.W. Gender differences in cortical complexity. *Nat. Neurosci.* **2004**, *7*, 799–800. [CrossRef]
68. Luders, E.; Thompson, P.M.; Narr, K.L.; Toga, A.W.; Jancke, L.; Gaser, C. A curvature-based approach to estimate local gyrification on the cortical surface. *NeuroImage* **2006**, *29*, 1224–1230. [CrossRef]
69. Salinas, J.; Mills, E.D.; Conrad, A.L.; Koscik, T.; Andreasen, N.C.; Nopoulos, P. Sex differences in parietal lobe structure and development. *Gend. Med.* **2012**, *9*, 44–55. [CrossRef]
70. Wierenga, L.M.; Langen, M.; Oranje, B.; Durston, S. Unique developmental trajectories of cortical thickness and surface area. *NeuroImage* **2014**, *87*, 120–126. [CrossRef]
71. Raznahan, A.; Shaw, P.; Lalonde, F.; Stockman, M.; Wallace, G.L.; Greenstein, D.; Clasen, L.; Gogtay, N.; Giedd, J.N. How does your cortex grow? *J. Neurosci.* **2011**, *31*, 7174–7177. [CrossRef]
72. Dupuy, O.; Gauthier, C.J.; Fraser, S.A.; Desjardins-Crepeau, L.; Desjardins, M.; Mekary, S.; Lesage, F.; Hoge, R.D.; Pouliot, P.; Bherer, L. Higher levels of cardiovascular fitness are associated with better executive function and prefrontal oxygenation in younger and older women. *Front. Hum. Neurosci.* **2015**, *9*, 66. [CrossRef]
73. Killgore, W.D.; Schwab, Z.J. Sex differences in the association between physical exercise and IQ. *Percept. Mot. Ski.* **2012**, *115*, 605–617. [CrossRef] [PubMed]
74. Thomas, S.; Reading, J.; Shephard, R.J. Revision of the Physical Activity Readiness Questionnaire (PAR-Q). *Can. J. Sport Sci.* **1992**, *17*, 338–345. [PubMed]
75. Fogelholm, M.; Malmberg, J.; Suni, J.; Santtila, M.; Kyrolainen, H.; Mantysaari, M.; Oja, P. International Physical Activity Questionnaire: Validity against fitness. *Med. Sci. Sports Exerc* **2006**, *38*, 753–760. [CrossRef] [PubMed]
76. Wallace, A.L.; Wade, N.E.; Hatcher, K.F.; Lisdahl, K.M. Effects of Cannabis Use and Subclinical ADHD Symptomology on Attention Based Tasks in Adolescents and Young Adults. *Arch. Clin. Neuropsychol.* **2018**, *34*, 700–705. [CrossRef] [PubMed]
77. Meier, M.H.; Caspi, A.; Ambler, A.; Harrington, H.; Houts, R.; Keefe, R.S.; McDonald, K.; Ward, A.; Poulton, R.; Moffitt, T.E. Persistent cannabis users show neuropsychological decline from childhood to midlife. *Proc. Natl. Acad. Sci. USA* **2012**, *109*, E2657–E2664. [CrossRef] [PubMed]
78. Sobell, L.C.; Sobell, M.B. Timeline Follow-Back. In *Measuring Alcohol Consumption*; Litten, R.Z., Allen, J.P., Eds.; Humana Press: Totowa, NJ, USA, 1992; pp. 41–72. [CrossRef]
79. Pescatello, L.S. *ACSM's Guidelines for Exercise Testing and Prescription*, 9th ed.; Wolters Kluwer/Lippincott Williams & Wilkins Health: Philadelphia, PA, USA, 2014.
80. Fitness Ranking by the KU Alzheimers Disease Center. Available online: http://www.kumc.edu/fitness-ranking.html (accessed on 10 November 2019).
81. Wallace, A.L.; Wade, N.E.; Lisdahl, K.M. Impact of Two-Weeks of Monitored Abstinence on Cognition in Adolescent and Young Adult Cannabis Users. *J. Int. Neuropsychol. Soc.* **2020**, in press.
82. Brown, S.A.; Myers, M.G.; Lippke, L.; Tapert, S.F.; Stewart, D.G.; Vik, P.W. Psychometric evaluation of the Customary Drinking and Drug Use Record (CDDR): A measure of adolescent alcohol and drug involvement. *J. Stud. Alcohol* **1998**, *59*, 427–438. [CrossRef]

83. Sheehan, D.V.; Lecrubier, Y.; Sheehan, K.H.; Amorim, P.; Janavs, J.; Weiller, E.; Hergueta, T.; Baker, R.; Dunbar, G.C. The Mini-International Neuropsychiatric Interview (M.I.N.I.): The development and validation of a structured diagnostic psychiatric interview for DSM-IV and ICD-10. *J. Clin. Psychiatry* **1998**, *59*, 22–23.
84. Sheehan, D.V.; Sheehan, K.H.; Shytle, R.D.; Janavs, J.; Bannon, Y.; Rogers, J.E.; Milo, K.M.; Stock, S.L.; Wilkinson, B. Reliability and validity of the Mini International Neuropsychiatric Interview for Children and Adolescents (MINI-KID). *J. Clin. Psychiatry* **2010**, *71*, 313–326. [CrossRef]
85. Howley, E.T.; Bassett, D.R., Jr.; Welch, H.G. Criteria for maximal oxygen uptake: Review and commentary. *Med. Sci. Sports Exerc.* **1995**, *27*, 1292–1301. [CrossRef]
86. Schaer, M.; Cuadra, M.B.; Schmansky, N.; Fischl, B.; Thiran, J.P.; Eliez, S. How to measure cortical folding from MR images: A step-by-step tutorial to compute local gyrification index. *J. Vis. Exp.* **2012**, e3417. [CrossRef] [PubMed]
87. R Core Team. *R: A Language and Environment for Statistical Computing, 3.5.2*; R Found. for Statistical Computing: Vienna, Austria, 2010; Available online: https://www.R-project.org/ (accessed on 22 February 2020).
88. Dale, A.M.; Fischl, B.; Sereno, M.I. Cortical surface-based analysis. I. Segmentation and surface reconstruction. *NeuroImage* **1999**, *9*, 179–194. [CrossRef] [PubMed]
89. Schaer, M.; Cuadra, M.B.; Tamarit, L.; Lazeyras, F.; Eliez, S.; Thiran, J.P. A surface-based approach to quantify local cortical gyrification. *IEEE Trans. Med. Imaging* **2008**, *27*, 161–170. [CrossRef] [PubMed]
90. Desikan, R.S.; Segonne, F.; Fischl, B.; Quinn, B.T.; Dickerson, B.C.; Blacker, D.; Buckner, R.L.; Dale, A.M.; Maguire, R.P.; Hyman, B.T.; et al. An automated labeling system for subdividing the human cerebral cortex on MRI scans into gyral based regions of interest. *NeuroImage* **2006**, *31*, 968–980. [CrossRef] [PubMed]
91. Degenhardt, L.; Chiu, W.T.; Sampson, N.; Kessler, R.C.; Anthony, J.C.; Angermeyer, M.; Bruffaerts, R.; de Girolamo, G.; Gureje, O.; Huang, Y.; et al. Toward a global view of alcohol, tobacco, cannabis, and cocaine use: Findings from the WHO World Mental Health Surveys. *PLoS Med.* **2008**, *5*, e141. [CrossRef]
92. Brellenthin, A.G.; Koltyn, K.F. Exercise as an adjunctive treatment for cannabis use disorder. *Am. J. Drug Alcohol Abus.* **2016**, *42*, 481–489. [CrossRef]
93. Tomasi, D.; Volkow, N.D. Association between functional connectivity hubs and brain networks. *Cereb. Cortex* **2011**, *21*, 2003–2013. [CrossRef]
94. Eggan, S.M.; Lewis, D.A. Immunocytochemical distribution of the cannabinoid CB1 receptor in the primate neocortex: A regional and laminar analysis. *Cereb. Cortex* **2007**, *17*, 175–191. [CrossRef]
95. Vijayakumar, N.; Allen, N.B.; Youssef, G.; Dennison, M.; Yucel, M.; Simmons, J.G.; Whittle, S. Brain development during adolescence: A mixed-longitudinal investigation of cortical thickness, surface area, and volume. *Hum. Brain Mapp.* **2016**, *37*, 2027–2038. [CrossRef]
96. Tervo-Clemmens, B.; Simmonds, D.; Calabro, F.J.; Montez, D.F.; Lekht, J.A.; Day, N.L.; Richardson, G.A.; Luna, B. Early Cannabis Use and Neurocognitive Risk: A Prospective Functional Neuroimaging Study. *Biol. Psychiatry Cognit. Neurosci. Neuroimaging* **2018**, *3*, 713–725. [CrossRef]
97. Martin-Santos, R.; Fagundo, A.B.; Crippa, J.A.; Atakan, Z.; Bhattacharyya, S.; Allen, P.; Fusar-Poli, P.; Borgwardt, S.; Seal, M.; Busatto, G.F.; et al. Neuroimaging in cannabis use: A systematic review of the literature. *Psychol. Med.* **2010**, *40*, 383–398. [CrossRef] [PubMed]
98. Lowe, R.H.; Abraham, T.T.; Darwin, W.D.; Herning, R.; Cadet, J.L.; Huestis, M.A. Extended urinary Δ9-tetrahydrocannabinol excretion in chronic cannabis users precludes use as a biomarker of new drug exposure. *Drug Alcohol Depend.* **2009**, *105*, 24–32. [CrossRef] [PubMed]
99. Battistella, G.; Fornari, E.; Annoni, J.M.; Chtioui, H.; Dao, K.; Fabritius, M.; Favrat, B.; Mall, J.F.; Maeder, P.; Giroud, C. Long-term effects of cannabis on brain structure. *Neuropsychopharmacol. Off. Publ. Am. Coll. Neuropsychopharmacol.* **2014**, *39*, 2041–2048. [CrossRef] [PubMed]
100. Weiland, B.J.; Thayer, R.E.; Depue, B.E.; Sabbineni, A.; Bryan, A.D.; Hutchison, K.E. Daily marijuana use is not associated with brain morphometric measures in adolescents or adults. *J. Neurosci.* **2015**, *35*, 1505–1512. [CrossRef]
101. Lenroot, R.K.; Giedd, J.N. Sex differences in the adolescent brain. *Brain Cognit.* **2010**, *72*, 46–55. [CrossRef]
102. Scott, E.P.; Brennan, E.; Benitez, A. A Systematic Review of the Neurocognitive Effects of Cannabis Use in Older Adults. *Curr. Addict. Rep.* **2019**. [CrossRef]

103. Gonzalez, R.; Pacheco-Colon, I.; Duperrouzel, J.C.; Hawes, S.W. Does Cannabis Use Cause Declines in Neuropsychological Functioning? A Review of Longitudinal Studies. *J. Int. Neuropsychol. Soc.* **2017**, *23*, 893–902. [CrossRef]
104. Cuttler, C.; Mischley, L.K.; Sexton, M. Sex Differences in Cannabis Use and Effects: A Cross-Sectional Survey of Cannabis Users. *Cannabis Cannabinoid Res.* **2016**, *1*, 166–175. [CrossRef]
105. Burston, J.J.; Wiley, J.L.; Craig, A.A.; Selley, D.E.; Sim-Selley, L.J. Regional enhancement of cannabinoid CB1 receptor desensitization in female adolescent rats following repeated Delta-tetrahydrocannabinol exposure. *Br. J. Pharmacol.* **2010**, *161*, 103–112. [CrossRef]
106. Rubino, T.; Vigano, D.; Realini, N.; Guidali, C.; Braida, D.; Capurro, V.; Castiglioni, C.; Cherubino, F.; Romualdi, P.; Candeletti, S.; et al. Chronic delta 9-tetrahydrocannabinol during adolescence provokes sex-dependent changes in the emotional profile in adult rats: Behavioral and biochemical correlates. *Neuropsychopharmacol. Off. Publ. Am. Coll. Neuropsychopharmacol.* **2008**, *33*, 2760–2771. [CrossRef]
107. Lenroot, R.K.; Gogtay, N.; Greenstein, D.K.; Wells, E.M.; Wallace, G.L.; Clasen, L.S.; Blumenthal, J.D.; Lerch, J.; Zijdenbos, A.P.; Evans, A.C.; et al. Sexual dimorphism of brain developmental trajectories during childhood and adolescence. *NeuroImage* **2007**, *36*, 1065–1073. [CrossRef] [PubMed]
108. Giedd, J.N.; Blumenthal, J.; Jeffries, N.O.; Castellanos, F.X.; Liu, H.; Zijdenbos, A.; Paus, T.; Evans, A.C.; Rapoport, J.L. Brain development during childhood and adolescence: A longitudinal MRI study. *Nat. Neurosci.* **1999**, *2*, 861–863. [CrossRef] [PubMed]
109. Wittfeld, K.; Jochem, C.; Dorr, M.; Schminke, U.; Glaser, S.; Bahls, M.; Markus, M.R.P.; Felix, S.B.; Leitzmann, M.F.; Ewert, R.; et al. Cardiorespiratory Fitness and Gray Matter Volume in the Temporal, Frontal, and Cerebellar Regions in the General Population. *Mayo Clin. Proc.* **2020**, *95*, 44–56. [CrossRef] [PubMed]
110. Herting, M.M.; Keenan, M.F. Exercise and the Developing Brain in Children and Adolescents. In *Physical Activity and the Aging Brain*; Watson, R.R., Ed.; Academic Press: London, UK, 2017; pp. 13–19. [CrossRef]
111. Haeger, A.; Costa, A.S.; Schulz, J.B.; Reetz, K. Cerebral changes improved by physical activity during cognitive decline: A systematic review on MRI studies. *NeuroImage Clin.* **2019**, *23*, 101933. [CrossRef]
112. Fleenor, B.S.; Marshall, K.D.; Durrant, J.R.; Lesniewski, L.A.; Seals, D.R. Arterial stiffening with ageing is associated with transforming growth factor-beta1-related changes in adventitial collagen: Reversal by aerobic exercise. *J. Physiol.* **2010**, *588*, 3971–3982. [CrossRef]
113. Lopez-Lopez, C.; LeRoith, D.; Torres-Aleman, I. Insulin-like growth factor I is required for vessel remodeling in the adult brain. *Proc. Natl. Acad. Sci. USA* **2004**, *101*, 9833–9838. [CrossRef]
114. Nokia, M.S.; Lensu, S.; Ahtiainen, J.P.; Johansson, P.P.; Koch, L.G.; Britton, S.L.; Kainulainen, H. Physical exercise increases adult hippocampal neurogenesis in male rats provided it is aerobic and sustained. *J. Physiol.* **2016**, *594*, 1855–1873. [CrossRef]
115. Waters, R.P.; Emerson, A.J.; Watt, M.J.; Forster, G.L.; Swallow, J.G.; Summers, C.H. Stress induces rapid changes in central catecholaminergic activity in Anolis carolinensis: Restraint and forced physical activity. *Brain Res. Bull.* **2005**, *67*, 210–218. [CrossRef]
116. Frodl, T.; Strehl, K.; Carballedo, A.; Tozzi, L.; Doyle, M.; Amico, F.; Gormley, J.; Lavelle, G.; O'Keane, V. Aerobic exercise increases hippocampal subfield volumes in younger adults and prevents volume decline in the elderly. *Brain Imaging Behav.* **2019**. [CrossRef]
117. Henchoz, Y.; Dupuis, M.; Deline, S.; Studer, J.; Baggio, S.; N'Goran, A.A.; Daeppen, J.B.; Gmel, G. Associations of physical activity and sport and exercise with at-risk substance use in young men: A longitudinal study. *Prev. Med.* **2014**, *64*, 27–31. [CrossRef]
118. Westin, A.A.; Mjones, G.; Burchardt, O.; Fuskevag, O.M.; Slordal, L. Can physical exercise or food deprivation cause release of fat-stored cannabinoids? *Basic Clin. Pharm. Toxicol.* **2014**, *115*, 467–471. [CrossRef] [PubMed]
119. Wong, A.; Montebello, M.E.; Norberg, M.M.; Rooney, K.; Lintzeris, N.; Bruno, R.; Booth, J.; Arnold, J.C.; McGregor, I.S. Exercise increases plasma THC concentrations in regular cannabis users. *Drug Alcohol Depend.* **2013**, *133*, 763–767. [CrossRef] [PubMed]
120. Meyer, J.D.; Crombie, K.M.; Cook, D.B.; Hillard, C.J.; Koltyn, K.F. Serum Endocannabinoid and Mood Changes after Exercise in Major Depressive Disorder. *Med. Sci. Sports Exerc.* **2019**, *51*, 1909–1917. [CrossRef] [PubMed]
121. Watkins, B.A. Endocannabinoids, exercise, pain, and a path to health with aging. *Mol. Asp. Med.* **2018**, *64*, 68–78. [CrossRef]

122. Hillard, C.J. Circulating Endocannabinoids: From Whence Do They Come and where are They Going? *Neuropsychopharmacol. Off. Publ. Am. Coll. Neuropsychopharmacol.* **2018**, *43*, 155–172. [CrossRef]
123. Gao, Y.; Riklin-Raviv, T.; Bouix, S. Shape analysis, a field in need of careful validation. *Hum. Brain Mapp.* **2014**, *35*, 4965–4978. [CrossRef]
124. Lee, J.K.; Lee, J.M.; Kim, J.S.; Kim, I.Y.; Evans, A.C.; Kim, S.I. A novel quantitative cross-validation of different cortical surface reconstruction algorithms using MRI phantom. *NeuroImage* **2006**, *31*, 572–584. [CrossRef]
125. Segonne, F.; Pacheco, J.; Fischl, B. Geometrically accurate topology-correction of cortical surfaces using nonseparating loops. *IEEE Trans. Med. Imaging* **2007**, *26*, 518–529. [CrossRef]
126. Verweij, K.J.; Zietsch, B.P.; Lynskey, M.T.; Medland, S.E.; Neale, M.C.; Martin, N.G.; Boomsma, D.I.; Vink, J.M. Genetic and environmental influences on cannabis use initiation and problematic use: A meta-analysis of twin studies. *Addiction* **2010**, *105*, 417–430. [CrossRef]
127. Shollenbarger, S.G.; Price, J.; Wieser, J.; Lisdahl, K. Poorer frontolimbic white matter integrity is associated with chronic cannabis use, FAAH genotype, and increased depressive and apathy symptoms in adolescents and young adults. *NeuroImage Clin.* **2015**, *8*, 117–125. [CrossRef]
128. Filbey, F.M.; Schacht, J.P.; Myers, U.S.; Chavez, R.S.; Hutchison, K.E. Individual and additive effects of the CNR1 and FAAH genes on brain response to marijuana cues. *Neuropsychopharmacol. Off. Publ. Am. Coll. Neuropsychopharmacol.* **2010**, *35*, 967–975. [CrossRef] [PubMed]
129. Moore, T.H.M.; Zammit, S.; Lingford-Hughes, A.; Barnes, T.R.E.; Jones, P.B.; Burke, M.; Lewis, G. Cannabis use and risk of psychotic or affective mental health outcomes: A systematic review. *Lancet* **2007**, *370*, 319–328. [CrossRef]
130. Crippa, J.A.; Zuardi, A.W.; Martin-Santos, R.; Bhattacharyya, S.; Atakan, Z.; McGuire, P.; Fusar-Poli, P. Cannabis and anxiety: A critical review of the evidence. *Hum. Psychopharmacol.* **2009**, *24*, 515–523. [CrossRef] [PubMed]
131. Lev-Ran, S.; Roerecke, M.; Le Foll, B.; George, T.P.; McKenzie, K.; Rehm, J. The association between cannabis use and depression: A systematic review and meta-analysis of longitudinal studies. *Psychol. Med.* **2014**, *44*, 797–810. [CrossRef] [PubMed]
132. Lisdahl, K.M.; Sher, K.J.; Conway, K.P.; Gonzalez, R.; Feldstein Ewing, S.W.; Nixon, S.J.; Tapert, S.; Bartsch, H.; Goldstein, R.Z.; Heitzeg, M. Adolescent brain cognitive development (ABCD) study: Overview of substance use assessment methods. *Dev. Cognit. Neurosci.* **2018**, *32*, 80–96. [CrossRef] [PubMed]

Perspective

The Endocannabinoid System as a Potential Mechanism through which Exercise Influences Episodic Memory Function

Paul D. Loprinzi [1,*], **Liye Zou** [2] and **Hong Li** [3,4,5,*]

1 Exercise & Memory Laboratory, Department of Health, Exercise Science and Recreation Management, The University of Mississippi, Oxford, MS 38677, USA

2 Lifestyle (Mind-Body Movement) Research Center, College of Psychology and Sociology, Shenzhen University, Shenzhen 518060, China; liyezou123@gmail.com

3 Shenzhen Key Laboratory of Affective and Social Cognitive Science, College of Psychology and Sociology, Shenzhen University, Shenzhen 518060, China

4 Research Centre of Brain Function and Psychological Science, Shenzhen University, Shenzhen 518060, China

5 Shenzhen Institute of Neuroscience, Shenzhen University, Shenzhen 518060, China

* Correspondence: pdloprin@olemiss.edu (P.D.L.); lihongszu@szu.edu.cn (H.L.); Tel.: +1-662-915-5561 (P.D.L.); +86-139-0297-3932 (H.L.)

Received: 4 May 2019; Accepted: 14 May 2019; Published: 16 May 2019

Abstract: Emerging research demonstrates that exercise, including both acute and chronic exercise, may influence episodic memory function. To date, mechanistic explanations of this effect are often attributed to alterations in long-term potentiation, neurotrophic production, angiogenesis, and neurogenesis. Herein, we discuss a complementary mechanistic model, suggesting that the endocannabinoid system may, in part, influence the effects of exercise on memory function. We discuss the role of the endocannabinoid system on memory function as well as the effects of exercise on endocannabinoid alterations. This is an exciting line of inquiry that should help delineate new insights into the mechanistic role of exercise on memory function.

Keywords: BDNF; CB1; CB2; episodic memory; exercise

1. Introduction

The purpose of the present review, written in a brief format, is to discuss a new potential mechanistic paradigm (endocannabinoid system) to elucidate the effects of exercise on episodic memory. This review is structured by first discussing the effects of exercise on memory; then briefly discussing the endocannabinoid system; then indicating the role of the endocannabinoid system on memory function; then how exercise may alter the function of the endocannabinoid system; and then lastly, introducing a hypothetical model indicating the potential moderational role of the endocannabinoid system on the exercise-memory interaction. This review is not meant to be an exhaustive review of the literature. Rather, the goal is to discuss a new mechanistic model and then succinctly provide support for the pathways within our model (Figure 1). Ultimately, the goal of this paper is to discuss a new mechanistic insight to help spawn the development of additional work in this important area of research.

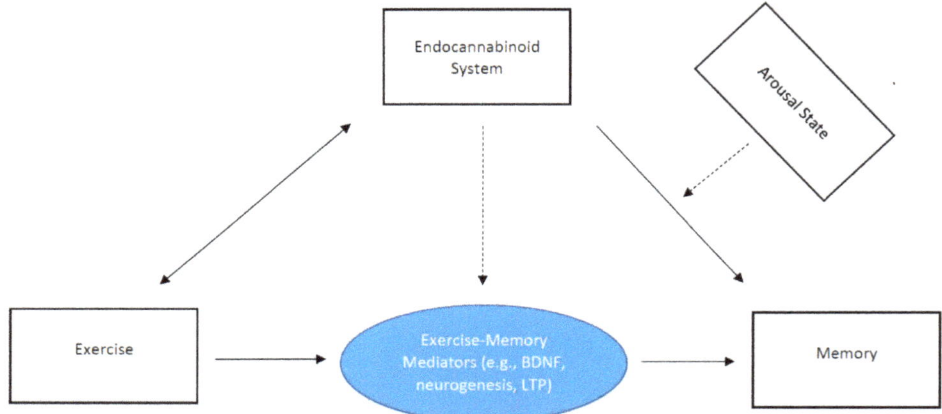

Figure 1. Schematic depicting the role of the endocannabinoid system on the exercise-memory interaction. The dashed lines indicate a moderation effect.

2. Effects of Exercise on Memory

Emerging research from our laboratory demonstrates that exercise, including both acute and chronic exercise, may be effective behaviors in enhancing memory function [1–9]. Various mediators of this exercise-memory interaction have been proposed [10–12]. From a chronic exercise perspective, potential mechanisms may occur at multiple levels, including molecular, cellular, and structural levels. At the molecular level, and as we have thoroughly detailed elsewhere [13–16], chronic exercise may increase levels of brain-derived neurotrophic factor (BDNF) [15,16], vascular endothelial growth factor (VEGF), insulin-like growth factor-1 (IGF-1) [14], and astrocytes [13]. These molecular alterations may induce cellular changes, including gliogenesis, neurogenesis, synaptogenesis, and angiogenesis. These cellular changes, in turn, may alter structural and functional adaptations, including increased white matter, gray matter, receptor activity, neural activity, and cerebral blood flow. Collectively, these molecular, cellular and structural/functional adaptations may improve behavioral performance in memory function.

From an acute exercise perspective, which we have discussed in detail elsewhere [10,11,15,17], various exercise-induced alterations may help facilitate long-term potentiation, a cellular correlate of episodic memory [18]. Acute exercise, via, for example, muscle spindle activation, may increase neuronal excitability in key memory-related brain structures (e.g., hippocampus). This increased neuronal excitability may increase central levels of BDNF, which may help upregulate the expression and function of NMDA receptors. Downstream of this BDNF/TrkB signaling pathway, activation of the PI3K/AKT pathway may contribute to the maintenance of long-term potentiation via NMDA activity [19].

The present paper builds on our previous discussions of potential mechanisms through which exercise influences memory. That is, here we discuss a unique role of the endocannabinoid system in influencing the effects of exercise on memory function.

3. The Endocannabinoid System

Detailed information on the endocannabinoid system can be found elsewhere [20,21]. The cannabinoid system contains two notable subtypes of G protein-coupled receptors, namely CB1 and CB2. The role of endocannabinoids on cognitive processes has mainly focused on CB1 receptors, which are widely distributed throughout the brain and body. CB1 receptors are distributed in the CNS (brainstem, cortex, nucleus, accumbens, hypothalamus, cerebellum, hippocampus, amygdala, spinal cord) and periphery (immune system, liver, bone marrow, pancreas, lungs, vascular system, muscles, GI tract, and reproductive organs) [22]. CB2 receptors are also distributed in the CNS (brainstem, glial cells) and periphery (immune system, liver, bone marrow, pancreas, spleen, bones, skin) [22].

4. The Endocannabinoid System and Memory Function

Previous reviews have detailed the role of the endocannabinoid system on memory function [22–24]. The influence of cannabinoids in memory function can be traced back to early work showing that marijuana intoxication (delta-9-tetrahydrocannabinol, THC) disrupts short-term memory function [25]. Such effects of THC on memory impairment appear to occur in a dose-dependent manner [26,27], with this disruption occurring primarily in the dentate gyrus, where high densities of cannabinoid receptors exist [28], and exist mainly in GABA-ergic inhibitory neurons. Further, memory impairment effects from marijuana may occur, in part, from its detrimental effects on information processing and reduced blood flow to the temporal lobe [24].

Acute systemic administration of CB1 agonists has been shown to impair acquisition of memory across multiple memory tasks, including the Morris water maze task [29]. Similar results have also been observed with intra-cranial administration of CB1 agonists [30]. Conversely, administration of antagonists of CB1 receptors has been shown to facilitate memory consolidation [31]. Blockage of CB1 receptors increases the release of acetylcholine (ACh) [32], a neurotransmitter essential for memory and learning.

Cannabinoid receptor activation may impair memory through various pathways. For example, activation of CB1 receptors is connected with inhibition of adenyl cyclase as well as calcium channels and leads to the activation of potassium channels [22]. As a result, this leads to short-term depression of neurotransmitter release. More specifically, CB1 receptor activation may inhibit cAMP accumulation within neurons, inhibit glutamate release, and inhibit voltage-activated calcium currents [33–35], of which may reduce the excitability of hippocampal neurons, and in turn, reduce neural transmission [32]. Further, cannabinoid agonists may interfere with long-term potentiation [36]. Notably, however, previous work has shown that pre-incubation of adult rat hippocampal slices with THC can either inhibit or potentiate long-term potentiation, depending on the concentration used [37]. Possible explanations for contrasting results for THC on memory is that cannabinoid receptors are expressed at both glutamatergic and GABA-ergic synapses, which often exert opposite effects on memory [38]. For example, CB1 activation from low doses impacts glutamatergic transmission, whereas higher doses affect GABA-ergic transmission [39]. Relatedly, a chronic low dose of THC has been shown to reverse age-related decline in cognitive performance, via enhanced expression of synaptic markers and increased hippocampal spine density [40]. Further, cannabinoid-induced depression of synaptic transmission is switched to stimulation when dopaminergic tone is increased [41]. Although less investigated than CB1 receptors, recent work suggests an important role of CB2 receptors in memory function [42]. Chronic activation of CB2 receptors in the hippocampus for 7–10 days has been shown to increase excitatory synaptic transmission [43]. Similarly, other related work demonstrates that CB2 receptors play an important role in the modulation of memory consolidation for aversive experiences [44]. Further, CB2 receptor agonists reduce neurodegeneration, neuroinflammation, and attenuates spatial memory impairment in an Alzheimer's disease model [45]. Relatedly, CB2 knockout has been shown to impair contextual long-term memory [46]. In addition to direct activation of CB2 receptors, other work also demonstrates the important role of key enzymes (e.g., fatty acid amide hydrolase, FAAH) that are responsible for the metabolism of key endocannabinoids (e.g., anandamide) [47]. For example, recent work has shown that FAAH inhibition modulates hippocampal microglial recruitment and activation that is associated with improved hippocampal-dependent memory [48]. Relatedly, FAAH inhibitor (URB597) infusion, which selectively increased anandamide levels at active synapses, enhanced emotional memory via consolidation-based processes [49]. Treatment with URB597 has also been shown to restore age-related decreases in long-term potentiation in the dentate gyrus [50].

The conflicting findings of the endocannabinoid system on memory function may also be context-dependent. As thoroughly detailed elsewhere [44], the endocannabinoid system may shape how environmental stimuli influence emotional responses. In a low arousal state, endocannabinoid activation was not associated with memory in rats, which was in contrast to their findings in a high arousal state, showing that short-term memory was enhanced when endocannabinoid activation occurred during the early memory consolidation stage [45]. Thus, environmental or behavioral events that influence different levels of stress and arousal may shape the responses to the memory effects of the cannabinoid system. As detailed elsewhere [44], emotionally arousing experiences, such as stress and physical exercise, increase stress hormones (e.g., cortisol and epinephrine), which bind to metabotropic receptors within the basolateral complex of the amygdala, activating the cAMP/PKA pathway to induce endocannabinoid synthesis. Endocannabinoids are then released, bind to GABAergic terminals, inhibits GABA release, and in turn, increase noradrenergic activation of postsynaptic β-adrenoceptors, ultimately facilitating memory consolidation of emotional/arousing events. These effects may, in part, help explain the potential beneficial effects of exercise on memory function. This may be particularly true for studies evaluating the effects of exercise on emotional memory. As we demonstrated recently [51], when exercise occurs during the memory consolidation stage, emotional memory is enhanced, whereas when it occurs prior to memory encoding, it remains unaffected [52].

5. Exercise and the Endocannabinoid System

Several studies have demonstrated that endocannabinoid levels may be altered with exercise [53,54], with their effects acting both centrally and peripherally [54]. Exercise has been shown to enhance CB1 receptor sensitivity [55]. Sparling et al. [56] demonstrated that higher levels of physical activity were associated with greater anandamide (an endogenous agonist of the cannabinoid CB1 and CB2 receptors) levels. Among rodents, Hill et al. [57] showed that 8 days of exercise increased anandamide levels. Further, Raichlen et al. [58] showed an intensity-dependent effect of exercise on anandamide levels, with moderate-intensity exercise enhancing anandamide levels. Fuss et al. [59] showed that wheel running increases endocannabinoid levels and ablation of CB1 receptors on GABAergic neurons inhibits running-induced anxiolysis. Notably, however, a bi-directional relationship may also exist, as research demonstrates that stimulation of CB1 receptors is a prerequisite for voluntary running in mice [60–62]. For example, CB1 activation on VTA (ventral tegmental area) GABAergic neurons may trigger disinhibition of VTA dopamine [60], implicated in reward-directed processes.

6. Hypothetical Model

Emerging work has started to evaluate the potential role of the cannabinoid system on subserving the exercise-memory relationship. Research demonstrates that exercise-induced hippocampal cell proliferation and neurogenesis depends on CB1 receptor signaling [57,63,64]. Notably, CB1 receptors have widespread expression over the entire dentate gyrus and voluntary wheel running has been shown to increase CB1 receptor mRNA in the hippocampus [63]. CB1 receptors specifically affect the stages of adult neurogenesis and the survival and maturation of new neurons [63].

Relatedly, research demonstrates that treadmill running improves spatial memory in mice, which is prevented by simultaneous treatment of a CB1 receptor antagonist [65]. Such exercise-related effects may be attributed to exercise-induced increases in CB1 receptor activation and BDNF expression in the hippocampus [65]. Thus, exercise-induced enhancement of memory function may, in part, be due to a number of endocannabinoid signaling mechanisms related to long-term potentiation, production of neurotrophic factors, and cellular neurogenesis. This is schematically illustrated in Figure 1. That is, there exists a bi-directional relationship between exercise and endocannabinoid levels. The endocannabinoid system may play an important role in episodic memory function, and as demonstrated previously, this may be moderated by arousal state. Further, key exercise-induced mechanisms (e.g., neurogenesis) that influence episodic memory function may be moderated by the endocannabinoid system. Key insight and support of this model have been demonstrated recently.

Bosch et al. [66] evaluated the effects of acute exercise intensity on memory function, with considerations of AEA (anandamide) and BDNF in mediating this relationship. Their results demonstrated consistent evidence of moderate-intensity acute exercise enhancing associative memory. They also demonstrated that increased AEA after moderate-intensity exercise correlated with neural activation of the right hippocampus [66].

7. Model Evaluation

Future work is needed to evaluate this model and, when appropriate, make necessary revisions. Such work should employ both acute and chronic exercise paradigms. From an acute exercise perspective, future within-subject experimental designs should employ multiple exercise intensities (e.g., control, moderate, and vigorous), and when doing so, carefully consider the temporal effects of acute exercise on memory function [4]. That is, consider integrating the acute bout of exercise prior to memory encoding and across different phases of memory consolidation. In human models, blood samples to assess endocannabinoid levels should be measured at multiple time points (e.g., before and after exercise; prior to memory encoding and retrieval, and during memory consolidation). Similarly, key mediators (e.g., BDNF, LTP) through which the endocannabinoid system may influence the effects of acute exercise on memory will need to be assessed at these time points. In human work, novel methodologies to assess LTP will need to be considered. For example, evaluating LTP-like responses, such as visually-evoked event-related potentials, is worth considering [67]. Further, the memory assessments should be carefully considered, and, for example, include hippocampal-dependent memory tasks and emotional memory tasks (given the abundance of CB1 receptors in the limbic system).

Among human models, chronic exercise training studies should carefully design the study to ensure that any potential effects are due to the chronic training stimulus, as opposed to a potential acute exercise response. Rarely do chronic training studies indicate whether participants avoided exercise shortly before the post-training memory assessment, and as such, it is challenging to determine whether post-training outcomes are from chronic adaptations from exercise, or rather, are an artifact of an acute exercise response. These chronic training studies should evaluate other potential mediators through which the endocannabinoid system may influence, such as neurogenesis, which can be measured from magnetic resonance imaging [68]. Lastly, animal studies should continue to design experimental studies that evaluate whether exercise activates the endocannabinoid system, whether this activation is associated with memory function, and whether blocking the endocannabinoid system prevents a direct effect of exercise on memory function.

8. Summary

In conclusion, this brief narrative review highlights the potential role of the cannabinoid system on the exercise-memory relationship. Future research is needed to fully test out this potential mechanistic paradigm. Such work should also delineate whether the site of CB1 activation (e.g., GABA-ergic, glutamatergic) moderates this relationship. This is an exciting line of inquiry that should help delineate new insights into the mechanistic role of exercise on memory function.

Author Contributions: P.D.L. conceptualized the model and drafted the manuscript. L.Z. and H.L. provided insights and feedback in revising the manuscript.

Funding: This research project is supported by both Guangdong-Government Funding for Scientific Research (2016KZDXM009) and Shenzhen-Government Research Grants Programme in Basic Sciences (JCYJ20150729104249783).

Conflicts of Interest: The authors declare no conflict of interest.

References

1. Loprinzi, P.D.; Frith, E.; Edwards, M.K.; Sng, E.; Ashpole, N. The Effects of Exercise on Memory Function Among Young to Middle-Aged Adults: Systematic Review and Recommendations for Future Research. *Am. J. Health Promot.* **2018**, *32*, 691–704. [CrossRef]
2. Iv, J.T.H.; Frith, E.; Sng, E.; Loprinzi, P.D. Experimental Effects of Acute Exercise on Episodic Memory Function: Considerations for the Timing of Exercise. *Psychol. Rep.* **2018**, 33294118786688.
3. Johnson, L.; Loprinzi, P.D. The effects of acute exercise on episodic memory function among young University students: Moderation considerations by biological sex. *Health Promot. Perspect.* **2019**.
4. Loprinzi, P.D.; Blough, J.; Crawford, L.; Ryu, S.; Zou, L.; Li, H. The Temporal Effects of Acute Exercise on Episodic Memory Function: Systematic Review with Meta-Analysis. *Brain Sci.* **2019**, *9*, 87. [CrossRef] [PubMed]
5. Loprinzi, P.D.; Frith, E.; Edwards, M.K. Resistance exercise and episodic memory function: a systematic review. *Clin. Physiol. Funct. Imaging* **2018**, *38*, 923–929. [CrossRef] [PubMed]
6. Ponce, P.; Loprinzi, P.D. A bi-directional model of exercise and episodic memory function. *Med. Hypotheses* **2018**, *117*, 3–6. [CrossRef]
7. Siddiqui, A.; Loprinzi, P.D. Experimental Investigation of the Time Course Effects of Acute Exercise on False Episodic Memory. *J. Clin. Med.* **2018**, *7*, 157. [CrossRef] [PubMed]
8. Sng, E.; Frith, E.; Loprinzi, P.D. Experimental effects of acute exercise on episodic memory acquisition: Decomposition of multi-trial gains and losses. *Physiol. Behav.* **2018**, *186*, 82–84. [CrossRef]
9. Loprinzi, P.D.; Scott, T.M.; Ikuta, T.; Addoh, O.; Tucker, K.L. Association of physical activity on changes in cognitive function: Boston Puerto Rican Health Study. *Physician Sportsmed.* **2018**, *47*, 227–231. [CrossRef]
10. Loprinzi, P.D.; Edwards, M.K.; Frith, E. Potential avenues for exercise to activate episodic memory-related pathways: a narrative review. *Eur. J. Neurosci.* **2017**, *46*, 2067–2077. [CrossRef]
11. Frith, E.; Loprinzi, P.D. Physical activity and individual cognitive funcion parameters: unique exercise-induced mechansims. *J. Cognit. Behav. Psychother. Res.* **2018**, *7*, 92–106. [CrossRef]
12. El-Sayes, J.; Harasym, D.; Turco, C.V.; Locke, M.B.; Nelson, A.J. Exercise-Induced Neuroplasticity: A Mechanistic Model and Prospects for Promoting Plasticity. *Neuroscientist* **2019**, *25*, 65–85. [CrossRef] [PubMed]
13. Loprinzi, P. The role of astrocytes on the effects of exercise on episodic memory function. *Physiol. Int.* **2019**, *106*, 21–28. [CrossRef]
14. Loprinzi, P.D. IGF-1 in exercise-induced enhancement of episodic memory. *Acta Physiol.* **2018**, *226*, e13154. [CrossRef] [PubMed]
15. Loprinzi, P.D.; Frith, E. A brief primer on the mediational role of BDNF in the exercise-memory link. *Clin. Physiol. Funct. Imaging* **2019**, *39*, 9–14. [CrossRef]
16. Loprinzi, P.D. Does brain-derived neurotrophic factor mediate the effects of exercise on memory? *Physician Sportsmed.* **2019**, 1–11. [CrossRef] [PubMed]
17. Loprinzi, P.; Ponce, P.; Frith, E. Hypothesized mechanisms through which acute exercise influences episodic memory. *Physiol. Int.* **2018**, *105*, 285–297. [CrossRef]
18. Poo, M.-M.; Pignatelli, M.; Ryan, T.J.; Tonegawa, S.; Bonhoeffer, T.; Martin, K.C.; Rudenko, A.; Tsai, L.-H.; Tsien, R.W.; Fishell, G.; et al. What is memory? The present state of the engram. *BMC Boil.* **2016**, *14*, 1133. [CrossRef]
19. Nakai, T.; Nagai, T.; Tanaka, M.; Itoh, N.; Asai, N.; Enomoto, A.; Asai, M.; Yamada, S.; Saifullah, A.B.; Sokabe, M.; et al. Girdin Phosphorylation Is Crucial for Synaptic Plasticity and Memory: A Potential Role in the Interaction of BDNF/TrkB/Akt Signaling with NMDA Receptor. *J. Neurosci.* **2014**, *34*, 14995–15008. [CrossRef]
20. Zou, S.; Kumar, U. Cannabinoid Receptors and the Endocannabinoid System: Signaling and Function in the Central Nervous System. *Int. J. Mol. Sci.* **2018**, *19*, 833.
21. Mechoulam, R.; Parker, L.A. The Endocannabinoid System and the Brain. *Annu. Psychol.* **2013**, *64*, 21–47. [CrossRef] [PubMed]
22. Kruk-Slomka, M.; Dzik, A.; Budzynska, B.; Biala, G. Endocannabinoid System: The Direct and Indirect Involvement in the Memory and Learning Processes-a Short Review. *Mol. Neurobiol.* **2017**, *54*, 8332–8347. [CrossRef] [PubMed]

23. Hampson, R.E.; Deadwyler, S.A. Role of Cannabinoid Receptors in Memory Storage. *Neurobiol. Dis.* **1998**, *5*, 474–482. [CrossRef] [PubMed]
24. Riedel, G.; Davies, S.N. Cannabinoid Function in Learning, Memory and Plasticity. *Cannabinoids* **2005**, *168*, 445–477.
25. Miller, L.L.; Branconnier, R.J. Cannabis: Effects on memory and the cholinergic limbic system. *Psychol. Bull.* **1983**, *93*, 441–456. [CrossRef]
26. Campbell, K.A.; Foster, T.C.; Hampson, R.E.; Deadwyler, S.A. Effects of delta 9-tetrahydrocannabinol on sensory-evoked discharges of granule cells in the dentate gyrus of behaving rats. *J. Pharmacol. Exp. Ther.* **1986**, *239*, 941–945. [PubMed]
27. Campbell, K.A.; Foster, T.C.; Hampson, R.E.; Deadwyler, S.A. delta 9-Tetrahydrocannabinol differentially affects sensory-evoked potentials in the rat dentate gyrus. *J. Pharmacol. Exp. Ther.* **1986**, *239*, 936–940. [PubMed]
28. Deadwyler, S.A.; Hampson, R.E.; Childers, S.R. Functional significance of cannabinoid receptors in brain. In *Cannabinoid Receptors*; Pertwee, R.G., Ed.; Academic Press: London, UK, 1995; pp. 205–231.
29. Wise, L.E.; Long, K.A.; Abdullah, R.A.; Long, J.Z.; Cravatt, B.F.; Lichtman, A.H. Dual Fatty Acid Amide Hydrolase and Monoacylglycerol Lipase Blockade Produces THC-Like Morris Water Maze Deficits in Mice. *ACS Chem. Neurosci.* **2012**, *3*, 369–378. [CrossRef]
30. Abush, H.; Akirav, I. Cannabinoids modulate hippocampal memory and plasticity. *Hippocampus* **2010**, *20*, 1126–1138. [CrossRef]
31. Takahashi, R.N.; Pamplona, F.A.; Fernandes, M.S. The cannabinoid antagonist SR141716A facilitates memory acquisition and consolidation in the mouse elevated T-maze. *Neurosci. Lett.* **2005**, *380*, 270–275. [CrossRef]
32. Kruk-Slomka, M.; Biala, G. CB1 receptors in the formation of the different phases of memory-related processes in the inhibitory avoidance test in mice. *Behav. Brain* **2016**, *301*, 84–95. [CrossRef] [PubMed]
33. Howlett, A.C.; Johnson, M.R.; Melvin, L.S.; Milne, G.M. Nonclassical cannabinoid analgetics inhibit adenylate cyclase: development of a cannabinoid receptor model. *Mol. Pharmacol.* **1988**, *33*, 297–302. [PubMed]
34. Twitchell, W.; Brown, S.; Mackie, K. Cannabinoids Inhibit N- and P/Q-Type Calcium Channels in Cultured Rat Hippocampal Neurons. *J. Neurophysiol.* **1997**, *78*, 43–50. [CrossRef] [PubMed]
35. Shen, M.; Piser, T.M.; Seybold, V.S.; Thayer, S.A. Cannabinoid Receptor Agonists Inhibit Glutamatergic Synaptic Transmission in Rat Hippocampal Cultures. *J. Neurosci.* **1996**, *16*, 4322–4334. [CrossRef]
36. Stella, N.; Schweitzer, P.; Piomelli, D. A second endogenous cannabinoid that modulates long-term potentiation. *Nat. Cell Boil.* **1997**, *388*, 773–778. [CrossRef] [PubMed]
37. Nowicky, A.V.; Teyler, T.J.; Vardaris, R.M. The modulation of long-term potentiation by delta-9-tetrahydrocannabinol in the rat hippocampus, in vitro. *Brain Res. Bull.* **1987**, *19*, 663–672. [CrossRef]
38. Ruehle, S.; Rey, A.A.; Remmers, F.; Lutz, B. The endocannabinoid system in anxiety, fear memory and habituation. *J. Psychopharmacol* **2012**, *26*, 23–39. [CrossRef]
39. Busquets-Garcia, A.; Bains, J.; Marsicano, G. CB1 Receptor Signaling in the Brain: Extracting Specificity from Ubiquity. *Neuropsychopharmacolo.* **2018**, *43*, 4–20. [CrossRef]
40. Bilkei-Gorzo, A.; Albayram, O.; Draffehn, A.; Michel, K.; Piyanova, A.; Oppenheimer, H.; Dvir-Ginzberg, M.; Rácz, I.; Ulas, T.; Imbeault, S.; et al. A chronic low dose of Δ9-tetrahydrocannabinol (THC) restores cognitive function in old mice. *Nat. Med.* **2017**, *23*, 782–787. [CrossRef]
41. Caballero-Florán, R.N.; Conde-Rojas, I.; Chávez, A.O.; Cortes-Calleja, H.; Lopez-Santiago, L.F.; Isom, L.L.; Aceves, J.; Erlij, D.; Florán, B. Cannabinoid-induced depression of synaptic transmission is switched to stimulation when dopaminergic tone is increased in the globus pallidus of the rodent. *Neuropharmacology* **2016**, *110*, 407–418. [CrossRef] [PubMed]
42. Ratano, P.; Palmery, M.; Trezza, V.; Campolongo, P. Cannabinoid Modulation of Memory Consolidation in Rats: Beyond the Role of Cannabinoid Receptor Subtype 1. *Front. Pharmacol.* **2017**, *8*, 45. [CrossRef]
43. Kim, J.; Li, Y. Chronic activation of CB2 cannabinoid receptors in the hippocampus increases excitatory synaptic transmission. *J. Physiol.* **2015**, *593*, 871–886. [CrossRef]
44. Ratano, P.; Petrella, C.; Forti, F.; Passeri, P.P.; Morena, M.; Palmery, M.; Trezza, V.; Severini, C.; Campolongo, P. Pharmacological inhibition of 2-arachidonoilglycerol hydrolysis enhances memory consolidation in rats through CB2 receptor activation and mTOR signaling modulation. *Neuropharmacology* **2018**, *138*, 210–218. [CrossRef]

45. Çakır, M.; Tekin, S.; Doğanyiğit, Z.; Erden, Y.; Soytürk, M.; Çiğremiş, Y.; Sandal, S. Cannabinoid type 2 receptor agonist JWH-133, attenuates Okadaic acid induced spatial memory impairment and neurodegeneration in rats. *Life Sci.* **2019**, *217*, 25–33. [CrossRef]
46. Li, Y.; Kim, J. CB2 Cannabinoid Receptor Knockout in Mice Impairs Contextual Long-Term Memory and Enhances Spatial Working Memory. *Neural Plast.* **2016**, *2016*, 9817089. [CrossRef]
47. Ahn, K.; Johnson, D.S.; Cravatt, B.F. Fatty acid amide hydrolase as a potential therapeutic target for the treatment of pain and CNS disorders. *Expert Opin. Drug Discov.* **2009**, *4*, 763–784. [CrossRef]
48. Rivera, P.; Fernández-Arjona, M.D.M.; Silva-Peña, D.; Blanco, E.; Vargas, A.; López-Ávalos, M.D.; Grondona, J.M.; Serrano, A.; Pavón, F.J.; De Fonseca, F.R.; et al. Pharmacological blockade of fatty acid amide hydrolase (FAAH) by URB597 improves memory and changes the phenotype of hippocampal microglia despite ethanol exposure. *Biochem. Pharmacol.* **2018**, *157*, 244–257. [CrossRef]
49. Morena, M.; Roozendaal, B.; Trezza, V.; Ratano, P.; Peloso, A.; Hauer, D.; Atsak, P.; Trabace, L.; Cuomo, V.; McGaugh, J.L.; et al. Endogenous cannabinoid release within prefrontal-limbic pathways affects memory consolidation of emotional training. *Proc. Acad. Sci.* **2014**, *111*, 18333–18338. [CrossRef] [PubMed]
50. Murphy, N.; Cowley, T.R.; Blau, C.W.; Dempsey, C.N.; Noonan, J.; Gowran, A.; Tanveer, R.; Olango, W.M.; Finn, D.P.; Campbell, V.; et al. The fatty acid amide hydrolase inhibitor URB597 exerts anti-inflammatory effects in hippocampus of aged rats and restores an age-related deficit in long-term potentiation. *J. Neuroinflammation* **2012**, *9*, 79. [CrossRef] [PubMed]
51. Loprinzi, P.D.; Frith, E.; Edwards, M.K. Exercise and Emotional Memory: a Systematic Review. *J. Cogn. Enhanc.* **2018**, *3*, 94–103. [CrossRef]
52. Wade, B.; Loprinzi, P.D. The Experimental Effects of Acute Exercise on Long-Term Emotional Memory. *J. Clin. Med.* **2018**, *7*, 486. [CrossRef] [PubMed]
53. Thompson, Z.; Argueta, D.; Garland, T., Jr.; DiPatrizio, N. Circulating levels of endocannabinoids respond acutely to voluntary exercise, are altered in mice selectively bred for high voluntary wheel running, and differ between the sexes. *Physiol. Behav.* **2017**, *170*, 141–150. [CrossRef] [PubMed]
54. Dietrich, A.; McDaniel, W.F. Endocannabinoids and exercise. *Br. J. Sports Med.* **2004**, *38*, 536–541. [CrossRef] [PubMed]
55. De Chiara, V.; Errico, F.; Musella, A.; Rossi, S.; Mataluni, G.; Sacchetti, L.; Siracusano, A.; Castelli, M.; Cavasinni, F.; Bernardi, G. Voluntary exercise and sucrose consumption enhance cannabinoid CB1 receptor sensitivity in the striatum. *Neuropsychopharmacology* **2010**, *35*, 374–387. [CrossRef] [PubMed]
56. Sparling, P.B.; Giuffrida, A.; Piomelli, D.; Rosskopf, L.; Dietrich, A. Exercise activates the endocannabinoid system. *NeuroReport* **2003**, *14*, 2209–2211. [CrossRef]
57. Hill, M.N.; McLaughlin, R.J.; Bingham, B.; Shrestha, L.; Lee, T.T.Y.; Gray, J.M.; Hillard, C.J.; Gorzalka, B.B.; Viau, V. Endogenous cannabinoid signaling is essential for stress adaptation. *Proc. Acad. Sci.* **2010**, *107*, 9406–9411. [CrossRef]
58. Raichlen, D.A.; Foster, A.D.; Seillier, A.; Giuffrida, A.; Gerdeman, G.L. Exercise-induced endocannabinoid signaling is modulated by intensity. *Eur. J. Appl. Physiol.* **2013**, *113*, 869–875. [CrossRef]
59. Fuss, J.; Steinle, J.; Bindila, L.; Auer, M.K.; Kirchherr, H.; Lutz, B.; Gass, P. A runner's high depends on cannabinoid receptors in mice. *Proc. Acad. Sci.* **2015**, *112*, 13105–13108. [CrossRef] [PubMed]
60. Dubreucq, S.; Durand, A.; Matias, I.; Benard, G.; Richard, E.; Soria-Gómez, E.; Glangetas, C.; Groc, L.; Wadleigh, A.; Massa, F.; et al. Ventral Tegmental Area Cannabinoid Type-1 Receptors Control Voluntary Exercise Performance. *Boil. Psychiatry* **2013**, *73*, 895–903. [CrossRef]
61. Chaouloff, F.; Dubreucq, S.; Bellocchio, L.; Marsicano, G. Endocannabinoids and Motor Behavior: CB1 Receptors Also Control Running Activity. *Physiology* **2011**, *26*, 76–77. [CrossRef] [PubMed]
62. Fuss, J.; Gass, P. Endocannabinoids and voluntary activity in mice: Runner's high and long-term consequences in emotional behaviors. *Exp. Neurol.* **2010**, *224*, 103–105. [CrossRef] [PubMed]
63. Wolf, S.A.; Bick-Sander, A.; Fabel, K.; Leal-Galicia, P.; Tauber, S.; Ramírez-Rodríguez, G.; Müller, A.; Melnik, A.; Waltinger, T.P.; Ullrich, O.; et al. Cannabinoid receptor CB1 mediates baseline and activity-induced survival of new neurons in adult hippocampal neurogenesis. *Cell Commun. Signal.* **2010**, *8*, 12. [CrossRef] [PubMed]
64. Hill, M.N.; Titterness, A.K.; Morrish, A.C.; Carrier, E.; Lee, T.; Gil-Mohapel, J.; Gorzalka, B.; Hillard, C.; Christie, B. Endogenous cannabinoid signaling is required for voluntary exercise-induced enhancement of progenitor cell proliferation in the hippocampus. *Hippocampus* **2010**, *20*, 513–523. [CrossRef]

65. Ferreira-Vieira, T.H.; Bastos, C.P.; Pereira, G.S.; Moreira, F.A.; Massensini, A.R. A role for the endocannabinoid system in exercise-induced spatial memory enhancement in mice. *Hippocampus* **2014**, *24*, 79–88. [CrossRef] [PubMed]
66. Bosch, B.M.; Bringard, A.; Logrieco, M.G.; Lauer, E.; Imobersteg, N.; Ferretti, G.; Thomas, A.; Schwartz, S.; Igloi, K. Acute physical exercise of moderate intensity improves memory consolidation in humans via BDNF and endocannabinoid signaling. *bioRxiv* **2019**. [CrossRef]
67. Smallwood, N.; Spriggs, M.J.; Thompson, C.S.; Wu, C.C.; Hamm, J.P.; Moreau, D.; Kirk, I.J. Influence of Physical Activity on Human Sensory Long-Term Potentiation. *J. Int. Neuropsychol. Soc.* **2015**, *21*, 831–840. [CrossRef]
68. Sierra, A.; Encinas, J.M.; Maletic-Savatic, M. Adult Human Neurogenesis: From Microscopy to Magnetic Resonance Imaging. *Front. Neurosci.* **2011**, *5*, 47. [CrossRef] [PubMed]

Editorial

Cannabis and Cognition: Connecting the Dots towards the Understanding of the Relationship

Marco Colizzi [1,2,*], Sarah Tosato [1] and Mirella Ruggeri [1]

[1] Section of Psychiatry, Department of Neurosciences, Biomedicine and Movement Sciences, University of Verona, 37134 Verona, Italy; sarah.tosato@univr.it (S.T.); mirella.ruggeri@univr.it (M.R.)

[2] Department of Psychosis Studies, Institute of Psychiatry, Psychology and Neuroscience, King's College London, London SE5 8AF, UK

* Correspondence: marco.colizzi@univr.it; Tel.: +39-045-812-6832

Received: 23 February 2020; Accepted: 25 February 2020; Published: 27 February 2020

Abstract: Several studies have advanced the understanding of the effects of cannabis on cognitive function. A comprehensive reappraisal of such literature may help in drawing conclusions about the potential risks associated with cannabis use. In summary, the evidence suggests that earlier age of use, high-frequency and high-potency cannabis use, as well as sustained use over time and use of synthetic cannabinoids, are all correlated with a higher likelihood of developing potentially severe and persistent executive function impairments. While the exact mechanisms underlying the adverse effects of cannabis on cognition are not completely clear, Magnetic Resonance Imaging (MRI) studies support the presence of both structural and functional alterations associated with cannabis use. Cognitive dysfunction is also a core feature of many neuropsychiatric disorders and care must be taken regarding the effects of cannabis use in these patient populations. Cognitive impairments affect patients' daily functions, sociability, and long-term outcome, posing elevated economic, social, and clinical burdens. There is, thus, a compelling case for implementing behavioral and cognitive rehabilitation therapies for these patients, as well as investigating the endocannabinoid system in the development of new psychopharmacological treatments.

Keywords: delta-9-tetrahydrocannabinol; endocannabinoid system; executive functions

---

With around 200 million users worldwide, cannabis takes the lead when it comes to the number of people using a drug for recreational purposes [1]. The growing popularity of cannabis has seen a parallel increase of the public interest into its safety. Accumulating evidence associates cannabis use with several adverse behavioral, physiological, and neural effects [2], with acute challenge studies implying a causal relationship for such associations [3]. Indeed, studies of the long-term impact of cannabis suggest the development of tolerance [2] and dependence [4] upon sustained use. However, the harmful effects of cannabis are still debated, especially their severity and whether they are of a long-lasting nature. Interestingly, in a nine-category matrix of physical and social harm of both illicit and legal drugs, cannabis did not score in the top 10, while alcohol and tobacco did [5]. Cognitive function is one of the domains mostly investigated with reference to cannabis use, but also one of those generating the most conflicting results, with not all studies indicating poorer cognitive performance in otherwise healthy individuals or patients with a severe mental disorder and even some evidence of better performance in cannabis-using psychosis patients [6]. Studies of the effects of cannabis on cognition conducted over the last five decades have progressively unfolded a relationship of a complex nature, where several factors come into play. First, evidence indicates non-uniform disrupting effects of cannabis across different cognitive domains [7]. Second, genetic background may determine different individual susceptibility to cannabis-induced cognitive impairments [8,9]. Third, cognition seems to be the domain most likely to demonstrate tolerance upon repeated exposure, with some evidence of full tolerance indicating a complete absence of acute effects [2,10,11]. Fourth, cannabis composition and patterns of use play a relevant role, with both high-potency cannabis varieties, i.e., cannabis high in concentration of the psychoactive component delta-9-tetrahydrocannabinol (Δ9-THC) [12], and frequent cannabis use, e.g., daily [13], being associated with more pronounced cognitive impairments, thus supporting a cumulative adverse effect of Δ9-THC. Fifth, synthetic cannabinoids, which act as more potent full agonists at the cannabinoid receptor type 1 than Δ9-THC, thus exerting a more severe disruption of the endocannabinoid system, have been shown to induce more evident cognitive impairments in healthy subjects, which are undistinguishable from those observed in psychosis [14]. Finally, the use of cannabis in adolescence may lead to more serious cognitive impairments, due to the drug interfering with brain maturation [15].

An interesting up-to-date review article, "The Effects of Cannabinoids on Executive Functions: Evidence from Cannabis and Synthetic Cannabinoids—A Systematic Review", published in *Brain Sciences*, brings together different lines of research about the effects of cannabis on cognition, including preclinical versus clinical evidence, acute versus long-term effects, occasional versus regular exposure and organic versus synthetic cannabinoids [16]. Such strategy emphasizes the importance of interpreting the available evidence altogether, to overcome the risks of interpreting the phenomenon based only on partial data [17]. Other merits of the review are that it applies rigorous inclusion criteria in terms of cognitive outcome measures, focusing only on objective measurements, as well as disentangles the effects of cannabis on each executive function sub-domain. High-level cognitive functions call on combinations of different component processes and there is evidence that changes in cognitive functioning, for instance, because of aging, are more likely to be masked when using more general cognitive measures compared to the use of more specific abilities [18]. It is, therefore, plausible that the same would happen with reference to the effects of cannabis use. Focusing on the three core executive functions, attention, working memory, and cognitive flexibility, separately [19], the authors make a noble attempt to deal with this potential issue. Moreover, in excluding studies performed on participants with psychiatric or substance use disorders, the review cut out two important arguments that could have hampered its conclusions; that is, the alternative explanation that the association between cannabis and cognitive impairments would be driven by use of other substances or coexisting psychopathological features, making cannabis users less proficient cognitively [20].

In the review by Cohen and Weinstein, one by one, all the apparent inconsistencies of the available literature find a possible explanation. Repeated exposure to cannabis is more clearly associated with the manifestation of executive function impairments. The evidence indicates a dose–response relationship for the effect of cannabis on executive functions, with frequent users and users of potent forms of cannabis presenting with more pronounced cognitive impairments. Exposure to synthetic cannabinoids is more clearly associated with long-lasting impairments. Exposure during adolescence increases the likelihood of such impairments being more severe and persisting in adulthood.

The exact mechanisms underlying the adverse effects of cannabis on cognition are not completely clear. However, implementing studies of the effect of cannabinoids on cognition in a Magnetic Resonance Imaging (MRI) design may help understanding the underlying neurobiological mechanisms [6]. Consistently, the evidence from structural MRI studies reviewed here support an association between chronic cannabis use and reduced gray matter volumes in brain regions relevant to cognitive processes, including the hippocampus and amygdala, with the extent of such alterations correlating with age of onset, frequency, and severity of cannabis use. Similarly, functional MRI studies indicate disputed brain activity in regions involved in the processing of several cognitive tasks as a function of cannabis use. Interestingly, some of this evidence suggests that, while performing a cognitive task, cannabis users' brain activity may be disrupted, even in the absence of a less proficient behavioral performance, reflecting an attempt to sustain performance by recruiting additional or different neural resources [21]. This would provide another possible explanation for the absence of the cannabis effect in those studies assessing exclusively the behavioral component of cognitive processing [22].

By affecting patients' daily function, sociability, and long-term outcome, cognitive impairments place important socioeconomic burdens on society and patients themselves, also posing significant challenges to healthcare practitioners [23]. As Cohen and Weinstein point out, understanding how different cannabinoids may modulate cognitive processes can shed new light into the neurobiological mechanisms that increase the risk of long-lasting cognitive impairments in regular cannabis users. Moreover, cannabis use can increase the risk of developing disabling neuropsychiatric disorders, such as psychosis [24], and cognitive dysfunction is a core feature of such disorders [23]. Interestingly, endocannabinoid alterations have been implied in the pathophysiology of psychosis, independent of cannabis use [25]. Based on this evidence, along with the implementation of behavioral and cognitive rehabilitation therapies for these patients, there is also a compelling case for investigating the endocannabinoid system in the development of new psychopharmacological treatments [26].

Conflicts of Interest: The authors declare no conflicts of interest.

References

1. National Academies of Sciences, Engineering, and Medicine. *The Health Effects of Cannabis and Cannabinoids: The Current State of Evidence and Recommendations for Research*; The National Academies Press: Washington, DC, USA, 2017.
2. Colizzi, M.; Bhattacharyya, S. Cannabis use and the development of tolerance: A systematic review of human evidence. *Neurosci. Biobehav. Rev.* **2018**, *93*, 1–25. [CrossRef]
3. Colizzi, M.; Weltens, N.; McGuire, P.; Lythgoe, D.; Williams, S.; Van Oudenhove, L.; Bhattacharyya, S. Delta-9-tetrahydrocannabinol increases striatal glutamate levels in healthy individuals: Implications for psychosis. *Mol. Psychiatry* **2019**, 1. [CrossRef]
4. Degenhardt, L.; Ferrari, A.J.; Calabria, B.; Hall, W.D.; Norman, R.E.; McGrath, J.J.; Flaxman, A.D.; Engell, R.E.; Freedman, G.D.; Whiteford, H.A.; et al. The Global Epidemiology and Contribution of Cannabis Use and Dependence to the Global Burden of Disease: Results from the GBD 2010 Study. *PLoS ONE* **2013**, *8*, e76635. [CrossRef] [PubMed]
5. Nutt, D.J.; A King, L.; Saulsbury, W.; Blakemore, C. Development of a rational scale to assess the harm of drugs of potential misuse. *Lancet* **2007**, *369*, 1047–1053. [CrossRef]
6. Colizzi, M.; Bhattacharyya, S. Neurocognitive effects of cannabis: Lessons learned from human experimental studies. *Prog. Brain Res.* **2018**, *242*, 179–216. [PubMed]
7. Lovell, M.; Akhurst, J.; Padgett, C.; Garry, M.I.; Matthews, A. Cognitive outcomes associated with long-term, regular, recreational cannabis use in adults: A meta-analysis. *Exp. Clin. Psychopharmacol.* **2019**. [CrossRef] [PubMed]
8. Taurisano, P.; Antonucci, L.A.; Fazio, L.; Rampino, A.; Romano, R.; Porcelli, A.; Masellis, R.; Colizzi, M.; Quarto, T.; Torretta, S.; et al. Prefrontal activity during working memory is modulated by the interaction of variation in CB1 and COX2 coding genes and correlates with frequency of cannabis use. *Cortex* **2016**, *81*, 231–238. [CrossRef] [PubMed]
9. Colizzi, M.; Fazio, L.; Ferranti, L.; Porcelli, A.; Masellis, R.; Marvulli, D.; Bonvino, A.; Ursini, G.; Blasi, G.; Bertolino, A. Functional genetic variation of the cannabinoid receptor I and cannabis use interact on prefrontal connectivity and related working memory behavior. *Neuropsychopharmacology* **2015**, *40*, 640–649. [CrossRef]
10. Colizzi, M.; McGuire, P.; Giampietro, V.; Williams, S.; Brammer, M.; Bhattacharyya, S. Previous cannabis exposure modulates the acute effects of delta-9-tetrahydrocannabinol on attentional salience and fear processing. *Exp. Clin. Psychopharmacol.* **2018**, *26*, 582–598. [CrossRef]
11. Colizzi, M.; McGuire, P.; Giampietro, V.; Williams, S.C.; Brammer, M.; Bhattacharyya, S. Modulation of acute effects of delta-9-tetrahydrocannabinol on psychotomimetic effects, cognition and brain function by previous cannabis exposure. *Eur. Neuropsychopharmacol.* **2018**, *28*, 850–862. [CrossRef]
12. Colizzi, M.; Bhattacharyya, S. Does Cannabis Composition Matter? Differential Effects of Delta-9-tetrahydrocannabinol and Cannabidiol on Human Cognition. *Curr. Addict. Rep.* **2017**, *4*, 62–74. [CrossRef] [PubMed]

13. Meier, M.H.; Caspi, A.; Ambler, A.; Harrington, H.; Houts, R.; Keefe, R.S.E.; McDonald, K.; Ward, A.; Poulton, R.; Moffitt, T. Persistent cannabis users show neuropsychological decline from childhood to midlife. *Proc. Natl. Acad. Sci. USA* **2012**, *109*, E2657–E2664. [CrossRef] [PubMed]
14. Altıntaş, M.; Inanc, L.; Oruc, G.A.; Arpacioglu, S.; Gulec, H. Clinical characteristics of synthetic cannabinoid-induced psychosis in relation to schizophrenia: A single-center cross-sectional analysis of concurrently hospitalized patients. *Neuropsychiatr. Dis. Treat.* **2016**, *12*, 1893–1900. [CrossRef] [PubMed]
15. Hurd, Y.L.; Manzoni, O.J.; Pletnikov, M.V.; Lee, F.S.; Bhattacharyya, S.; Melis, M. Cannabis and the Developing Brain: Insights into Its Long-Lasting Effects. *J. Neurosci.* **2019**, *39*, 8250–8258. [CrossRef] [PubMed]
16. Cohen, K.; Weinstein, A.M. The Effects of Cannabinoids on Executive Functions: Evidence from Cannabis and Synthetic Cannabinoids—A Systematic Review. *Brain Sci.* **2018**, *8*, 40. [CrossRef] [PubMed]
17. Loke, Y.K.; Price, D.; Herxheimer, A. Systematic reviews of adverse effects: Framework for a structured approach. *BMC Med. Res. Methodol.* **2007**, *7*, 32. [CrossRef]
18. Harada, C.N.; Love, M.N.; Triebel, K. Normal cognitive aging. *Clin. Geriatr. Med.* **2013**, *29*, 737–752. [CrossRef]
19. Diamond, A. Executive functions. *Annu. Rev. Psychol.* **2013**, *64*, 135–168. [CrossRef]
20. Haney, M.; Evins, A.E. Does Cannabis Cause, Exacerbate or Ameliorate Psychiatric Disorders? An Oversimplified Debate Discussed. *Neuropsychopharmacology* **2015**, *41*, 393–401. [CrossRef]
21. Bossong, M.G.; Jager, G.; Bhattacharyya, S.; Allen, P. Acute and non-acute effects of cannabis on human memory function: A critical review of neuroimaging studies. *Curr. Pharm. Des.* **2014**, *20*, 2114–2125. [CrossRef]
22. Pope, H.G.; Gruber, A.J.; Yurgelun-Todd, D. Residual neuropsychologic effects of cannabis. *Curr. Psychiatry Rep.* **2001**, *3*, 507–512. [CrossRef]
23. Stuchlik, A.; Sumiyoshi, T. Cognitive Deficits in Schizophrenia and Other Neuropsychiatric Disorders: Convergence of Preclinical and Clinical Evidence. *Front. Behav. Neurosci.* **2014**, *8*, 444. [CrossRef] [PubMed]
24. Colizzi, M.; Murray, R. Cannabis and psychosis: What do we know and what should we do? *Br. J. Psychiatry* **2018**, *212*, 195–196. [CrossRef] [PubMed]
25. Appiah-Kusi, E.; Wilson, R.; Colizzi, M.; Foglia, E.; Klamerus, E.; Caldwell, A.; Bossong, M.G.; McGuire, P.; Bhattacharyya, S. Childhood trauma and being at-risk for psychosis are associated with higher peripheral endocannabinoids. *Psychol. Med.* **2019**, 1–10. [CrossRef] [PubMed]
26. O'Neill, A.; Wilson, R.; Blest-Hopley, G.; Annibale, L.; Colizzi, M.; Brammer, M.; Giampietro, V.; Bhattacharyya, S. Normalization of mediotemporal and prefrontal activity, and mediotemporal-striatal connectivity, may underlie antipsychotic effects of cannabidiol in psychosis. *Psychol. Med.* **2020**, *2020*, 1–11. [CrossRef] [PubMed]

Article

Descriptive Psychopathology of the Acute Effects of Intravenous Delta-9-Tetrahydrocannabinol Administration in Humans

Marco Colizzi [1,*], Nathalie Weltens [2], Philip McGuire [1], Lukas Van Oudenhove [2] and Sagnik Bhattacharyya [1]

1 Department of Psychosis Studies, Institute of Psychiatry, Psychology and Neuroscience, King's College London, London SE5 8AF, UK; philip.mcguire@kcl.ac.uk (P.M.); sagnik.2.bhattacharyya@kcl.ac.uk (S.B.)

2 Translational Research Center for Gastrointestinal Disorders (TARGID), Department of Chronic Diseases, Metabolism and Ageing, University of Leuven, Leuven 3000, Belgium; nathalie.weltens@kuleuven.be (N.W.); lukas.vanoudenhove@kuleuven.be (L.V.O.)

* Correspondence: marco.v.colizzi@kcl.ac.uk; Tel.: +44-(0)20-7848-0049

Received: 3 April 2019; Accepted: 19 April 2019; Published: 25 April 2019

Abstract: Background: Cannabis use can increase the risk of psychosis, and the acute administration of its key psychoactive ingredient, delta-9-tetrahydrocannabinol (Δ9-THC), can induce transient psychotomimetic symptoms. Methods: A double-blind, randomized, placebo-controlled crossover design was used to investigate the symptomatic effects of acute intravenous administration of Δ9-THC (1.19 mg/2 mL) in 16 healthy participants (seven males) with modest previous cannabis exposure. Results: In the 20 min following acute Δ9-THC administration, symptomatic effects of at least mild severity were present in 94% of the cohort, with moderate to severe symptoms having a much lower prevalence (19%). Nearly one-third (31%) of the volunteers were still experiencing protracted mild symptomatic effects 2.5 h after exposure to Δ9-THC. Compared to the Δ9-THC challenge, most of the study participants did not experience any symptomatic effects following placebo administration (62%). Acute physical reactions were 2.5 times more frequent after Δ9-THC (31%) than placebo (12%). Male and female participants differed in terms of acute Δ9-THC effects, with some negative symptoms occurring more frequently in female (56% to 89%) than male participants (0% to 29%), and acute physical reactions occurring exclusively in the female gender (56%). Conclusions: These results have implications for future research, also in light of cannabis being the most widely used illicit drug.

Keywords: delta-9-tetrahydrocannabinol; placebo; cannabis-associated psychosis; schizophrenia

---

1. Introduction

Psychosis is a severe mental disorder resulting from a complex interplay between genetic and environmental determinants leading to a disruption of central nervous system function [1]. In order to better understand its pathophysiological mechanisms, different models of psychosis have been proposed [2]. Over the last two decades, there has been growing interest in the drug-induced model of psychosis, due to the potential of several pharmacological agents to elicit psychotomimetic symptoms that resemble those observed in psychosis patients [3]. In particular, in-human models of psychosis have become available involving the acute administration of dopaminergic [4], serotoninergic [5], glutamatergic [6], and cannabinoid compounds [7,8]. Compared to animal models, which have been implicated as not adequately modeling the complexity of the disorder [9], the transient symptoms induced by acute challenge with psychotomimetic drugs in healthy individuals are of interest, as they may share pathophysiological mechanisms with the full-blown disorder.

The administration of cannabis' key psychoactive ingredient delta-9-tetrahydrocannabinol (Δ9-THC) has been shown to induce transient psychosis-like symptoms in otherwise healthy individuals [10–13]. The association between cannabinoids and psychosis is further supported by several lines of research: (i) the evidence for a higher risk of psychosis in cannabis users [14–16], especially against a specific genetic background [17,18]; (ii) the evidence that cannabis use can exacerbate psychotic symptoms and cause relapse in patients with schizophrenia [19–23]; and (iii) the evidence that the endocannabinoid system might be disrupted in patients with schizophrenia both in the context of cannabis use and in its absence [24,25], as well as involved in modulating cognitive function in healthy individuals [26–28].

Although clinical research is needed to further understand psychosis in cannabis users, limited evidence from anecdotal studies has been published on the nature of the transient clinical manifestations of acute cannabis intoxication in healthy individuals [29–31]. In many respects, experimental studies examining the nature of the psychotomimetic effects of Δ9-THC may arguably be a priority because they can inform further studies of cannabis-associated psychosis, including aetiology, course, prognosis, and treatment. Previous studies that have assessed the acute psychotomimetic effects of Δ9-THC have reported them as summary measure using the PANSS (Positive and Negative Syndrome Scale) [11,12,32–36], BPRS (Brief Psychiatric Rating Scale) [37], SSPS (State Social Paranoia Scale) [35], or self-report questionnaires [12,32,34]. A limited range of other effects has also been investigated using self-report questionnaires and visual analogue measures, including dissociation [12], affect and mood [11,12,32,34–37], sedation and intoxication [11,12,36,37], and anxiety and panic [11,12,36].

Also, evidence indicates that frequent cannabis users have a more blunted response to the acute psychotomimetic effects of Δ9-THC compared to a group of healthy controls, suggesting the potential development of tolerance [38,39]. Thus, studies conducted among frequent users may have limited usefulness in informing on the nature of the symptoms acutely induced by cannabis in healthy individuals.

Employing a placebo-controlled acute pharmacological challenge design, the aim of this study was to investigate the symptomatic effects of acute Δ9-THC administration under controlled experimental conditions in a group of healthy individuals with modest previous cannabis use.

2. Materials and Methods

This study employed a double-blind, randomized, placebo-controlled, repeated-measures, within-subject design, with a counterbalanced order of drug administration, using an established protocol [13,40]. Sixteen healthy participants (seven males) were assessed on two different occasions separated by at least a two-week interval, with each session preceded by intravenous administration of Δ9-THC (1.19 mg/2 mL) or placebo. All the subjects underwent structural Magnetic Resonance Imaging (MRI), functional MRI (fMRI) and proton magnetic resonance spectroscopy (1H-MRS) scanning in both sessions. The present report focuses on the psychopathological assessment.

2.1. Experimental Procedure

Prior to each study visit, participants were advised to get at least six to eight hours sleep overnight and to refrain from smoking for four hours, taking caffeine for 12 h, and consuming alcohol for 24 h. Also, subjects had been abstinent from cannabis for at least six months before the first study visit, and were advised to abstain from using any substance throughout the duration of the study. On arrival at the study center in the morning, participants had a light standardized breakfast after an overnight fast. All the subjects had a negative urinary drug screen for amphetamines, benzodiazepines, cocaine, opiates, and Δ9-THC, and were tested on each study day using immunometric assay kits. All the female participants had a negative pregnancy test; also, all of them were consistently using a reliable contraceptive method, apart from a single subject who underwent both study visits in the first week of the menstrual cycle. After a physical examination performed by a medical doctor, an indwelling intravenous line in the non-dominant arm was placed by a trained nurse. This cannula was used for the intravenous administration of Δ9-THC (1.19 mg/ 2 mL, ≥99% pure; THC-Pharm, Frankfurt, Germany, http://biochem.thc-pharm.de; pharmaceutical formulation at the Barts Health NHS Trust pharmacy according to previous work [41]) or placebo as well as blood collection a different time points before and after drug challenge. A dose of 1.19 mg was used, as previous work has suggested that an intravenous dose range between 0.015–0.03 mg/kg is consistently associated with an induction of psychotomimetic symptoms [42]. Heart rate and blood pressure were monitored via a digital recorder and an automated arm cuff for the entire duration of the study.

2.2. Subjects

Sixteen healthy, English-speaking, right-handed individuals participated in this study. Demographic information such as age, ethnicity, and level of education was recorded. All the subjects gave written, informed consent, and completed all of the components of the study. Personal or family history of psychiatric illness in first-degree relatives represented an exclusion criterion. None of the subjects included in the study had used more than 21 units/week of alcohol on a regular basis. Only three subjects had a regular smoking habit (two of them smoking <10 cigarettes/day and one smoking two cigarettes/week), six had smoked occasionally/experimentally, and seven had never smoked. Apart from three subjects who had a single experimental use of 3,4-Methylenedioxymethamphetamine (MDMA), all the remaining participants had never used any other substance. Regarding previous lifetime cannabis exposure, nine subjects had used cannabis ≤5 times, three subjects ≤10 times, two subjects ≤20 times, one subject 20 times, and one subject 60 times.

2.3. Psychopathological Assessment

All the participants were interviewed by a psychiatrist with a specific expertise in Diagnostic and Statistical Manual of Mental Disorders, 5th edition (DSM-5) schizophrenia and other psychotic disorders as well as substance use disorders [43], using the Structured Clinical Interview for DSM-5 (SCID-5) as a guide for the assessment of the psychotic spectrum [44]. Assessments were carried out immediately before and at 20 min and 2.5 h after drug administration, and clinically discussed with a senior psychiatrist at the end of each study visit. Psychopathological ratings were recorded using the Positive and Negative Syndrome Scale [45] (PANSS), which is a well-established scale that is used for measuring the symptom severity of individuals with psychosis. Verbatim quotations from participants were also recorded, as research evidence indicates that the inclusion of excerpts from transcripts might help clarify links between data, interpretation, and conclusion [46]. Participants were contacted the day after each study visit for a health check as part of the study standard operating procedure (SOP). Putative symptoms lasting longer than expected or occurring after the end of the study visit were also recorded.

2.4. Ethics Approval

The study was approved by the Joint South London and Maudsley (SLaM) and Institute of Psychiatry, Psychology & Neuroscience (IoPPN) National Health Service Research Ethics Committee (PNM/13/14-38), and the investigators had a license to use Δ9-THC for research purposes.

3. Results

3.1. Demographic Information

Study participants had a mean age of 24.44 (standard deviation, SD: 4.29) years. All except three (with self-described mixed ethnic origin) of the volunteers were white Europeans. They had 16.94 ± 2.84 years (mean, M ± SD) of education.

The effects of Δ9-THC administration on blood pressure and heart rate and related statistics as well as Δ9-THC plasma levels have been previously reported [40].

3.2. Prevalence and Severity of Symptoms: Results at a Glance

3.2.1. Following Acute Δ9-THC Administration

Apart from one participant with minimal and questionable symptoms, who did not score more than two on any PANSS item, the entire study cohort reported at least mild and clearly detectable symptomatic effects (94%; ≥3 on at least one PANSS item) within 20 min following acute challenge with Δ9-THC. More severe symptomatic effects were experienced by a smaller proportion of participants, with 10 volunteers reporting at least moderate symptoms (62%; ≥4 on at least one PANSS item) and three of them reporting moderate to severe symptoms (19%; ≥5 on at least one PANSS item). Acute physical reactions, including effects on movement, blood pressure, heart rate, skin vascularity, and vagal response, occurred on five occasions (31%).

Two hours and 30 minutes after the intravenous administration of Δ9-THC, five (31%) and three (19%) participants were still experiencing mild (= 3 on at least one PANSS item) and minimal (= 2 on at least one PANSS item) symptoms, respectively. In contrast, by this time, no physical reaction was evident. Upon telephone follow-up, six participants (37%) reported long-lasting effects of the drug, which faded away by the end of the study day or the subsequent morning. These symptoms were mainly included fatigue and food craving. In one case, these effects included psychosis-related symptoms such as suspiciousness, hostility, tension, and poor impulse control (6%).

3.2.2. Following Placebo Administration

Differently from the Δ9-THC condition, most of the study participants did not experience any symptomatic effects following placebo administration (n = 10; 62%). Three volunteers (19%) reported minimal and questionable symptoms (= 2 on at least one PANSS item) and, interestingly, only three subjects (19%) had detectable symptoms of mild severity (= 3 on at least one PANSS item). Acute physical reactions, including effects on heart rate and skin vascularity, were present in two occasions (12%), occurring at a lower rate compared to the Δ9-THC condition.

Two hours and 30 minutes after the intravenous administration of placebo, only one participant (6%) was experiencing minimal and questionable psychotomimetic symptoms (= 2 on at least one PANSS item). Also, similarly to the Δ9-THC condition, no physical reaction was evident at that time point. Finally, differently from the Δ9-THC condition, only one participant (6%) reported long-lasting effects at the telephone follow-up after placebo administration, which faded away by the end of the day. However, these effects included psychosis-related symptoms such as suspiciousness, which was totally overlapping with the frequency of long-lasting psychosis-related symptoms following acute challenge with Δ9-THC (6%).

3.3. Symptoms Description

3.3.1. Psychosis-Related Positive Symptoms and Disorganization

The effects of Δ9-THC administration on the PANSS positive symptom subscale and related statistics have been previously reported [40].

Conceptual disorganization was the most frequently observed symptom in the ~20 min following the acute administration of Δ9-THC, with all the participants reporting such symptoms in a minimal to severe form (2 ≤ PANSS-related item ≤ 6). Further frequent symptoms (≥2 on PANSS-related item) included hallucinatory behavior (62%), excitement (62%), and suspiciousness/persecution (56%). A lower percentage of participants also reported symptoms of grandiosity (25%), hostility (19%), and delusions (19%).

Some symptoms were still detectable 2.5 h after the injection (≥2 on PANSS-related item), even if in a more attenuated form, with conceptual disorganization being the most frequent symptom (37%), followed by hallucinatory behavior (6%) and excitement (6%). Volunteers showing a more severe conceptual disorganization immediately after the intravenous administration of Δ9-THC were more likely to still experience such symptom 2.5 h after the injection, with five out of nine participants experiencing moderate to severe conceptual disorganization (4 ≤ PANSS-related item ≤ 6) versus one out of seven participants with minimal to mild conceptual disorganization (≤3 on PANSS-related item).

In the ~20 min following the acute administration of placebo, positive symptoms were reported by four participants (≥2 on PANSS-related item) and only two of them had clearly detectable symptoms (mild severity, ≥3 on PANSS-related item; 12%), which was a percentage that was 7.5 times smaller than that observed in participants under the influence of Δ9-THC. In both cases, these symptoms were within the conceptual disorganization domain. Also, only one participant was still experiencing a disorganized process of thinking 2.5 h after the injection of placebo.

Overall, Δ9-THC-induced excitement and grandiosity were more frequent in male (86% and 43% respectively) than female participants (44% and 11% respectively). Instead, hostility was observable only in a percentage of female participants (33%) (Table 1).

3.3.2. Psychosis-Related Negative Symptoms

The effects of Δ9-THC administration on the PANSS negative symptom subscale and related statistics have been previously reported [40].

A lack of spontaneity and reduced flow of conversation was the most frequently observed symptom in the ~20 min following the acute administration of Δ9-THC, with 13 participants reporting such symptoms in a minimal to moderately severe form (2 ≤ PANSS-related item ≤ 5; 81%). Further frequent symptoms (≥2 on PANSS-related item) included stereotyped thinking (69%), blunted affect (62%), poor rapport (62%), and difficulty in abstract thinking (50%). A lower percentage of participants also reported emotional (44%) and social withdrawal (31%). In only three participants (19%), some symptoms were still detectable 2.5 h after the injection, even if in a more attenuated form (lack of spontaneity and reduced flow of conversation, 12%; blunted affect, 6%; difficulty in abstract thinking, 6%; stereotyped thinking, 6%).

In the ~20 min following the acute administration of placebo, negative symptoms were detectable only in two participants (mild severity, = 3 on PANSS-related item, 12%), which was a percentage that was 7.5 times smaller than that observed in participants under the influence of Δ9-THC. In these cases, symptoms included the poor rapport (6%) and/or the lack of spontaneity (12%) domains. Also, only one participant was still experiencing a lack of spontaneity and reduced flow of conversation 2.5 h after the injection of placebo.

Overall, Δ9-THC-induced poor rapport, emotional withdrawal, and social withdrawal were more frequent in female participants (56% to 89%) than male participants (0% to 29%) (Table 2).

Table 1. Psychosis-related positive symptoms and disorganization.

Study Participant	Drug	Delusions	Conceptual Disorganization	Hallucinatory Behavior	Excitement	Grandiosity	Suspiciousness/Persecution	Hostility
male 1	Δ9-THC	×	**mild**	×	minimal	×	mild	×
male 2	Δ9-THC	×	*textsurd* mild	×	*textsurd* mild	*textsurd* minimal	×	×
male 3	Δ9-THC	×	***textsurd* severe**	*textsurd* mild	***textsurd* moderate**	*textsurd* moderate	*textsurd* mild	×
male 4	Δ9-THC	×	*textsurd* minimal	×	*textsurd* minimal	×	×	×
male 5	Δ9-THC	×	*textsurd* moderate	*textsurd* mild	×	×	*textsurd* minimal	×
male 6	Δ9-THC	*textsurd* mild	*textsurd* moderate	*textsurd* mild	*textsurd* moderate	*textsurd* moderate	×	×
male 7	Δ9-THC	×	***textsurd* moderate**	***textsurd* mild**	*textsurd* mild	×	*textsurd* minimal	×
female 1	Δ9-THC	×	*textsurd* mild	×	×	×	×	*textsurd* mild
female 2	Δ9-THC	×	*textsurd* mild	*textsurd* mild	×	×	×	×
female 3	Δ9-THC	×	*textsurd* mild	×	*textsurd* mild	×	×	×
female 4	Δ9-THC	×	***textsurd* moderate**	*textsurd* minimal	*textsurd* mild	×	*textsurd* mild	*textsurd* mild
female 5	Δ9-THC	×	*textsurd* mild	*textsurd* mild	*textsurd* mild	×	*textsurd* mild	×
female 6	Δ9-THC	×	***textsurd* moderate**	*textsurd* mild	*textsurd* minimal	×	×	×
female 7	Δ9-THC	×	*textsurd* moderate	×	×	×	*textsurd* minimal	×
female 8	Δ9-THC	*textsurd* moderate	***textsurd* moderate/severe**	*textsurd* severe	×	*textsurd* moderate	*textsurd* minimal	×
female 9	Δ9-THC	*textsurd* minimal	*textsurd* moderate	*textsurd* moderate	×	×	*textsurd* moderate	*textsurd* mild
male 1	placebo	×	×	×	×	×	×	×
male 2	placebo	×	×	×	×	×	×	×
male 3	placebo	×	×	×	*textsurd* minimal	×	×	×
male 4	placebo	×	×	×	×	×	×	×
male 5	placebo	×	*textsurd* mild	×	×	×	×	×
male 6	placebo	×	×	×	×	×	×	×
male 7	placebo	×	***textsurd* mild**	***textsurd* minimal**	×	×	×	×
female 1	placebo	×	×	×	×	×	×	×
female 2	placebo	×	×	×	×	×	×	×
female 3	placebo	×	×	×	minimal	×	×	×
female 4	placebo	×	×	×	×	×	×	×
female 5	placebo	×	×	×	×	×	×	×
female 6	placebo	×	×	×	×	×	×	×
female 7	placebo	×	×	×	×	×	×	×
female 8	placebo	×	×	×	×	×	×	×
female 9	placebo	×	×	×	×	×	×	×

Δ9-THC, (−)-trans-Δ9-tetrahydrocannabinol; symptoms highlighted in bold were still observable 2.5 h after the injection.

Table 2. Psychosis-related negative symptoms.

Study Participant	Drug	Blunted Affect	Emotional Withdrawal	Poor Rapport	Social Withdrawal	Difficulty in Abstract Thinking	Lack of Spontaneity and Flow of Conversation	Stereotyped Thinking
male 1	Δ9-THC	***textsurd* minimal**	×	×	×	×	***textsurd* mild**	***textsurd* minimal**
male 2	Δ9-THC	*textsurd* mild	×	×	×	*textsurd* mild	*textsurd* mild	*textsurd* minimal
male 3	Δ9-THC	×	×	*textsurd* mild	×	***textsurd* moderate**	*textsurd* minimal	*textsurd* mild
male 4	Δ9-THC	×	×	×	×	×	×	×
male 5	Δ9-THC	×	×	×	×	×	*textsurd* moderate	*textsurd* minimal
male 6	Δ9-THC	×	×	×	×	*textsurd* mild	×	*textsurd* minimal
male 7	Δ9-THC	*textsurd* mild	×	*textsurd* moderate	×	×	*textsurd* moderate	*textsurd* mild
female 1	Δ9-THC	*textsurd* minimal	*textsurd* mild	*textsurd* mild	*textsurd* mild	*textsurd* mild	*textsurd* mild	*textsurd* mild
female 2	Δ9-THC	×	×	×	×	×	*textsurd* mild	×
female 3	Δ9-THC	×	*textsurd* mild	*textsurd* minimal	×	×	*textsurd* mild	×
female 4	Δ9-THC	*textsurd* mild	*textsurd* mild	*textsurd* mild	×	*textsurd* moderate	*textsurd* moderate	*textsurd* mild
female 5	Δ9-THC	*textsurd* mild	×	*textsurd* mild	*textsurd* mild	*textsurd* moderate	*textsurd* mild	*textsurd* mild
female 6	Δ9-THC	*textsurd* moderate	*textsurd* mild	*textsurd* mild	*textsurd* mild	×	*textsurd* moderate	×
female 7	Δ9-THC	*textsurd* mild	*textsurd* moderate	*textsurd* mild	*textsurd* mild	*textsurd* moderate/ severe	*textsurd* mild	×
female 8	Δ9-THC	*textsurd* moderate/severe	*textsurd* moderate/severe	*textsurd* severe	*textsurd* moderate	*textsurd* moderate	***textsurd* moderate/severe**	*textsurd* moderate
female 9	Δ9-THC	*textsurd* mild	*textsurd* minimal	*textsurd* mild	×	×	×	*textsurd* minimal
male 1	placebo	×	×	×	×	×	×	×
male 2	placebo	×	×	×	×	×	×	×
male 3	placebo	×	×	×	×	×	×	×
male 4	placebo	×	×	×	×	×	×	×
male 5	placebo	×	×	×	×	×	*textsurd* mild	×
male 6	placebo	×	×	×	×	×	×	×
male 7	placebo	×	×	*textsurd* mild	×	×	***textsurd* mild**	×
female 1	placebo	×	×	×	×	×	×	×
female 2	placebo	×	×	×	×	×	×	×
female 3	placebo	×	×	×	×	×	×	×
female 4	placebo	×	×	×	×	×	×	×
female 5	placebo	×	×	×	×	×	×	×
female 6	placebo	×	×	×	×	×	×	×
female 7	placebo	×	×	×	×	×	×	×
female 8	placebo	×	×	×	×	×	×	×
female 9	placebo	×	×	×	×	×	×	×

Δ9-THC, (−)-trans-Δ9-tetrahydrocannabinol; symptoms highlighted in bold were still observable 2.5 h after the injection.

3.3.3. Psychosis-Related General Psychopathology

The effects of Δ9-THC administration on the PANSS general psychopathology subscale and related statistics have been previously reported [40].

Poor attention was the most frequently observed symptom in the ~20 min following the acute administration of Δ9-THC, with 14 participants reporting such symptoms in a mild to moderate form ($3 \leq$ PANSS-related item ≤ 4; 87%). Most of the participants also experienced a disturbance of volition (75%), disorientation (69%), and poor impulse control (62%). Further frequent symptoms (≥ 2 on PANSS-related item) included somatic concern (50%), preoccupation (50%), motor retardation (50%), mannerisms and posturing (50%), unusual thought content (50%), tension (44%), and active social avoidance (44%). A lower percentage of participants also reported a lack of judgment and insight (37%), symptoms of anxiety and depression (31%), uncooperativeness (31%), and feelings of guilt (12%). One participant reported feeling less depressed after the acute challenge with Δ9-THC.

Some symptoms were still detectable 2.5 h after the injection ($\geq$2 on PANSS-related item), even if in a more attenuated form, with poor attention and disorientation being the most frequent symptom (31%), followed by motor retardation (12%), somatic concern (6%), preoccupation (6%), anxiety (6%), tension (6%), and uncooperativeness (6%).

In the ~20 min following the acute administration of placebo, general psychopathological symptoms were observable in six participants ($\geq$2 on PANSS-related item) and only three of them had clearly detectable symptoms (mild severity, = 3 on PANSS-related item; 19%), which was a percentage that was five times smaller than that observed in participants under the influence of Δ9-THC. In these cases, symptoms included poor attention (12%), somatic concern and preoccupation (6%), tension (6%), disorientation (6%), disturbance of volition (6%), and poor impulse control (6%). Also, only one participant was still experiencing a disturbance of volition 2.5 h after the injection of placebo. Male and female participants were similar in terms of the prevalence of Δ9-THC-induced general psychopathology (Table 3).

3.4. Subjects' Quotes

Table 4 reports the most relevant symptoms experienced by the participants from a narrative perspective. The quality of symptoms showed similarity to the psychotic symptoms reported by schizophrenia patients. When under the acute effect of Δ9-THC, individuals reported both positive and negative symptoms. The most relevant positive symptoms induced by Δ9-THC included suspiciousness, paranoid and grandiose ideas/delusions, conceptual disorganization, and perceptual alterations. Negative symptoms included reduced rapport, a lack of spontaneity, emotional withdrawal, and concrete thinking. The administration of Δ9-THC also induced altered body perception, depersonalization/derealization, slowing of time, euphoria, and anxiety.

3.5. Additional Symptoms

Additional symptoms not necessarily related to psychosis also occurred during the trial. In particular, five female participants (56%) had an acute physical reaction to the Δ9-THC administration, including generalized tremors, vagal reaction, paleness, orthostatic hypotension, sick feeling, flushing, and symptoms of fainting. In contrast, no male participant experienced any physical reaction after the drug challenge. Less intense physical reactions also occurred during the placebo session in two occasions (Table 5).

After the intravenous administration of Δ9-THC, eight volunteers (50%) showed difficulty in motor coordination and indecisiveness with different levels of severity (Table 5; the most severe occurrence is shown in Figure 1). A participant had a protracted posture alteration. Also, two participants showed over-inclusive thinking and protracted internal absorption, respectively (Table 5).

Table 3. Psychosis-related general psychopathology.

Study Participant	Drug	Somatic Concern	Anxiety	Guilt Feelings	Tension	Mannerism and Posturing	Depression	Motor Retardation	Uncooperative	Unusual Thought Content	Disorientation	Poor Attention	Lack of Judgment & Insight	Disturbance of Volition	Poor Impulse Control	Preoccupation	Active Social Avoidance
male 1	Δ9-THC	*textsurd*minimal	×	×	×	*textsurd*minimal	↓	×	×	×	***textsurd*minimal**	***textsurd*mild**	×	×	*textsurd*minimal	***textsurd*mild**	×
male 2	Δ9-THC	×	×	×	×	*textsurd*minimal	×	*textsurd*mild	×	×	*textsurd*mild	*textsurd*mild	×	*textsurd*mild	×	*textsurd*mild	*textsurd*mild
male 3	Δ9-THC	*textsurd*mild	×	*textsurd* mild	×	×	*textsurd*minimal	×	×	*textsurd*mild	***textsurd*mild**	***textsurd*moderate**	×	*textsurd*moderate	*textsurd*mild	*textsurd*mild	*textsurd*minimal
male 4	Δ9-THC	×	×	×	×	×	×	×	×	×	×	×	×	×	*textsurd*minimal	×	×
male 5	Δ9-THC	*textsurd*mild	*textsurd*mild	×	*textsurd*mild	×	*textsurd*mild	×	×	×	*textsurd*mild	*textsurd*moderate	×	*textsurd*mild	*textsurd*mild	*textsurd*mild	×
male 6	Δ9-THC	×	×	×	×	*textsurd*moderate	×	×	×	*textsurd*mild	*textsurd*minimal	*textsurd*mild	*textsurd*mild	*textsurd*mild	*textsurd*mild	×	×
male 7	Δ9-THC	*textsurd*mild	*textsurd*minimal	×	*textsurd*minimal	×	×	*textsurd*mild	*textsurd*minimal	*textsurd*minimal	***textsurd*mild**	***textsurd*moderate**	*textsurd*mild	*textsurd*moderate	×	×	×
female 1	Δ9-THC	×	×	*textsurd* minimal	*textsurd*minimal	×	*textsurd*minimal	*textsurd*mild	*textsurd*mild	*textsurd*minimal	*textsurd*minimal	*textsurd*mild	×	*textsurd*mild	*textsurd*mild	×	*textsurd*minimal
female 2	Δ9-THC	×	×	×	×	×	×	*textsurd*minimal	×	×	×	×	×	×	*textsurd*minimal	×	×
female 3	Δ9-THC	*textsurd*mild	×	×	*textsurd*mild	×	×	×	×	×	×	*textsurd*mild	×	*textsurd*mild	×	*textsurd*mild	×
female 4	Δ9-THC	×	×	×	×	*textsurd*minimal	×	*textsurd*mild	*textsurd*minimal	*textsurd*mild	*textsurd*mild	*textsurd*moderate	*textsurd*mild	*textsurd*mild	*textsurd*mild	×	*textsurd*minimal
female 5	Δ9-THC	***textsurd*mild**	***textsurd*moderate**	×	***textsurd*mild**	*textsurd*minimal	*textsurd*minimal	×	×	*textsurd*mild	*textsurd*mild	*textsurd*mild	×	*textsurd*mild	×	*textsurd*moderate	×
female 6	Δ9-THC	×	×	×	×	*textsurd*mild	*textsurd*mild	*textsurd*moderate	×	×	***textsurd*mild**	*textsurd*moderate	*textsurd*minimal	*textsurd*mild	×	×	*textsurd*mild
female 7	Δ9-THC	*textsurd*mild	*textsurd*mild	×	*textsurd*mild	×	×	***textsurd*mild**	×	×	×	***textsurd*mild**	×	*textsurd*mild	*textsurd*minimal	*textsurd*minimal	*textsurd*mild
female 8	Δ9-THC	*textsurd*mild	*textsurd*mild	×	*textsurd*mild	*textsurd*mild	×	***textsurd*mod./severe**	***textsurd*minimal**	*textsurd*moderate	***textsurd*mild**	***textsurd*moderate**	*textsurd*moderate	*textsurd*moderate	×	*textsurd*mod./severe	*textsurd*mild
female 9	Δ9-THC	×	×	×	×	*textsurd*minimal	×	×	*textsurd*mild	*textsurd*mild	×	*textsurd*moderate	*textsurd*mild	×	*textsurd*minimal	×	×
male 1	placebo	×	×	×	×	×	×	×	×	×	×	×	×	×	×	×	×
male 2	placebo	×	×	×	×	×	×	×	×	×	×	×	×	×	×	×	×
male 3	placebo	×	*textsurd*minimal	×	×	×	×	×	×	×	*textsurd*minimal	×	×	×	×	×	×
male 4	placebo	×	×	×	×	×	×	×	×	×	×	×	×	×	×	×	×
male 5	placebo	×	×	×	×	×	×	×	×	×	*textsurd*mild	*textsurd*mild	×	×	*textsurd*mild	×	×
male 6	placebo	×	*textsurd*minimal	×	*textsurd*minimal	×	×	×	×	×	×	×	×	×	×	×	×
male 7	placebo	×	×	×	×	×	×	×	×	×	×	*textsurd*mild	×	***textsurd*mild**	×	×	×
female 1	placebo	*textsurd*mild	*textsurd*minimal	×	*textsurd*mild	*textsurd*minimal	×	×	×	×	×	×	×	×	×	*textsurd*mild	×
female 2	placebo	×	×	×	×	×	×	×	×	×	×	×	×	×	×	×	×
female 3	placebo	×	×	×	×	×	×	×	×	×	×	*textsurd*minimal	×	×	×	×	×
female 4	placebo	×	×	×	×	×	×	×	×	×	×	×	×	×	×	×	×
female 5	placebo	×	×	×	×	×	×	×	×	×	×	×	×	×	×	×	×
female 6	placebo	×	×	×	×	×	×	×	×	×	×	×	×	×	×	×	×
female 7	placebo	×	×	×	×	×	×	×	×	×	×	×	×	×	×	×	×
female 8	placebo	×	×	×	×	×	×	×	×	×	×	×	×	×	×	×	×
female 9	placebo	×	×	×	×	×	×	×	×	×	×	×	×	×	×	×	×

Δ9-THC, (−)-trans-Δ9-tetrahydrocannabinol; symptoms highlighted in bold were still observable 2.5 h after the injection.

Table 4. Subjects' quotes.

Study Participant	Δ9-THC	Symptom
male 1	'I was so stressed, irritable, and prone to anger that I started an argument with my partner the afternoon after the study'	Hostility, irritability, mood lability
male 2	'I feel I am all over the place and can't stop laughing, thinking you will expose me, I will say something stupid or strange'	Conceptual disorganization, thought disorder, loosening of associations, excitement
male 3	'I can't follow my thoughts, I am not able to think'	Conceptual disorganization, racing thoughts
male 3	'I can understand things better and look for details, I am superior to others'	Grandiosity
male 3	'I might have done something wrong, not willing to say'	Feelings of guilt
male 3	'I am thinking about death and cemeteries'	Unusual thought content
male 6	'The injection changed me into someone with increased abilities'	Grandiosity
male 7	'I am feeling quite confused and disoriented, like time is passing slower and the space is different, like from a camera zoom'	Conceptual disorganization, disorientation
female 1	'I am not interested and I am not willing to talk, I don't care, I want just to go... I feel down, under the weather; maybe I am depressed'	Negative symptoms, depression
female 4	'What have you done to me? I understand, you want to make me paranoid with brainwashing questions...'	Suspiciousness/paranoia with loss of insight
female 4	'What an apple and a ball have in common...You can eat the apple, but not the ball'	Difficulty in abstract thinking
female 5	'I thought you were going to attack me, people are entering the room to check on me'	Suspiciousness/paranoia, ideas of reference
female 7	'What an apple and a ball have in common...You can put the apple in the ball'	Difficulty in abstract thinking
female 8	'My mind went blank, empty, with no thoughts'	Thought blocking
female 8	'The ventilator's noise is louder...This noise is actually rain, it's raining inside the room, I can see and feel it, there is a black sky with seven blue drops, I can count them, someone has opened the ceiling to let the rain in, and put my bed closer to the ceiling...Maybe someone is projecting a sky in front of me'	Inability to 'filter' out irrelevant background stimuli, hallucinatory behavior, secondary delusions
female 9	'Is this real? Is this a fake interview made by a fake doctor, like a Truman show?'	Depersonalization/derealization
female 9	'Colors are brighter, noises louder, and I have something making a noise in the back of my head'	Hallucinatory behavior
male 3, 7; female 5, 7	'I think I am going to chock up...something is wrong with my body...the heart is racing'	Preoccupation/somatic concern, anxiety/tension
Study Participant	**Placebo**	**Symptom**
male 7	'I felt spaced out, a little bit paranoid and upset after a conversation with someone who had a strange facial expression'	Suspiciousness/paranoia

Δ9-THC, (−)-trans-Δ9-tetrahydrocannabinol.

Table 5. Additional symptoms.

Study Participant	Δ9-THC
	physical reactions
female 2	generalized tremors, vagal reaction
female 3	about to faint
female 5	about to faint, paleness
female 6	about to faint, orthostatic hypotension, sick feeling
female 8	flushing
	observed symptoms
males 2, 3, 6; females 1, 2, 3, 4, 8	Different handwriting, errors and corrections in filling out the questionnaires (still present at 2.5 h after the injection for males 3 and 6)
male 6	The participant kept the arm in a distinctively awkward position for 30 min
female 4	The participant committed errors when asked four times to recall words related to a memory task (night instead of black 2/4, vegetable instead of apple 3/4, crisis instead of cries 4/4)
female 8	The participant was internally absorbed and didn't engage at all with a cognitive task
	reported long-lasting symptoms (telephone follow-up)
male 1	suspiciousness, hostility, tension, and poor impulse control until afternoon
male 1	headache, sick and weak feeling, fatigue, exhaustion, physical and mental strain until day after
male 2	tiredness, sleepiness, postprandial somnolence
male 6	disorientation and tiredness until evening
female 5	tiredness and cravings for savory foods until afternoon
female 6	tiredness, sleepiness, and hunger until the end of the day
female 9	sleepiness, thirst, and hungry
females 1 and 4	Symptoms reported during the trial were the same experienced in the past when using recreational cannabis
Study participant	**Placebo**
	physical reactions
female 1	flushing, drowsiness, stomach pain
female 6	flushing, increase in heart rate, heavy chest feeling
	reported long-lasting symptoms (telephone follow-up)
male 7	suspiciousness until afternoon

Δ9-THC, (−)-trans-Δ9-tetrahydrocannabinol.

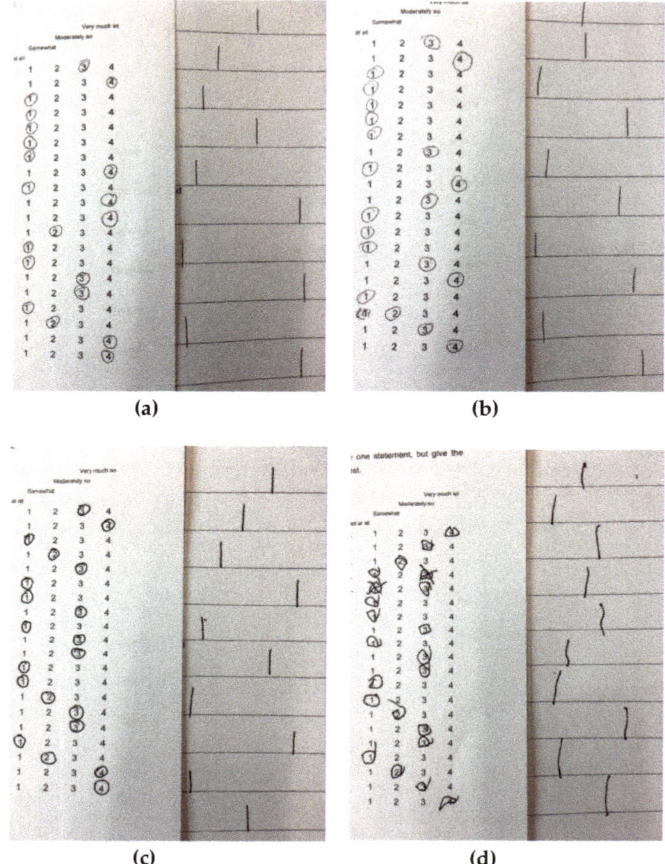

Figure 1. Difficulty in motor coordination and indecisiveness. (**a**) Before placebo administration; (**b**) After placebo administration; (**c**) Before delta-9-tetrahydrocannabinol administration; (**d**) After delta-9-tetrahydrocannabinol administration.

4. Discussion

The purpose of this clinical investigation was to systematically assess the transient psychotic reaction to the intravenous administration of pure Δ9-THC in healthy subjects in a controlled setting, which was in line with the evidence that this cannabinoid represents a valid pharmacologic model for psychosis [10–12]. Results from this study indicate: (i) detectable acute Δ9-THC-induced symptomatic effects in over 90% of the cohort, with moderate to severe symptoms having a lower prevalence (<20%); (ii) protracted minimal to mild Δ9-THC-induced symptomatic effects in 50% of the cohort (~2.5 h after the exposure); (iii) acute physical reactions to Δ9-THC in about 30% of the cohort and only in female participants; (iv) long-lasting Δ9-THC-induced physical symptoms and psychosis-related symptomatic effects in less than 40% and 6% of the cohort, respectively; (v) detectable and mild symptomatic effects after placebo administration in less than 20% of the cohort; (vi) protracted minimal and questionable symptomatic effects after placebo administration in 6% of the cohort (~2.5 h after the exposure); (vii) acute physical reactions to placebo in about 12% of the cohort and only in female participants; and (viii) long-lasting symptomatic effects of placebo in only 6% of the cohort.

The constellation of symptomatic effects induced by Δ9-THC resembled several dimensions of psychotic disorders and overlapped with evidence from previous acute challenge studies with Δ9-THC [12]. However, in order to better understand the extent of its detrimental effects, this investigation took into account the potential nonspecific effects of the drug administration, the so-called placebo effects when they are beneficial, and nocebo effects when they are harmful [47]. Study participants reported a number of symptoms and signs when administered placebo, indicating a nocebo effect. Both psychological (conditioning, negative expectations) and neurobiological (cholecystokinin, endogenous opioids, and dopamine) mechanisms might explain the nocebo effects observed in this study [48]. When controlling for the prevalence, quality, and severity of the subjective and objective changes occurring under placebo, the manifestation of symptomatic effects following Δ9-THC administration remained significantly higher and of greater severity, with most of the transient psychosis-like symptoms occurring only under Δ9-THC. Also, psychotomimetic symptoms lasted significantly longer under Δ9-THC compared to the placebo condition. Similarly, some objective protracted symptoms such as poor motor coordination, posture alteration, over-inclusive thinking, and internal absorption occurred only under Δ9-THC.

Relatively longer-lasting (<24 h) self-reported effects such as tiredness, sleepiness, and increased appetite occurred only under Δ9-THC. Acute physical reactions to the intravenous administration of the drug were more prevalent and clinically more severe in participants under the influence of Δ9-THC than under placebo. Also, they appeared to be gender-specific, with only female participants showing such reactions. Physical and somatic effects were not unexpected, as Δ9-THC has been shown to acutely induce sedation and intoxication [40].

Upon comparing results from this study with previous research, several factors need to be considered, including, but not limited to, the degree of current cannabis use (tolerance effect) and lifetime cannabis exposure (residual effect) of the study samples, and study design. Some research evidence indicates less prominent acute behavioral effects of Δ9-THC in current cannabis users [49], individuals with a past history of frequent cannabis use [38], and when administering Δ9-THC orally [50], as also recently reviewed [39]. Further evidence suggests that the development of tolerance may be explained by the less marked effects of acute Δ9-THC administration on brain function [51,52]. Participants taking part in our intravenous Δ9-THC challenge had been abstinent from cannabis for at least six months. Apart from one subject, the study cohort had also modest previous cannabis exposure. Altogether, previous evidence and our findings suggest that healthy subjects with modest previous cannabis exposure and a proper abstinent period might be more reliable to study the psychotropic effect of Δ9-THC and its underlying mechanisms.

Only individuals with negligible use of other substances (alcohol, tobacco, and other illicit drugs) were invited to take part in the study. Therefore, we can reasonably rule out the possibility that some of the results observed could be attributed to the effects of other substance use. Moreover, this study observed an interval between the two study visits of at least 14 days. This allows us to exclude the possibility of carryover effects from the first session, in light of evidence that Δ9-THC has an elimination half-life of 18 h to 4.3 days [53]. Furthermore, all the participants' urine samples collected at each study visit baseline were negative for the presence of Δ9-THC.

For the purpose of the study and due to ethical reasons, individuals with cannabis dependence or previous negative response to cannabis were excluded from the study. While this allowed us to examine a more homogeneous sample, this might have limited the application of the present results to the general population. Also, caution should be used when making inferences to the general population, as this experiment was conducted in a relatively small sample. The intravenous route of administration was used to allow much more consistent Δ9-THC blood levels across participants and potentially reduce inter-individual variability in drug response [12]. For instance, absorption is slower when cannabinoids are ingested, with Δ9-THC peak concentrations that are lower and more delayed [54], and absorption may also considerably vary between subjects [55]. Similarly, another line of research suggests that cannabinoid levels following cannabis smoking may vary depending on how intensively and efficiently people smoke [56]. However, the intravenous route of administration might have affected the generalizability of the results to the effects of recreational cannabis use. Future studies are needed to assess in the same individuals the effects of acute cannabis challenge using different routes of administration.

5. Conclusions

In conclusion, these results provide further evidence of the psychoactive properties of Δ9-THC and have implications for research in this area. Acute psychosis can be secondary to cannabis use, with some patients relapsing with a similar presentation, and a proportion developing a long-lasting psychotic disorder [57]. More research is needed into the chronology and components of the onset of cannabis-associated psychosis. Acute Δ9-THC challenge studies may help elucidate the nature of psychotic symptom development in cannabis users, ultimately enhancing our understanding of the onset, course, and outcome of cannabis-associated psychosis.

Author Contributions: Conceptualization, M.C., N.W., P.M., L.V.O. and S.B.; Methodology, M.C., N.W., P.M., L.V.O. and S.B.; Validation, P.M., L.V.O. and S.B.; Formal analysis, M.C. and S.B.; Investigation, M.C. and N.W.; Resources, M.C., N.W., P.M., L.V.O. and S.B.; Data curation, M.C. and S.B.; Writing—original draft preparation, M.C. and S.B.; Writing—review and editing, M.C., N.W., P.M., L.V.O., and S.B.; Visualization, M.C., N.W., P.M., L.V.O. and S.B.; Supervision, P.M., L.V.O. and S.B.; Project administration, P.M., L.V.O., and S.B.; Funding acquisition, N.W., L.V.O. and S.B.

Funding: This research was funded by the People Programme (Marie Curie Actions) of the European Union's Seventh Framework Programme (FP7/2007-2013) under REA grant agreement no. 608765. S.B. has been supported by a NIHR Clinician Scientist Award (NIHR CS-11-001) and the MRC (MR/J012149/1). Lukas Van Oudenhove has been supported by a KU Leuven Special Research Fund (BOF, Bijzonder Onderzoeksfonds). Nathalie Weltens has been supported by a KU Leuven BOF Post-Doctoral Mandate.

Acknowledgments: The authors acknowledge infrastructure support from the NIHR Mental Health Biomedical Research Centre at South London and Maudsley NHS Foundation Trust and King's College London. The views expressed are those of the author(s) and not necessarily those of the NHS, the NIHR or the Department of Health.

Conflicts of Interest: The authors declare no conflicts of interest. The funders had no role in the design of the study; in the collection, analyses, or interpretation of data; in the writing of the manuscript, or in the decision to publish the results.

References

1. Tsuang, M.T.; Bar, J.L.; Stone, W.S.; Faraone, S.V. Gene-environment interactions in mental disorders. *World Psychiatry* **2004**, *3*, 73–83. [PubMed]
2. Rapoport, J.L.; Giedd, J.N.; Gogtay, N. Neurodevelopmental model of schizophrenia: Update 2012. *Mol. Psychiatry* **2012**, *17*, 1228–1238. [CrossRef]
3. Gouzoulis-Mayfrank, E.; Hermle, L.; Thelen, B.; Sass, H. History, rationale and potential of human experimental hallucinogenic drug research in psychiatry. *Pharmacopsychiatry* **1998**, *31*, 63–68. [CrossRef]
4. Strakowski, S.M.; Sax, K.W.; Setters, M.J.; Keck, P.E., Jr. Enhanced response to repeated d-amphetamine challenge: Evidence for behavioral sensitization in humans. *Biol. Psychiatry* **1996**, *40*, 872–880. [CrossRef]
5. Vollenweider, F.X.; Vollenweider-Scherpenhuyzen, M.F.; Babler, A.; Vogel, H.; Hell, D. Psilocybin induces schizophrenia-like psychosis in humans via a serotonin-2 agonist action. *Neuroreport* **1998**, *9*, 3897–3902. [CrossRef]
6. Krystal, J.H.; Karper, L.P.; Seibyl, J.P.; Freeman, G.K.; Delaney, R.; Bremner, J.D.; Heninger, G.R.; Bowers, M.B., Jr.; Charney, D.S. Subanesthetic effects of the noncompetitive NMDA antagonist, ketamine, in humans. Psychotomimetic, perceptual, cognitive, and neuroendocrine responses. *Arch. Gen. Psychiatry* **1994**, *51*, 199–214. [CrossRef] [PubMed]
7. Sewell, R.A.; Skosnik, P.D.; Garcia-Sosa, I.; Ranganathan, M.; D'Souza, D.C. Behavioral, cognitive and psychophysiological effects of cannabinoids: Relevance to psychosis and schizophrenia. *Rev. Bras. Psiquiatr.* **2010**, *32*, 15–30.
8. Bhattacharyya, S.; Crippa, J.A.; Martin-Santos, R.; Winton-Brown, T.; Fusar-Poli, P. Imaging the neural effects of cannabinoids: Current status and future opportunities for psychopharmacology. *Curr. Pharm. Des.* **2009**, *15*, 2603–2614. [CrossRef] [PubMed]

9. Nestler, E.J.; Hyman, S.E. Animal models of neuropsychiatric disorders. *Nat. Neurosci.* **2010**, *13*, 1161–1169. [CrossRef]
10. Bhattacharyya, S.; Fusar-Poli, P.; Borgwardt, S.; Martin-Santos, R.; Nosarti, C.; O'Carroll, C.; Allen, P.; Seal, M.L.; Fletcher, P.C.; Crippa, J.A.; et al. Modulation of Mediotemporal and Ventrostriatal Function in Humans by Delta 9-Tetrahydrocannabinol A Neural Basis for the Effects of Cannabis sativa on Learning and Psychosis. *Arch. Gen. Psychiatry* **2009**, *66*, 442–451. [CrossRef]
11. Bhattacharyya, S.; Crippa, J.A.; Allen, P.; Martin-Santos, R.; Borgwardt, S.; Fusar-Poli, P.; Rubia, K.; Kambeitz, J.; O'Carroll, C.; Seal, M.L.; et al. Induction of psychosis by Delta9-tetrahydrocannabinol reflects modulation of prefrontal and striatal function during attentional salience processing. *Arch. Gen. Psychiatry* **2012**, *69*, 27–36. [CrossRef]
12. D'Souza, D.C.; Perry, E.; MacDougall, L.; Ammerman, Y.; Cooper, T.; Wu, Y.T.; Braley, G.; Gueorguieva, R.; Krystal, J.H. The psychotomimetic effects of intravenous delta-9-tetrahydrocannabinol in healthy individuals: Implications for psychosis. *Neuropsychopharmacology* **2004**, *29*, 1558–1572. [CrossRef] [PubMed]
13. Bhattacharyya, S.; Atakan, Z.; Martin-Santos, R.; Crippa, J.A.; Kambeitz, J.; Prata, D.; Williams, S.; Brammer, M.; Collier, D.A.; McGuire, P.K. Preliminary report of biological basis of sensitivity to the effects of cannabis on psychosis: AKT1 and DAT1 genotype modulates the effects of delta-9-tetrahydrocannabinol on midbrain and striatal function. *Mol. Psychiatry* **2012**, *17*, 1152–1155. [CrossRef]
14. Moore, T.H.; Zammit, S.; Lingford-Hughes, A.; Barnes, T.R.; Jones, P.B.; Burke, M.; Lewis, G. Cannabis use and risk of psychotic or affective mental health outcomes: A systematic review. *Lancet* **2007**, *370*, 319–328. [CrossRef]
15. Morrison, P.D.; Bhattacharyya, S.; Murray, R. Recreational cannabis: The risk of schizophrenia. In *Handbook of Cannabis*; Pertwee, R.G., Ed.; Oxford University Press: Oxford, UK, 2015; pp. 661–673.
16. Colizzi, M.; Murray, R. Cannabis and psychosis: What do we know and what should we do? *Br. J. Psychiatry* **2018**, *212*, 195–196. [CrossRef] [PubMed]
17. Colizzi, M.; Iyegbe, C.; Powell, J.; Ursini, G.; Porcelli, A.; Bonvino, A.; Taurisano, P.; Romano, R.; Masellis, R.; Blasi, G.; et al. Interaction Between Functional Genetic Variation of DRD2 and Cannabis Use on Risk of Psychosis. *Schizophr. Bull.* **2015**, *41*, 1171–1182. [CrossRef]
18. Colizzi, M.; Iyegbe, C.; Powell, J.; Blasi, G.; Bertolino, A.; Murray, R.M.; Di Forti, M. Interaction between DRD2 and AKT1 genetic variations on risk of psychosis in cannabis users: A case–control study. *Npj Schizophr.* **2015**, *1*, 15025. [CrossRef]
19. Colizzi, M.; Carra, E.; Fraietta, S.; Lally J.; Quattrone, D.; Bonaccorso, S.; Mondelli, V.; Ajnakina, O.; Dazzan, P.; Trotta, A.; et al. Substance use, medication adherence and outcome one year following a first episode of psychosis. *Schizophr. Res.* **2016**, *170*, 311–317. [CrossRef]
20. Patel, R.; Wilson, R.; Jackson, R.; Ball, M.; Shetty, H.; Broadbent, M.; Stewart, R.; McGuire, P.; Bhattacharyya, S. Association of cannabis use with hospital admission and antipsychotic treatment failure in first episode psychosis: An observational study. *BMJ Open* **2016**, *6*, e009888. [CrossRef] [PubMed]
21. Schoeler, T.; Petros, N.; Di Forti, M.; Klamerus, E.; Foglia, E.; Ajnakina, O.; Gayer-Anderson, C.; Colizzi, M.; Quattrone, D.; Behlke, I.; et al. Effects of continuation, frequency, and type of cannabis use on relapse in the first 2 years after onset of psychosis: An observational study. *Lancet Psychiatry* **2016**, *3*, 947–953. [CrossRef]
22. Schoeler, T.; Petros, N.; Di Forti, M.; Pingault, J.B.; Klamerus, E.; Foglia, E.; Small, A.; Murray, R.; Bhattacharyya, S. Association Between Continued Cannabis Use and Risk of Relapse in First-Episode Psychosis: A Quasi-Experimental Investigation Within an Observational Study. *JAMA Psychiatry* **2016**, *73*, 1173–1179. [CrossRef]
23. Schoeler, T.; Monk, A.; Sami, M.B.; Klamerus, E.; Foglia, E.; Brown, R.; Camuri, G.; Altamura, A.C.; Murray, R.; Bhattacharyya, S. Continued versus discontinued cannabis use in patients with psychosis: A systematic review and meta-analysis. *Lancet Psychiatry* **2016**, *3*, 215–225. [CrossRef]
24. Appiah-Kusi, E.; Leyden, E.; Parmar, S.; Mondelli, V.; McGuire, P.; Bhattacharyya, S. Abnormalities in neuroendocrine stress response in psychosis: The role of endocannabinoids. *Psychol. Med.* **2016**, *46*, 27–45. [CrossRef]
25. Leweke, F.M.; Giuffrida, A.; Koethe, D.; Schreiber, D.; Nolden, B.M.; Kranaster, L.; Neatby, M.A.; Schneider, M.; Gerth, C.W.; Hellmich, M.; et al. Anandamide levels in cerebrospinal fluid of first-episode schizophrenic patients: Impact of cannabis use. *Schizophr. Res.* **2007**, *94*, 29–36. [CrossRef]

26. Taurisano, P.; Antonucci, L.A.; Fazio, L.; Rampino, A.; Romano, R.; Porcelli, A.; Masellis, R.; Colizzi, M.; Quarto, T.; Torretta, S.; et al. Prefrontal activity during working memory is modulated by the interaction of variation in CB1 and COX2 coding genes and correlates with frequency of cannabis use. *Cortex* **2016**, *81*, 231–238. [CrossRef]
27. Colizzi, M.; Bhattacharyya, S. Does Cannabis Composition Matter? Differential Effects of Delta-9-tetrahydrocannabinol and Cannabidiol on Human Cognition. *Curr. Addict. Rep.* **2017**, *4*, 62–74. [CrossRef]
28. Colizzi, M.; Fazio, L.; Ferranti, L.; Porcelli, A.; Masellis, R.; Marvulli, D.; Bonvino, A.; Ursini, G.; Blasi, G.; Bertolino, A. Functional genetic variation of the cannabinoid receptor 1 and cannabis use interact on prefrontal connectivity and related working memory behavior. *Neuropsychopharmacology* **2015**, *40*, 640–649. [CrossRef]
29. Chopra, G.S.; Smith, J.W. Psychotic Reactions Following Cannabis Use in East Indians. *Arch. Gen. Psychiatry* **1974**, *30*, 24–27. [CrossRef] [PubMed]
30. Talbott, J.A.; Teague, J.W. Marihuana Psychosis—Acute Toxic Psychosis Associated with Use of Cannabis Derivatives. *J. Amer. Med. Assoc.* **1969**, *210*, 299–302. [CrossRef]
31. Keeler, M.H.; Moore, E. Paranoid reactions while using marijuana. *Dis. Nerv. Syst.* **1974**, *35*, 535–536.
32. Morrison, P.D.; Zois, V.; McKeown, D.A.; Lee, T.D.; Holt, D.W.; Powell, J.F.; Kapur, S.; Murray, R.M. The acute effects of synthetic intravenous Delta(9)-tetrahydrocannabinol on psychosis, mood and cognitive functioning. *Psychol. Med.* **2009**, *39*, 1607–1616. [CrossRef]
33. Barkus, E.; Morrison, P.D.; Vuletic, D.; Dickson, J.C.; Ell, P.J.; Pilowsky, L.S.; Brenneisen, R.; Holt, D.W.; Powell, J.; Kapur, S.; et al. Does intravenous Δ9-tetrahydrocannabinol increase dopamine release? A SPET study. *J. Psychopharmacol.* **2011**, *25*, 1462–1468. [CrossRef]
34. Morrison, P.D.; Stone, J.M. Synthetic delta-9-tetrahydrocannabinol elicits schizophrenia-like negative symptoms which are distinct from sedation. *Hum. Psychopharmacol.* **2011**, *26*, 77–80. [CrossRef]
35. Englund, A.; Morrison, P.D.; Nottage, J.; Hague, D.; Kane, F.; Bonaccorso, S.; Stone, J.M.; Reichenberg, A.; Brenneisen, R.; Holt, D.; et al. Cannabidiol inhibits THC-elicited paranoid symptoms and hippocampal-dependent memory impairment. *J. Psychopharmacol.* **2013**, *27*, 19–27. [CrossRef]
36. Martin-Santos, R.; Crippa, J.A.; Batalla, A.; Bhattacharyya, S.; Atakan, Z.; Borgwardt, S.; Allen, P.; Seal, M.; Langohr, K.; Farre, M.; et al. Acute Effects of a Single, Oral dose of d9-tetrahydrocannabinol (THC) and Cannabidiol (CBD) Administration in Healthy Volunteers. *Curr. Pharm. Des.* **2012**, *18*, 4966–4979. [CrossRef] [PubMed]
37. Kaufmann, R.M.; Kraft, B.; Frey, R.; Winkler, D.; Weiszenbichler, S.; Backer, C.; Kasper, S.; Kress, H.G. Acute psychotropic effects of oral cannabis extract with a defined content of Delta9-tetrahydrocannabinol (THC) in healthy volunteers. *Pharmacopsychiatry* **2010**, *43*, 24–32. [CrossRef]
38. D'Souza, D.C.; Ranganathan, M.; Braley, G.; Gueorguieva, R.; Zimolo, Z.; Cooper, T.; Perry, E.; Krystal, J. Blunted psychotomimetic and amnestic effects of delta-9-tetrahydrocannabinol in frequent users of cannabis. *Neuropsychopharmacology* **2008**, *33*, 2505–2516. [CrossRef] [PubMed]
39. Colizzi, M.; Bhattacharyya, S. Cannabis use and the development of tolerance: A systematic review of human evidence. *Neurosci. Biobehav. Rev.* **2018**, *93*, 1–25. [CrossRef] [PubMed]
40. Colizzi, M.; Weltens, N.; McGuire, P.; Lythgoe, D.; Williams, S.; Van Oudenhove, L.; Bhattacharyya, S. Delta-9-tetrahydrocannabinol increases striatal glutamate levels in healthy individuals: Implications for psychosis. *Mol. Psychiatry* **2019**. [CrossRef] [PubMed]
41. Naef, M.; Russmann, S.; Petersen-Felix, S.; Brenneisen, R. Development and pharmacokinetic characterization of pulmonal and intravenous delta-9-tetrahydrocannabinol (THC) in humans. *J. Pharm. Sci.* **2004**, *93*, 1176–1184. [CrossRef]
42. Radhakrishnan, R.; Wilkinson, S.T.; D'Souza, D.C. Gone to pot—A review of the association between cannabis and psychosis. *Front. Psychiatry* **2014**, *5*. [CrossRef]
43. APA. *Diagnostic and Statistical Manual of Mental Disorders*, 5th ed.; American Psychiatric Publishing: Arlington, VA, USA, 2013.
44. First, M.B.; Williams, J.B.W.; Karg, R.S.; Spitzer, R.L. *Structured Clinical Interview for DSM-5 Disorders, Clinician Version (SCID-5-CV)*; American Psychiatric Association: Arlington, VA, USA, 2015.
45. Kay, S.R.; Fiszbein, A.; Opler, L.A. The positive and negative syndrome scale (PANSS) for schizophrenia. *Schizophr. Bull.* **1987**, *13*, 261–276. [CrossRef]

46. Spencer, L.; Ritchie, J.; Lewis, J.; Dillon, L. *Quality in Qualitative Evaluation: A Framework for Assessing Research Evidence*; Cabinet Office: London, UK, 2003.
47. Hauser, W.; Sarzi-Puttini, P.; Tolle, T.R.; Wolfe, F. Placebo and nocebo responses in randomised controlled trials of drugs applying for approval for fibromyalgia syndrome treatment: Systematic review and meta-analysis. *Clin. Exp. Rheumatol.* **2012**, *30*, S78–S87.
48. Planes, S.; Villier, C.; Mallaret, M. The nocebo effect of drugs. *Pharmacol. Res. Perspect.* **2016**, *4*, e00208. [CrossRef]
49. Hart, C.L.; van Gorp, W.; Haney, M.; Foltin, R.W.; Fischman, M.W. Effects of acute smoked marijuana on complex cognitive performance. *Neuropsychopharmacology* **2001**, *25*, 757–765. [CrossRef]
50. Curran, H.V.; Brignell, C.; Fletcher, S.; Middleton, P.; Henry, J. Cognitive and subjective dose-response effects of acute oral Delta(9)-tetrahydrocannabinol (THC) in infrequent cannabis users. *Psychopharmacology* **2002**, *164*, 61–70. [CrossRef]
51. Colizzi, M.; McGuire, P.; Giampietro, V.; Williams, S.; Brammer, M.; Bhattacharyya, S. Previous cannabis exposure modulates the acute effects of delta-9-tetrahydrocannabinol on attentional salience and fear processing. *Exp. Clin. Psychopharm.* **2018**, *26*, 582–598. [CrossRef]
52. Colizzi, M.; McGuire, P.; Giampietro, V.; Williams, S.; Brammer, M.; Bhattacharyya, S. Modulation of acute effects of delta-9-tetrahydrocannabinol on psychotomimetic effects, cognition and brain function by previous cannabis exposure. *Eur. Neuropsychopharm.* **2018**, *28*, 850–862. [CrossRef]
53. Kelly, P.; Jones, R.T. Metabolism of Tetrahydrocannabinol in Frequent and Infrequent Marijuana Users. *J. Anal. Toxicol.* **1992**, *16*, 228–235. [CrossRef]
54. Law, B.; Mason, P.A.; Moffat, A.C.; Gleadle, R.I.; King, L.J. Forensic aspects of the metabolism and excretion of cannabinoids following oral ingestion of cannabis resin. *J. Pharm. Pharmacol.* **1984**, *36*, 289–294. [CrossRef] [PubMed]
55. Huestis, M.A.; Gustafson, R.A.; Moolchan, E.T.; Barnes, A.; Bourland, J.A.; Sweeney, S.A.; Hayes, E.F.; Carpenter, P.M.; Smith, M.L. Cannabinoid concentrations in hair from documented cannabis users. *Forensic Sci. Int.* **2007**, *169*, 129–136. [CrossRef]
56. Colizzi, M.; Bhattacharyya, S. Neurocognitive effects of cannabis: Lessons learned from human experimental studies. *Prog. Brain. Res.* **2018**, *242*, 179–216. [CrossRef] [PubMed]
57. Semple, D.M.; McIntosh, A.M.; Lawrie, S.M. Cannabis as a risk factor for psychosis: Systematic review. *J. Psychopharmacol.* **2005**, *19*, 187–194. [CrossRef] [PubMed]

Review

Factors Moderating the Association between Cannabis Use and Psychosis Risk: A Systematic Review

Sanne J. van der Steur, Albert Batalla and Matthijs G. Bossong *

Department of Psychiatry, University Medical Center Utrecht Brain Center, Utrecht University, 3584CX Utrecht, The Netherlands; s.j.vandersteur@umcutrecht.nl (S.J.v.d.S.); A.BatallaCases@umcutrecht.nl (A.B.)

* Correspondence: m.bossong@umcutrecht.nl; Tel.: +31-88-7556369

Received: 21 January 2020; Accepted: 10 February 2020; Published: 12 February 2020 

Abstract: Increasing evidence indicates a relationship between cannabis use and psychosis risk. Specific factors, such as determinants of cannabis use or the genetic profile of cannabis users, appear to moderate this association. The present systematic review presents a detailed and up-to-date literature overview on factors that influence the relationship between cannabis use and psychosis risk. A systematic search was performed according to the PRISMA guidelines in MEDLINE and Embase, and 56 studies were included. The results show that, in particular, frequent cannabis use, especially daily use, and the consumption of high-potency cannabis are associated with a higher risk of developing psychosis. Moreover, several genotypes moderate the impact of cannabis use on psychosis risk, particularly those involved in the dopamine function, such as AKT1. Finally, cannabis use is associated with an earlier psychosis onset and increased risk of transition in individuals at a clinical high risk of psychosis. These findings indicate that changing cannabis use behavior could be a harm reduction strategy employed to lower the risk of developing psychosis. Future research should aim to further develop specific biomarkers and genetic profiles for psychosis, thereby contributing to the identification of individuals at the highest risk of developing a psychotic disorder.

Keywords: cannabis use; psychotic disorder; genetics; age of onset; clinical high risk

---

1. Introduction

Schizophrenia and other psychotic disorders are a burdensome group of disorders, with occurrence of psychosis as an overlapping phenomenon and a lifetime morbid risk of about 7 per 1000 [1,2]. Characteristics of psychosis include positive symptoms, such as hallucinations and delusions; negative and affective symptoms, such as a lack of motivation and depression; and neurocognitive alterations [2]. Although one of the most robust pathophysiological features of psychosis is an increase in the striatal dopamine function, accumulating evidence indicates abnormalities in the endocannabinoid system of patients with a psychotic disorder [3–5]. For example, patients exhibit enhanced levels of endogenous cannabinoid ligands in cerebrospinal fluid [6,7], as well as increased post-mortem CB1 receptor densities [8,9] and in vivo CB1 receptor availability in the brain [10,11].

A convincing amount of evidence indicates an association between cannabis use and the risk of psychosis. First, the administration of Δ9-tetrahydrocannabinol (THC), which is the main psychoactive component of cannabis, to healthy individuals can induce transient psychotic-like experiences [12–14] (see for a review, [15]). In addition, early epidemiological studies showed a relationship between cannabis use and the development of psychotic symptoms or a psychotic disorder later in life [16–22]. Systematic reviews and meta-analyses of these data have confirmed cannabis use as a risk factor in the development of psychosis or schizophrenia [23–26]. In particular, Moore et al. concluded that the risk of any psychotic outcome in individuals who had ever used cannabis was increased by approximately 1.5 times [26].

Specific factors appear to moderate this association between cannabis use and the development of psychosis, in particular, determinants of cannabis use. Several studies have shown that a higher frequency of cannabis use is related to a higher risk of psychosis [18,19] (see for reviews and meta-analysis, [26–28]). Similarly, a lower age of onset of cannabis use [17,19,21] and the use of more potent types of cannabis, with higher THC and lower cannabidiol (CBD) concentrations [22,28], have been shown to further increase the risk of psychosis. CBD is a non-intoxicating cannabinoid compound that may attenuate some of the negative effects associated with cannabis use [29–31]. The genetic profile of cannabis users has also been implicated as a moderator of the association between cannabis use and the development of psychosis [32–34]. For example, the catechol-O-methyltransferase (COMT) Val158Met genotype, involved in dopamine turnover in the prefrontal cortex, was shown to interact with the impact of cannabis use on psychosis risk [35,36]. Initial studies also suggested that cannabis use is associated with an earlier age of psychosis onset [37–40]. Finally, cannabis use by individuals at a clinical high risk (CHR) for psychosis may exacerbate psychotic symptoms and drive an earlier transition to psychosis [41], although contrasting results have also been reported [42].

The present systematic review aims to give an up-to-date, detailed overview of the most recent literature (2009–2019) on factors that influence the relationship between recreational cannabis use and the risk of psychosis. The reviewed articles investigated how this relationship was affected by (1) patterns of cannabis use (e.g., dose and frequency); (2) age of initiation of cannabis use; (3) type of cannabis used; and (4) the individual genetic profile. Studies on the age of onset of psychosis and the influence of cannabis use on the transition to psychosis in people at CHR are also discussed.

2. Materials and Methods

2.1. Literature Search and Selection Procedures

A systematic search was performed in two databases—MEDLINE and Embase—to identify relevant studies conforming to the Preferred Reporting Items for Systematic Reviews and Meta-Analyses (PRISMA) guidelines [43]. The final systematic search was performed on July 23rd, 2019. The search was limited to human studies published between 2009 and 2019 and was run through titles and abstracts. The exact search term used was as follows:

```
((((("Cannabis"[Mesh]) OR ((Cannabis[Title/Abstract] OR
Marihuana*[Title/Abstract] OR Marijuana*[Title/Abstract] OR
Hashish*[Title/Abstract] OR Hemp[Title/Abstract])))) AND
(("Psychotic Disorders"[Mesh]) OR ((psychotic
disorder*[Title/Abstract] OR psychosis[Title/Abstract] OR
psychoses[Title/Abstract] OR psychotic[Title/Abstract]))))) NOT
(animals[MeSH Terms] NOT humans[MeSH Terms]).
```

Further publications were found by screening reference lists from included articles and relevant reviews. The PRISMA flowchart presented in Figure 1 shows the selection procedure employed to identify relevant studies. A total of 3348 records were identified through the search. An additional 12 records were found through screening reference lists of relevant articles. After the removal of duplicates, 2433 articles were screened and 2340 records were excluded. In total, 93 records were assessed for eligibility, leaving 56 studies to be included. Please see Table S1 in the Supplementary Materials for an overview of included studies.

2.2. Selection Criteria

Publications were screened by two researchers. Only published, peer-reviewed, and observational studies investigating the relationship between cannabis use and psychosis were considered and were selected when they examined one of the following moderating factors: (1) patterns of cannabis use (e.g., dose and frequency); (2) age of initiation of cannabis use; (3) type of cannabis used; (4) the individual genetic profile; (5) cannabis use related to the age of onset of psychosis; and (6) the influence of cannabis use on the transition to psychosis in individuals at CHR.

If cannabis was not the only substance being investigated, studies were included only if cannabis was the most frequently used illicit substance, when each analysis was conducted separately for each substance, or when other substance use was controlled for. Studies that exclusively reported measures of lifetime cannabis use (ever vs. never), that only examined other potential risk factors for psychosis (e.g., childhood trauma), or that reported data from overlapping cohorts were excluded.

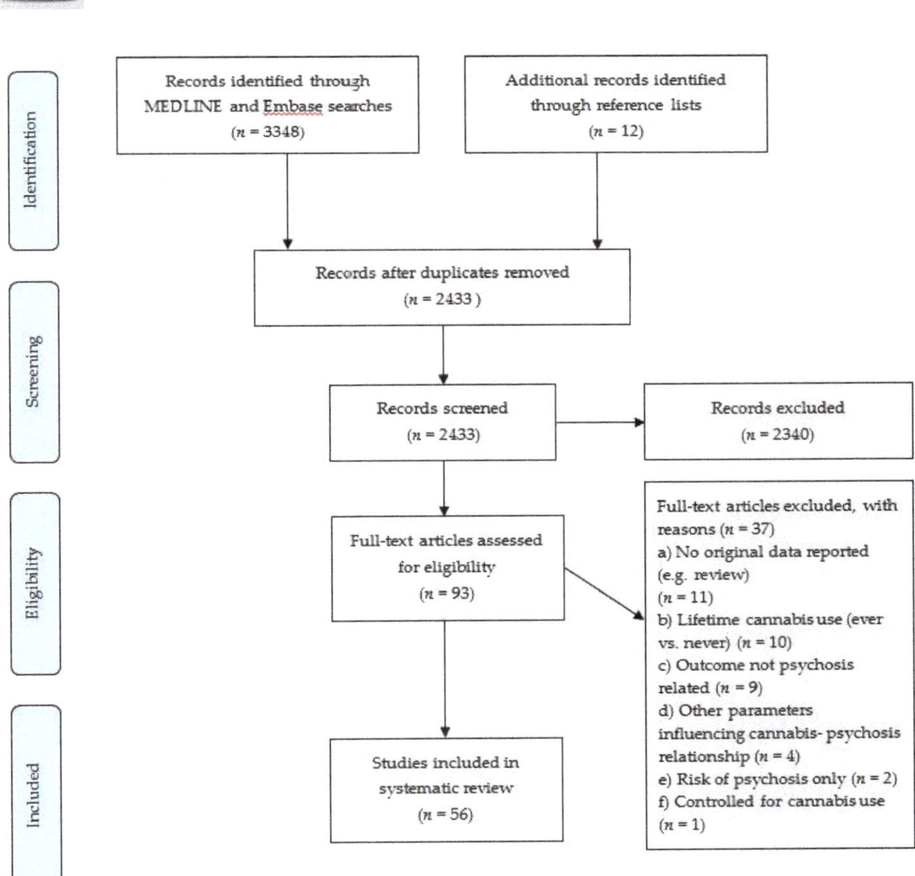

Figure 1. Literature search and selection of studies adapted from the Preferred Reporting Items for Systematic Reviews and Meta-Analyses (PRISMA) flowchart (http://www.prisma-statement.org/).

2.3. Statistical Outcome Measures

Because the included papers analysed and reported their results in different ways, an explanation of statistical outcome measures is provided in Table 1.

Table 1. Explanation of statistical outcome measures.

OR	Odds ratio	An OR is a statistic that quantifies the strength of the association between two events, for example, the use of cannabis and the development of a psychotic disorder. An OR greater than 1 indicates that the two events are associated.
HR	Hazard ratio	HR is a measure of an effect of an intervention on an outcome of interest over time, for example, daily cannabis use and the onset of psychosis. An HR of 1 indicates that event rates (e.g., onset of psychosis) are the same in both groups (e.g., daily cannabis use vs. no cannabis use).
Regression beta coefficient		A regression assesses whether predictor variables (e.g., age of onset of cannabis use) account for variability in a dependent variable (onset of psychosis). The beta coefficient is the degree of change in the outcome variable for every 1-unit of change in the predictor variable. If the beta coefficient is positive and significant, the interpretation is that for every 1-unit increase in the predictor variable, the outcome variable will increase by the beta coefficient value.

3. Results

3.1. Cannabis Use and the Development of Psychotic Disorders

3.1.1. Patterns of Cannabis Use

In the last decade, six studies have been published that have examined the role of patterns of cannabis use in the incidence of psychosis. In their study issued in 2009, Di Forti and colleagues found that first-episode psychosis (FEP) patients were more likely to be current cannabis users (odds ratio (OR) = 6.4, 95% Confidence Interval (CI) 3.2–28.6, $P < 0.05$) and to have used cannabis for over 5 years (OR = 2.1, 95% CI 0.9–8.4, $P < 0.05$) compared to a healthy control group [22]. In an international follow-up study published in 2019, Di Forti et al. took six measures of cannabis use across eleven European sites in more than 900 FEP patients and 1200 controls. Daily cannabis use increased the odds of psychotic disorder to 3.2 (fully adjusted OR, 95% CI 2.2–4.1, $P < 0.0001$) compared with never having used cannabis, irrespective of the type of cannabis that was used [44]. Compton et al. (2009) found that past cannabis use had no significant effect on the risk of the onset of psychosis. However, daily cannabis use did increase the risk of the onset of psychosis (hazard ratio (HR) = 1.997, $P < 0.05$) compared to non-cannabis use [45]. According to a 2019 study by Karcher et al., individuals who reported frequent lifetime cannabis use (defined as cannabis use >100 times lifetime) were diagnosed with a cannabis use disorder (CUD) and those who reported current cannabis use had an increased risk of 1.21 to 1.26 times of at least one psychotic-like experience. Moreover, psychotic-like experiences were associated with frequent cannabis use ($\beta = 0.11$, 95% CI 0.08–0.14), cannabis use disorder ($\beta = 0.13$, 95% CI 0.09–0.16), and current cannabis use ($\beta = 0.07$, 95% CI 0.04–0.10), even after adjustment for covariates ($P < 0.05$) [46]. Leadbeater and colleagues (2019) reported that more frequent cannabis use ($\beta = 0.13$, 95% CI 0.002–0.25, $P < 0.05$), as well as CUD ($\beta = 0.51$, 95% CI 0.01–1.01, $P < 0.05$), during adolescence were associated with psychotic symptoms at ages 22 and 23 [47]. Finally, Marconi and colleagues (2016) performed a meta-analysis of 18 studies reporting on the level of cannabis use that were published before 2014 and demonstrated that higher levels of cannabis use were associated with an increased risk for psychosis in all the included studies. A logistic regression model gave an OR of

3.90 (95% CI 2.84–5.34) for the risk of psychotic disorder among the heaviest cannabis users compared to the non-users [27].

In summary, all studies published in the last decade that investigated patterns of cannabis use indicate that frequent cannabis use, in particular daily use, is associated with an increased risk for both psychotic-like experiences and psychotic disorders.

3.1.2. Age of Onset of Cannabis Use

In the past decade, six studies have been published that have investigated the influence of age of initiation of cannabis use on the development of psychosis. Using an online version of the Community Assessment of Psychic Experiences (CAPE) questionnaire in a group of more than 17,000 adolescents, Schubart and colleagues (2011) showed that cannabis use at age 12 or younger was associated with an OR of 3.1 (95% CI 2.1–4.3, $P < 0.05$) for a top 10% score on the subscale measuring psychotic experiences. This significant association between cannabis use and the presence of psychotic experiences was absent when cannabis use was initiated at 15 years or older [48]. Van Gastel and colleagues (2012) found similar effects in a survey of secondary school children aged 12–16 ($N = 4552$), showing that an earlier age of initiation of heavy cannabis use was significantly related to the onset of psychotic experiences ($\beta = 0.065$, $P < 0.001$) [49]. In a cohort study of 1756 adolescents, Gage et al. (2014) investigated cannabis use at age 16 in relation to the emergence of psychotic experiences at age 18 and found that frequent cannabis use increased the odds of psychotic experiences (OR = 1.48, 95% CI 1.18–1.86, $P = 0.001$). Nevertheless, controlling for tobacco or other drugs attenuated the odds to 1.2 (95% CI 0.91–1.76, $P = 0.160$) and 1.25 (95% CI 0.91–1.73, $P = 0.165$), respectively, indicating that the use of other substances may be involved in the association [50]. In a cohort of 410 FEP patients, Di Forti et al. (2014) demonstrated that those who had started cannabis at age 15 or younger had an earlier onset of psychosis than those who had started after 15 years (27.0 ± 6.2 vs. 29.1 ± 8.5; HR = 1.40; 95% CI 1.06–1.84, $P = 0.05$) [51]. In the 2019 study by Di Forti and colleagues, they found that individuals who had started using cannabis when 15 years old or younger presented increased odds of psychosis (crude OR = 3.9, 95% CI 3.0–4.9, $P < 0.0001$) [44]. Finally, Setién-Suero and colleagues (2018) showed a significant correlation between age of cannabis use onset and age of onset of psychosis in FEP patients (rho = 0.441; $P \leq 0.001$). However, there were no significant differences in the age of psychosis onset between early and late cannabis users [52].

Overall, the above-mentioned studies indicate that the early initiation of cannabis use is related to an increased risk of developing psychosis. However, because only a limited number of studies have compared the odds of psychosis between early and late cannabis users, it is still unclear if young cannabis users are at a higher risk than late cannabis users. Confounding factors, such as the use of other substances or the cumulative amount of consumed cannabis, should be taken into account as these may explain part of the risk.

3.1.3. Type of Cannabis Used

Four studies have reported on the type of cannabis used in relation to the risk of psychosis. First, Di Forti and colleagues reported, in their study from 2009, that 78% of the cannabis-using FEP patients consumed high-potency cannabis (containing 12–18% THC) compared with 37% of the control group (OR = 6.8, 95% CI 2.6–25.4, $P < 0.05$) [22]. In a 2015 follow-up study by Di Forti et al., it was reported that the use of high-potency cannabis (containing >10% THC) increased the odds of a psychotic disorder to almost 3.0 compared to those who never used cannabis (OR = 2.92, 95% CI 1.52–3.45, $P = 0.001$). Importantly, everyday use of high-potency cannabis further increased the risk of psychotic disorder (OR = 5.4, 95% CI 2.81–11.31, $P = 0.002$) [53]. In their 2019 study, Di Forti and colleagues concluded that the use of high-potency cannabis (containing >10% THC) modestly increased the odds of a psychotic disorder (OR = 1.5, 95% CI 1.1–2.6), and correction for daily use did not change this. However, the daily use of high-potency cannabis did substantially increase the odds of psychotic disorder (OR = 4.8, 95% CI 2.5–6.3). There was no evidence of an interaction between the frequency of use and type of cannabis used [44]. Finally, based on the results from their 2011 online survey, Schubart et al. reported that the use of cannabis with a low THC:CBD ratio (<55) was related to fewer psychotic experiences than cannabis with a high THC:CBD ratio (>55) ($F(1,1877):14.577$; $P < 0.001$) [54].

In summary, these results suggest that the use of high-potency cannabis (i.e., cannabis with a high THC content or a high THC:CBD ratio) significantly increases the risk of a psychotic disorder.

3.1.4. Genetics

AKT1 Gene

Three studies have investigated the moderating effect of the *AKT1* genotype (rs2494732 locus) on the association between cannabis use and the risk of psychosis. AKT1 is a protein kinase involved in the dopamine signaling cascade downstream of the D2 dopamine receptor [55]. Van Winkel et al. (2011) investigated individuals affected by psychosis and their unaffected siblings, and showed that carriers of the C/C genotype had an approximately two-fold increased odds of being diagnosed with a psychotic disorder when using cannabis (OR = 2.08, 95% CI 0.92–4.67) [56]. Di Forti and colleagues found, in a 2012 study, that daily cannabis use of C/C genotype carriers severely increased the odds of a psychotic disorder (OR= 7.23, 95% CI 1.37–38.12) compared to T/T carriers [57]. Finally, Morgan and colleagues (2016) also showed an association between the *AKT1* genotype and cannabis use in the emergence of positive psychotomimetic symptoms, which increased with the C/T or C/C genotype ($\beta = 0.119$, $P = 0.0015$) [58].

COMT Gene

Nine studies have investigated the influence of the *COMT* Val158Met genotype (rs4680 locus) on the relationship between cannabis use and the incidence of psychosis. Variation in the *COMT* Val158Met polymorphism is associated with dopamine turnover in the prefrontal cortex, with Val/Val carriers having higher COMT enzyme activity and thus reduced dopamine levels compared to Met/Met carriers [59,60]. Henquet and colleagues found, in 2009, that individuals with the Val allele showed an increase in hallucinations after cannabis exposure compared to Met allele carriers [61]. The interaction between *COMT* and cannabis use ($F = 3.556$; $P = 0.007$; Eta Squared (η^2)= 0.044) had a significant effect on the age of psychosis onset in a 2010 study by Pelayo-Terán ($F = 3.816$; $P = 0.024$; $\eta^2 = 0.045$) [62]. Costas and colleagues (2011) demonstrated a doubled probability of lifetime cannabis use in individuals homozygous for the Met allele compared to the homozygous Val genotype (Mantel–Haenszel OR = 2.07, 95% CI 1.27–3.26). There was no significant difference between the different genotypes [63]. Nieman and colleagues (2016) reported an interaction effect between the *COMT* Val158Met genotype and cannabis use on subclinical psychotic symptoms in subjects at CHR for psychosis [64].

In contrast, cannabis consumption was not associated with the *COMT* Val158Met polymorphism according to Gutiérrez and colleagues (2009). However, for female regular cannabis users, the psychosis risk was highest when they were also homozygous carriers of the Val allele. Importantly, this seemingly dose-dependent interaction between cannabis consumption, risk for psychosis, and carrying the Val allele in females was not significant [65]. Similarly, Zammit et al. concluded from their 2011 study that there was no evidence of an interaction between cannabis use and six *COMT* single nucleotide polymorphisms (SNPs) in the risk of developing psychotic symptoms [66]. In accordance with these findings, a 2015 study on the rs4680 SNP in a Pakistani population concluded that there was no association between cannabis use and the polymorphism in schizophrenia [67]. Mané and colleagues (2017) studied the interaction between *COMT* and cannabis use with respect to FEP. They reported that early cannabis use and the presence of the *COMT* Val158Met polymorphism were not significantly associated with an earlier onset of psychosis [68]. Another study published in 2017 by Lodhi and colleagues failed to demonstrate a significant moderating effect of the COMT genotype on the age of psychosis onset in cannabis users that initiated use before 20 years of age ($P = 0.051$) [69].

NOS1AP, DRD2, BNDF, and FAAH

In the last decade, four other genes related to cannabis use and psychosis have been examined, namely, *NOS1AP*, *DRD2*, *BDNF*, and *FAAH*. First, Husted et al. (2012) demonstrated that the presence of the *NOS1AP* risk of the schizophrenia genotype did not influence schizophrenia expression associated with cannabis use [70]. Colizzi and colleagues (2015) investigated the dopamine receptor *D2* (*DRD2*) genotype rs1076560 locus and found that cannabis-using carriers of the T-allele had a three-fold increase in the odds of psychosis compared to GG carriers (OR = 3.07; 95% CI 1.22–7.63). Among daily cannabis users, T carriers showed a five-fold increased odds of psychosis (OR = 4.82; 95% CI 1.39–16.71) [71]. The presence of the Met allele on the brain-derived neurotrophic factor (*BDNF*) gene Val66Met polymorphism was not associated with early cannabis use according to a 2017 study by Mané et al. [68]. Bioque et al. (2019) found that T/T carriers of the Fatty Acid Amide Hydrolase (*FAAH)* rs2295633 SNP (encoding the FAAH enzyme which is involved in the reuptake and degradation of endogenous cannabinoid ligands) had a ten-fold increased probability of presenting with FEP (OR = 10.36, statistical power 0.78), if they used cannabis frequently [72].

Mendelian Randomization Studies

Mendelian randomization studies use genetic variants or polygenic scores as instrumental variables to control for gene–environment correlation while estimating the association between an exposure and outcome. In the last decade, three Mendelian randomization studies have been performed that have focused on the potential causal relationship between cannabis use and psychosis. First, Gage and colleagues (2017) found some evidence consistent with a causal effect of cannabis initiation on the risk of psychosis (OR = 1.04, 95% CI 1.01–1.07). However, there was strong evidence consistent with a causal effect of psychosis risk on the likelihood of cannabis initiation (OR = 1.10, 95% CI 1.05–1.14) [73]. A second study performed in 2018 by Vaucher et al. demonstrated that the use of cannabis was associated with an increased risk of psychosis (OR for users vs. nonusers of cannabis = 1.37, 95% CI 1.09–1.67), thereby supporting epidemiological studies arguing that the use of cannabis is causally related to psychosis risk [74]. Finally, Pasman and colleagues (2018) replicated the findings of Gage et al. by showing some weak (non-significant) evidence for a causal influence of lifetime cannabis use on psychosis risk (OR = 1.10, 95% CI 0.99–1.21, $P = 0.074$). They also found stronger evidence for a causal positive influence of psychosis risk on lifetime cannabis use (OR = 1.16, 95% CI 1.06–1.27, $P = 0.001$) [75]. Overall, Mendelian randomization studies suggest that the association between cannabis and psychosis may be bidirectional.

In summary, evidence exists for a moderating effect of the *AKT1* genotype (rs2494732 locus) on the association between cannabis use and the risk of psychosis, with the highest risk for C/C carriers. Several studies have shown the COMT Val158Met genotype to be a moderator of the relationship between cannabis use and the later development of a psychotic disorder, with the highest risk for Val carriers. However, a significant amount of studies have failed to show an impact on the association between cannabis and psychosis. Interestingly, both dopamine receptor *D2* and FAAH genotypes appear to modulate the effect of cannabis use on the development of psychosis, but these initial findings need to be replicated. Using genetic approaches, Mendelian randomization studies suggest that the association between cannabis and psychosis may be bidirectional.

3.2. Cannabis Use and the Age of Onset of Psychosis

Twenty-two studies have reported on the association between cannabis use and the age of onset of psychosis. In a 2009 study, lifetime cannabis abuse or dependence was associated with a significantly earlier onset of psychosis ($\beta = -3.11$, $t(198) = -3.54$; $P < 0.001$) by 3.1 years (95% CI = 1.4–4.8 years earlier). The age at onset was another 3 years earlier in individuals with lifetime cannabis abuse or dependence [76]. Furthermore, progression to daily cannabis use over time predicted an earlier onset of illness or prodromal symptoms (HR = 2.065, $P < 0.05$) [45]. In a 2010 study by Barrigón and colleagues, the age of onset of psychosis treatment was significantly associated with cannabis use; the earlier the age of onset of the heaviest use, the earlier the start of treatment, indicating a dose-response relationship (sex-adjusted log-rank $\chi^2(1) = 23.43$, $P < 0.001$) [77]. First-episode schizophrenia patients with CUD seemed to have an earlier onset of psychotic symptoms according to Sevy et al. (2010). This result did not hold in a multivariate analysis when additional variables related to CUD were taken into account [78]. In contrast, Schimmelmann and colleagues showed, in 2011, that CUD subjects did not have an earlier age of onset than non-CUD subjects ($F(1) = 3.4$; $P = 0.067$; $\eta_p^2 = 0.01$). However, early CUD (starting at age 14 or younger) was associated with an earlier onset of psychosis ($\beta = -0.49$, R^2-change = 0.25, $P < 0.001$) [79]. A systematic meta-analysis by Large and colleagues published in 2011 reported that cannabis use was associated with an earlier onset of psychosis, with the onset for cannabis users being 2.7 years earlier (32 months) (Standardized Mean Difference (SMD) = −0.414, effect size −2.70 years) [80]. Accordingly, cannabis use predicted an earlier age at onset (1.5 years) in schizophrenia patients [81], and cannabis users had a 1.5-year earlier onset of psychosis compared to non-drug users, as reported in 2012 by Dekker and colleagues (difference = 1.7, B = −1.7, Standard Error (SE) = 0.6, $t = -2.6$, $P = 0.009$) [82]. Furthermore, as reported by Estrada and colleagues (2011), there was a significant positive relationship between the age at first use of cannabis and the age of the onset of psychosis ($\beta = 1.66$, SE = 0.78, $P = 0.04$). They also investigated the role of the *COMT* Val158Met genotype in this association and found an interaction with cannabis use, showing an earlier psychosis onset for Val/Val carriers than for Val/Met and Met/Met carriers ($\beta = -2.72$, SE = 1.30, $P = 0.04$) [83]. Grech and colleagues (2012) found that the age at admission, as a proxy measure for the age of onset of psychosis, was significantly different for individuals who tested positive for cannabis in a urine test (age = 24.63) and negative testers (age = 44.63) (Mann–Whitney $P = 0.001$) [84]. Leeson et al. (2012) found that cannabis users had a significantly younger age at onset of psychosis than non-users ($F(1,98) = 9.43$, $P = 0.003$) [85]. Furthermore, in a 2012 study by Galvez-Buccollini, a significant interaction between age at the initiation of cannabis use and age of onset of psychosis was found ($\beta = 0.4$, 95% CI 0.1–0.7, $P = 0.008$) [86]. Allegri and colleagues demonstrated, in 2013, that users of only cannabis experienced psychotic symptoms 6.2 years earlier than individuals who did not use any drugs (24.2 ± 5.0 vs. 32.9 ± 9.8 years; $t(1) = 4.26$; $P < 0.001$) [87]. In addition, in subjects with schizophrenia spectrum disorder (SSD), age at the initiation of cannabis use and age at the onset of psychosis were significantly linearly associated after adjusting for confounders ($F(11,984) = 13.77$, $P < 0.001$) [88]. Similarly, after adjustment for diagnosis and gender, FEP patients who used cannabis had an earlier onset of psychosis than abstinent patients ($F(1,291) = 16.29$, $P< 0.001$) [89]. Di Forti and team (2014) also showed that daily users of high-potency cannabis had an earlier onset of psychosis compared to non-cannabis users of 6 years on average (HR = 1.99; 95% CI 1.50–2.65; $P < 0.0001$) [51]. Additionally, in a 2014 study by Donoghue and colleagues, use of cannabis before the onset of psychosis was associated with an earlier onset of symptoms. In the same study, it was shown that cannabis use interacting with gender was the most parsimonious model influencing the age of onset of symptoms [90]. Among 555 FEP patients, history of cannabis abuse was associated with a nearly 6-year earlier onset of psychosis in 2015 (22.8 vs. 28.7 years, Z = −5.9 years, $P < 0.001$) [91]. Finally, cannabis use was significantly associated with a 3-year earlier onset of SSD in a large multi-site sample ($N = 1119$) study published in 2016, while other factors did not influence the age of onset ($F(1,1003) = 31.66$, $P < 0.001$) [92].

In contrast to the above-mentioned findings, one study reported, in 2009, that the use of cannabis prior to the age of 18 was not associated with cannabis use in individuals with psychotic disorders [93]. DeRosse and colleagues (2010) also concluded that schizophrenia patients with CUD did not have an earlier onset compared to patients without CUD [94]. Finally, in a sample of 633 schizophrenia patients that was subdivided based on the presence of large, rare genetic deletions or large, rare duplications, only those without large duplications had an earlier age of onset related to cannabis abuse [95].

In summary, the vast majority of studies published in the last decade investigating the age of onset of psychosis in relation to cannabis use have found that cannabis use is related to an earlier onset of psychotic symptoms, psychotic experiences, or a psychotic disorder. A few studies have shown an association between an earlier onset of cannabis use and earlier onset of psychosis.

3.3. Cannabis Use and the Transition to Psychosis in CHR Subjects

Five studies have examined cannabis use and the transition to psychosis in CHR individuals. In a study published in 2012, lifetime cannabis abuse was not shown to be related to the transition to psychosis in a small sample of people at CHR for psychosis (N = 15) [96]. In a much bigger sample of 182 CHR individuals, Valmaggia and colleagues (2014) investigated the influence of cannabis use on the transition to psychosis and reported that transition was more prevalent in frequent compared to non-frequent users ($\chi^2(1) = 4.994$, $P = 0.025$). Furthermore, the prevalence was higher in early-onset users (i.e., <age 15) who continued to use frequently compared to late-users ($\chi^2(1) = 7.093$, $P = 0.008$). However, in the overall sample, transition rates were not significantly increased in cannabis users compared to non-users (χ^2 (1) = 1.061, P = 0.303) [97]. A study by Auther et al. (2015) reported a more prevalent and earlier transition in CHR individuals diagnosed with cannabis abuse or CUD compared to non-users (log-rank test – $\chi^2 = 4.67$, $P = 0.031$) and cannabis users without impairment (log-rank test – $\chi^2 = 4.92$, $P = 0.027$). However, adjusting for alcohol use weakened this relationship (HR = 1.875, CI 0.963–3.651, P = 0.064), which suggests that the association between cannabis misuse and the transition to psychosis was confounded by alcohol use [98]. McHugh and colleagues demonstrated, in 2017, that individuals with a history of cannabis-induced attenuated psychotic symptoms were 4.90 (95% CI 1.93–12.44) times more likely to develop a psychotic disorder. In these individuals, greater cannabis abuse also indicated a greater risk of transition [99]. Finally, a meta-analysis of seven studies published before 2016 by Kraan et al. (2017) concluded that current cannabis abuse or dependence in subjects at CHR for psychosis increased the odds of transition to 1.75 (95% CI 1.135–2.710) [100].

Taken together, these studies indicate that frequent cannabis use, lifetime cannabis abuse, or cannabis dependence increases the risk of the transition to psychosis in subjects at CHR. However, this effect may be most pronounced in heavy cannabis users and may be confounded by the use of alcohol or other drugs of abuse.

4. Discussion

The current systematic review presents a detailed and up-to-date literature overview on factors that influence the relationship between cannabis use and psychosis risk. Overall, the results show that, in particular, frequent cannabis use, especially daily use, and the consumption of high-potency cannabis (i.e., cannabis with high THC and low CBD concentrations) are associated with a higher risk of developing psychosis. Several genotypes have been shown to moderate the impact of cannabis use on psychosis risk, particularly those involved in the dopamine function, such as *AKT1*. Finally, the results indicate that cannabis use is associated with an earlier onset of psychosis and increased risk of transition in individuals at CHR of psychosis.

High cannabis exposure (i.e., more than weekly, especially daily use) and the use of high-potency cannabis are factors particularly associated with an increased risk of developing psychosis. Although more research is needed to clarify the specific effect of cannabis use in subjects at CHR for psychosis, most studies indicate that cannabis use increases the risk of the transition to psychosis. Since the genetic profile seems to modulate the risk cannabis use poses for the development of psychosis, genetic predisposition should be taken into account when exploring the risk of developing psychotic symptoms of illness. A better integration and understanding of genetic and environmental factors is needed in order to identify the individuals that are most sensitive or more resilient to the effects of cannabis on the risk of psychosis. Future studies should focus on developing methods to identify those individuals with either resilient or vulnerable genotypes, while considering cannabis use within the whole exposome [101]. In addition, the biological mechanisms underlying the influence of cannabis on psychosis risk remain largely unclear [3,30]. Study of the endocannabinoid system might help to find new markers of illness and recovery, and identify new routes for the development of novel treatments [30]. Importantly, CBD, a non-intoxicating compound of cannabis, has shown promising effects in individuals with psychotic symptoms [29,102,103]. Future studies could assess the effect of cannabis with a high CBD content on psychotic symptoms. In any case, there is enough evidence to set up educational programs to inform cannabis users about the risks for developing psychosis, especially in those with a family history of psychosis or at CHR.

When studying single candidate genes, *AKT1* and *COMT* have frequently been associated with cannabis use and an increased risk of psychosis. All recent studies that have investigated the moderating effect of the *AKT1* genotype on the relationship between cannabis and psychosis have concluded that C carriers of the rs2494732 SNP have an increased risk of developing psychosis after cannabis use [56–58]. Interestingly, *AKT1* C/C carriers showed an increased sensitivity to the psychotic effects of THC and to THC-induced memory-related activity in both the striatum and midbrain, which further suggests that psychotic effects of cannabis might be mediated by dopamine [104]. Regarding the *COMT* gene, several studies have shown the Val158Met genotype to be a moderator of the association between cannabis use and the later development of a psychotic disorder, with the highest risk for Val carriers [61–64]. However, a significant number of studies have failed to show an impact on the relationship between cannabis and psychosis [65–69]. As was also concluded by a recent meta-analysis [105], an association between the *COMT* Val158Met genotype, cannabis use, and psychosis cannot be proven at the moment, accounting for the dissimilarity in results. These controversial results probably reflect the limited value of the study of single genes for predicting mental health outcomes [106,107], since complex phenotypes like psychosis are associated with multiple genes [108]. Over the past decade, advanced genetic analyses that allow complex phenotypes to be predicted have been developed, such as polygenic risk scores [109]. Future studies should investigate the moderating effect of psychosis polygenic risk scores on the association between cannabis use and the development of psychosis, rather than examining single candidate genes. Mendelian randomization studies that use genetic variants or polygenic scores could shed further light on causality and the direction of the relationship between cannabis and psychosis.

A few factors should be taken into account when interpreting the results from the present systematic review. First, the influence of childhood trauma, the use of substances other than cannabis, and exposure to other environmental factors on the relationship between cannabis use and psychotic disorders was beyond the scope of the current review. However, several studies have demonstrated an impact of childhood trauma on the association between cannabis use and psychosis risk [110–112]. Additionally, cannabis use is related to the use of other substances (e.g., tobacco) that have independently been associated with psychosis [113], which may have an effect on the relationship between cannabis and psychosis. Therefore, future longitudinal studies may attempt to integrate these potential moderators when studying this relationship. Second, definitions of cannabis use parameters differed between studies. For example, frequent cannabis use was defined as weekly use in some studies, but as daily use in others. Likewise, the definition of high-potency cannabis varied between studies, and actual THC/CBD concentrations of cannabis samples were not assessed. Standardized measures for assessing cannabis use and determining THC/CBD concentrations in cannabis are highly needed for improving the comparison and interpretation of results [114,115]. Third, although the present systematic review provides an up-to-date and detailed overview of the most recent literature, possible differences in the quality of the included studies were not assessed in a systematic way. Therefore, variation in, for example, the study design and sample sizes, ranging from international multicenter studies with standardized assessments of thousands of patients to retrospective measures in smaller patient groups, may explain the differences in reported findings. Finally, the present review did not necessarily focus on confounders of the relationship between cannabis use and psychosis. A few studies have shown that confounding factors, such as age, gender, and the use of other substances, may attenuate the effect of cannabis use on the risk of psychosis [50]. Future studies or literature reviews could clarify which factors are important and could give more insight into the precise effect of cannabis use on the incidence of psychosis.

In conclusion, literature from the last decade shows that, in particular, frequent cannabis use and the consumption of high-potency cannabis increase the risk of psychosis. Furthermore, cannabis use lowers the age of onset of psychosis by 3 years, and increases the risk of transition in subjects at CHR for psychosis. Cannabis use is a risk factor—not necessary or sufficient for causing psychosis—that interplays with the genetic background, together with other environmental factors. These gene–environmental interactions are complex and still unclear. Further research should aim to clarify which genes are implicated in modulating the effect of modifiable environmental factors like cannabis use within the whole exposome. This might help to predict how changes in these environmental factors can positively impact the prognosis of patients with psychosis or at CHR by counterbalancing a vulnerable genetic profile. In addition, new studies should focus on the biological effect of cannabis on the endocannabinoid system and its relation with psychosis, as well as the potential pharmaceutical properties of CBD.

Supplementary Materials: The following are available online at http://www.mdpi.com/2076-3425/10/2/97/s1, Table S1: Overview of studies included in the present systematic review.

Author Contributions: S.J.v.d.S. performed the systematic search and drafted the manuscript, S.J.v.d.S. and M.G.B. screened potentially eligible publications, and A.B. and M.G.B. critically revised the manuscript for important intellectual content. All authors are listed and have significantly contributed to the manuscript. All authors have read and agreed to the published version of the manuscript.

Conflicts of Interest: The authors declare no conflict of interest.

References

1. McGrath, J.; Saha, S.; Chant, D.; Welham, J. Schizophrenia: A concise overview of incidence, prevalence, and mortality. *Epidemiol. Rev.* **2008**, *30*, 67–76. [CrossRef] [PubMed]
2. Van Os, J.; Kapur, S. Schizophrenia. *Lancet* **2009**, *374*, 635–645. [CrossRef]
3. Bossong, M.G.; Niesink, R.J.M. Adolescent brain maturation, the endogenous cannabinoid system and the neurobiology of cannabis-induced schizophrenia. *Prog. Neurobiol.* **2010**, *92*, 370–385. [CrossRef] [PubMed]
4. Bossong, M.G.; Jansma, J.M.; Bhattacharyya, S.; Ramsey, N.F. Role of the endocannabinoid system in brain functions relevant for schizophrenia: An overview of human challenge studies with cannabis or Δ9-tetrahydrocannabinol (THC). *Prog. Neuropsychopharmacol. Biol. Psychiatry* **2014**, *52*, 53–69. [CrossRef] [PubMed]
5. Minichino, A.; Senior, M.; Brondino, N.; Zhang, S.H.; Godwlewska, B.R.; Burnet, P.W.J.; Cipriani, A.; Lennox, B.R. Measuring Disturbance of the Endocannabinoid System in Psychosis: A Systematic Review and Meta-analysis. *JAMA Psychiatry* **2019**. [CrossRef] [PubMed]
6. Giuffrida, A.; Leweke, F.M.; Gerth, C.W.; Schreiber, D.; Koethe, D.; Faulhaber, J.; Klosterkötter, J.; Piomelli, D. Cerebrospinal anandamide levels are elevated in acute schizophrenia and are inversely correlated with psychotic symptoms. *Neuropsychopharmacology* **2004**, *29*, 2108–2114. [CrossRef]
7. Leweke, F.M.; Giuffrida, A.; Koethe, D.; Schreiber, D.; Nolden, B.M.; Kranaster, L.; Neatby, M.A.; Schneider, M.; Gerth, C.W.; Hellmich, M.; et al. Anandamide levels in cerebrospinal fluid of first-episode schizophrenic patients: Impact of cannabis use. *Schizophr. Res.* **2007**, *94*, 29–36. [CrossRef]
8. Dalton, V.S.; Long, L.E.; Weickert, C.S.; Zavitsanou, K. Paranoid schizophrenia is characterized by increased CB(1) receptor binding in the dorsolateral prefrontal cortex. *Neuropsychopharmacology* **2011**, *36*, 1620–1630. [CrossRef]
9. Jenko, K.J.; Hirvonen, J.; Henter, I.D.; Anderson, K.B.; Zoghbi, S.S.; Hyde, T.M.; Deep-Soboslay, A.; Innis, R.B.; Kleinman, J.E. Binding of a tritiated inverse agonist to cannabinoid CB1 receptors is increased in patients with schizophrenia. *Schizophr. Res.* **2012**, *141*, 185–188. [CrossRef]
10. Ceccarini, J.; De Hert, M.; Van Winkel, R.; Peuskens, J.; Bormans, G.; Kranaster, L.; Enning, F.; Koethe, D.; Leweke, F.M.; Van Laere, K. Increased ventral striatal CB1 receptor binding is related to negative symptoms in drug-free patients with schizophrenia. *Neuroimage* **2013**, *79*, 304–312. [CrossRef]
11. Wong, D.F.; Kuwabara, H.; Horti, A.G.; Raymont, V.; Brasic, J.; Guevara, M.; Ye, W.; Dannals, R.F.; Ravert, H.T.; Nandi, A.; et al. Quantification of cerebral cannabinoid receptors subtype 1 (CB1) in healthy subjects and schizophrenia by the novel PET radioligand [11C]OMAR. *Neuroimage* **2010**, *52*, 1505–1513. [CrossRef] [PubMed]

12. D'Souza, D.C.; Perry, E.; MacDougall, L.; Ammerman, Y.; Cooper, T.; Wu, Y.T.; Braley, G.; Gueorguieva, R.; Krystal, J.H. The psychotomimetic effects of intravenous delta-9-tetrahydrocannabinol in healthy individuals: Implications for psychosis. *Neuropsychopharmacology* **2004**, *29*, 1558–1572. [CrossRef] [PubMed]
13. Bossong, M.G.; Jansma, J.M.; van Hell, H.H.; Jager, G.; Kahn, R.S.; Ramsey, N.F. Default Mode Network in the Effects of Δ9-Tetrahydrocannabinol (THC) on Human Executive Function. *PLoS ONE* **2013**, *8*, e70074. [CrossRef] [PubMed]
14. Bhattacharyya, S.; Fusar-Poli, P.; Borgwardt, S.; Martin-Santos, R.; Nosarti, C.; O'Carrol, C.; Allen, P.; Seal, M.; Fletcher, P.; Crippa, J.A.; et al. Modulation of mediotemporal and ventrostriatal function in humans by Δ9-tetrahydrocannabinol a neural basis for the effects of cannabis sativa on learning and psychosis. *Arch. Gen. Psychiatry* **2009**, *66*, 442–451. [CrossRef]
15. Sherif, M.; Radhakrishnan, R.; D'Souza, D.C.; Ranganathan, M. Human laboratory studies on cannabinoids and psychosis. *Biol. Psychiatry* **2016**, *79*, 526–538. [CrossRef] [PubMed]
16. Tien, A.Y.; Anthony, J.C. Epidemiological analysis of alcohol and drug use as risk factors for psychotic experiences. *J. Nerv. Ment. Dis.* **1990**, *178*, 473–480. [CrossRef]
17. Arseneault, L.; Cannon, M.; Poulton, R.; Murray, R.; Caspi, A.; Moffitt, T.E. Cannabis use in adolescence and risk for adult psychosis: Longitudinal prospective study. *BMJ* **2002**, *325*, 1212–1213. [CrossRef]
18. Van Os, J.; Bak, M.; Hanssen, M.; Bijl, R.V.; De Graaf, R.; Verdoux, H. Cannabis use and psychosis: A longitudinal population-based study. *Am. J. Epidemiol.* **2002**, *156*, 319–327. [CrossRef]
19. Zammit, S.; Allebeck, P.; Andreasson, S.; Lundberg, I.; Lewis, G. Self reported cannabis use as a risk factor for schizophrenia in Swedish conscripts of 1969: Historical cohort study. *BMJ* **2002**, *325*, 1199. [CrossRef]
20. Fergusson, D.M.; Horwood, L.J.; Ridder, E.M. Tests of causal linkages between cannabis use and psychotic symptoms. *Addiction* **2005**, *100*, 354–366. [CrossRef]
21. Henquet, C.; Krabbendam, L.; Spauwen, J.; Kaplan, C.; Lieb, R.; Wittchen, H.U.; van Os, J. Prospective cohort study of cannabis use, predisposition for psychosis, and psychotic symptoms in young people. *BMJ* **2005**, *330*, 11. [CrossRef] [PubMed]
22. Di Forti, M.; Morgan, C.; Dazzan, P.; Pariante, C.; Mondelli, V.; Marques, T.R.; Handley, R.; Luzi, S.; Russo, M.; Paparelli, A.; et al. High-potency cannabis and the risk of psychosis. *Br. J. Psychiatry* **2009**, *195*, 488–491. [CrossRef] [PubMed]
23. Arseneault, L.; Cannon, M.; Witton, J.; Murray, R.M. Causal association between cannabis and psychosis: Examination of the evidence. *Br. J. Psychiatry* **2004**, *184*, 110–117. [CrossRef] [PubMed]
24. Smit, F.; Bolier, L.; Cuijpers, P. Cannabis use and the risk of later schizophrenia: A review. *Addiction* **2004**, *99*, 425–430. [CrossRef] [PubMed]
25. Semple, D.M.; McIntosh, A.M.; Lawrie, S.M. Cannabis as a risk factor for psychosis: Systematic review. *J. Psychopharmacol.* **2005**, *19*, 187–194. [CrossRef] [PubMed]
26. Moore, T.H.; Zammit, S.; Lingford-Hughes, A.; Barnes, T.R.; Jones, P.B.; Burke, M.; Lewis, G. Cannabis use and risk of psychotic or affective mental health outcomes: A systematic review. *Lancet* **2007**, *370*, 319–328. [CrossRef]
27. Marconi, A.; di Forti, M.; Lewis, C.M.; Murray, R.M.; Vassos, E. Meta-analysis of the association between the level of cannabis use and risk of psychosis. *Schizophr. Bull.* **2016**, *42*, 1262–1269. [CrossRef]
28. Murray, R.M.; Quigley, H.; Quattrone, D.; Englund, A.; Di Forti, M. Traditional marijuana, high-potency cannabis and synthetic cannabinoids: Increasing risk for psychosis. *World Psychiatry* **2016**, *15*, 195–204. [CrossRef]
29. Iseger, T.A.; Bossong, M.G. A systematic review of the antipsychotic properties of cannabidiol in humans. *Schizophr. Res.* **2015**, *162*, 153–161. [CrossRef]
30. Batalla, A.; Janssen, H.; Gangadin, S.S.; Bossong, M.G. The Potential of Cannabidiol as a Treatment for Psychosis and Addiction: Who Benefits Most? A Systematic Review. *J. Clin. Med.* **2019**, *8*, 1058. [CrossRef]
31. Freeman, A.M.; Petrilli, K.; Lees, R.; Hindocha, C.; Mokrysz, C.; Curran, H.V.; Saunders, R.; Freeman, T.P. How does cannabidiol (CBD) influence the acute effects of delta-9-tetrahydrocannabinol (THC) in humans? A systematic review. *Neurosci. Biobehav. Rev.* **2019**, *107*, 696–712. [CrossRef] [PubMed]
32. Pelayo-Teran, J.M.; Suarez-Pinilla, P.; Chadi, N.; Crespo-Facorro, B. Gene-Environment Interactions Underlying the Effect of Cannabis in First Episode Psychosis. *Curr. Pharm. Des.* **2012**, *18*, 5024–5035. [CrossRef] [PubMed]

33. Henquet, C.; Di Forti, M.; Morrison, P.; Kuepper, R.; Murray, R.M. Gene-environment interplay between cannabis and psychosis. *Schizophr. Bull.* **2008**, *34*, 1111–1121. [CrossRef] [PubMed]
34. Van Winkel, R.; Kuepper, R. Epidemiological, neurobiological, and genetic clues to the mechanisms linking cannabis use to risk for nonaffective psychosis. *Annu. Rev. Clin. Psychol.* **2014**, *10*, 767–791. [CrossRef] [PubMed]
35. Caspi, A.; Moffitt, T.E.; Cannon, M.; McClay, J.; Murray, R.; Harrington, H.; Taylor, A.; Arseneault, L.; Williams, B.; Braithwaite, A.; et al. Moderation of the effect of adolescent-onset cannabis use on adult psychosis by a functional polymorphism in the catechol-O-methyltransferase gene: Longitudinal evidence of a gene X environment interaction. *Biol. Psychiatry* **2005**, *57*, 1117–1127. [CrossRef]
36. Henquet, C.; Rosa, A.; Krabbendam, L.; Papiol, S.; Fananas, L.; Drukker, M.; Ramaekers, J.G.; van Os, J. An experimental study of catechol-o-methyltransferase Val158Met moderation of delta-9-tetrahydrocannabinol-induced effects on psychosis and cognition. *Neuropsychopharmacology* **2006**, *31*, 2748–2757. [CrossRef]
37. Veen, N.D.; Selten, J.-P.; van der Tweel, I.; Feller, W.G.; Hoek, H.W.; Kahn, R.S. Cannabis Use and Age at Onset of Schizophrenia. *Am. J. Psychiatry* **2004**, *161*, 501–506. [CrossRef]
38. Barnes, T.R.E.; Mutsatsa, S.H.; Hutton, S.B.; Watt, H.C.; Joyce, E.M. Comorbid substance use and age at onset of schizophrenia. *Br. J. Psychiatry* **2006**, *188*, 237–242. [CrossRef]
39. Pelayo-Terán, J.M.; Pérez-Iglesias, R.; Ramírez-Bonilla, M.; González-Blanch, C.; Martínez-García, O.; Pardo-García, G.; Rodríguez-Sánchez, J.M.; Roiz-Santiáñez, R.; Tordesillas-Gutiérrez, D.; Mata, I.; et al. Epidemiological factors associated with treated incidence of first-episode non-affective psychosis in Cantabria: Insights from the Clinical Programme on Early Phases of Psychosis. *Early Interv. Psychiatry* **2008**, *2*, 178–187. [CrossRef]
40. González-Pinto, A.; Vega, P.; Ibáñez, B.; Mosquera, F.; Barbeito, S.; Gutiérrez, M.; Ruiz de Azúa, S.; Ruiz, I.; Vieta, E. Impact of cannabis and other drugs on age at onset of psychosis. *J. Clin. Psychiatry* **2008**, *69*, 1210–1216. [CrossRef]
41. Kristensen, K.; Cadenhead, K.S. Cannabis abuse and risk for psychosis in a prodromal sample. *Psychiatry Res.* **2007**, *151*, 151–154. [CrossRef] [PubMed]
42. Phillips, L.J.; Curry, C.; Yung, A.R.; Yuen, H.P.; Adlard, S.; McGorry, P.D. Cannabis use is not associated with the development of psychosis in an 'ultra' high-risk group. *Aust. N. Z. J. Psychiatry* **2002**, *36*, 800–806. [CrossRef] [PubMed]
43. Moher, D.; Liberati, A.; Tetzlaff, J.; Altman, D.G. PRISMA 2009 Flow Diagram. *The PRISMA statement* **2009**, *6*, 1000097.
44. Di Forti, M.; Quattrone, D.; Freeman, T.P.; Tripoli, G.; Gayer-Anderson, C.; Quigley, H.; Rodriguez, V.; Jongsma, H.E.; Ferraro, L.; La Cascia, C.; et al. The contribution of cannabis use to variation in the incidence of psychotic disorder across Europe (EU-GEI): A multicentre case-control study. *Lancet Psychiatry* **2019**, *6*, 427–436. [CrossRef]
45. Compton, M.T.; Kelley, M.E.; Ramsay, C.E.; Pringle, M.; Goulding, S.M.; Esterberg, M.L.; Stewart, T.; Walker, E.F. Association of pre-onset cannabis, alcohol, and tobacco use with age at onset of prodrome and age at onset of psychosis in first-episode patients. *Am. J. Psychiatry* **2009**, *166*, 1251–1257. [CrossRef]
46. Karcher, N.R.; Barch, D.M.; Demers, C.H.; Baranger, D.A.A.; Heath, A.C.; Lynskey, M.T.; Agrawal, A. Genetic Predisposition vs Individual-Specific Processes in the Association between Psychotic-like Experiences and Cannabis Use. *JAMA Psychiatry* **2019**, *76*, 87–94. [CrossRef]
47. Leadbeater, B.J.; Ames, M.E.; Linden-Carmichael, A.N. Age-varying effects of cannabis use frequency and disorder on symptoms of psychosis, depression and anxiety in adolescents and adults. *Addiction* **2019**, *114*, 278–293. [CrossRef]
48. Schubart, C.D.; van Gastel, W.A.; Breetvelt, E.J.; Beetz, S.L.; Ophoff, R.A.; Sommer, I.E.; Kahn, R.S.; Boks, M.P. Cannabis use at a young age is associated with psychotic experiences. *Psychol. Med.* **2011**, *41*, 1301–1310. [CrossRef]
49. Van Gastel, W.A.; Wigman, J.T.; Monshouwer, K.; Kahn, R.S.; van Os, J.; Boks, M.P.; Vollebergh, W.A. Cannabis use and subclinical positive psychotic experiences in early adolescence: Findings from a Dutch survey. *Addiction* **2012**, *107*, 381–387. [CrossRef]

50. Gage, S.H.; Hickman, M.; Heron, J.; Munafò, M.R.; Lewis, G.; Macleod, J.; Zammit, S. Associations of cannabis and cigarette use with psychotic experiences at age 18: Findings from the Avon Longitudinal Study of Parents and Children. *Psychol. Med.* **2014**, *44*, 3435–3444. [CrossRef]
51. Di Forti, M.; Sallis, H.; Allegri, F.; Trotta, A.; Ferraro, L.; Stilo, S.A.; Marconi, A.; La Cascia, C.; Reis Marques, T.; Pariante, C.; et al. Daily use, especially of high-potency cannabis, drives the earlier onset of psychosis in cannabis users. *Schizophr. Bull.* **2014**, *40*, 1509–1517. [CrossRef] [PubMed]
52. Setién-Suero, E.; Martínez-García, O.; de la Foz, V.O.; Vázquez-Bourgon, J.; Correa-Ghisays, P.; Ferro, A.; Crespo-Facorro, B.; Ayesa-Arriola, R. Age of onset of Cannabis use and cognitive function in first-episode non-affective psychosis patients: Outcome at three-year follow-up. *Schizophr. Res.* **2018**, *201*, 159–166. [CrossRef]
53. Di Forti, M.; Marconi, A.; Carra, E.; Fraietta, S.; Trotta, A.; Bonomo, M.; Bianconi, F.; Gardner-Sood, P.; O'Connor, J.; Russo, M.; et al. Proportion of patients in south London with first-episode psychosis attributable to use of high potency cannabis: A case-control study. *Lancet Psychiatry* **2015**, *2*, 233–238. [CrossRef]
54. Schubart, C.D.; Sommer, I.E.C.; van Gastel, W.A.; Goetgebuer, R.L.; Kahn, R.S.; Boks, M.P.M. Cannabis with high cannabidiol content is associated with fewer psychotic experiences. *Schizophr. Res.* **2011**, *130*, 216–221. [CrossRef] [PubMed]
55. Ozaita, A.; Puighermanal, E.; Maldonado, R. Regulation of PI3K/Akt/GSK-3 pathway by cannabinoids in the brain. *J. Neurochem.* **2007**, *102*, 1105–1114. [CrossRef] [PubMed]
56. van Winkel, R. Genetic Risk and Outcome of Psychosis (GROUP) Investigators. Family-based analysis of genetic variation underlying psychosis-inducing effects of cannabis: Sibling analysis and proband follow-up. *Arch. Gen. Psychiatry* **2011**, *68*, 148–157. [CrossRef]
57. Di Forti, M.; Iyegbe, C.; Sallis, H.; Kolliakou, A.; Falcone, M.A.; Paparelli, A.; Sirianni, M.; La Cascia, C.; Stilo, S.A.; Marques, T.R.; et al. Confirmation that the AKT1 (rs2494732) genotype influences the risk of psychosis in cannabis users. *Biol. Psychiatry* **2012**, *72*, 811–816. [CrossRef]
58. Morgan, C.J.A.; Freeman, T.P.; Powell, J.; Curran, H.V. AKT1 genotype moderates the acute psychotomimetic effects of naturalistically smoked cannabis in young cannabis smokers. *Transl. Psychiatry* **2016**, *6*, e738. [CrossRef]
59. Chen, J.; Lipska, B.K.; Halim, N.; Ma, Q.D.; Matsumoto, M.; Melhem, S.; Kolachana, B.S.; Hyde, T.M.; Herman, M.M.; Apud, J.; et al. Functional analysis of genetic variation in catechol-O-methyltransferase (COMT): Effects on mRNA, protein, and enzyme activity in postmortem human brain. *Am. J. Hum. Genet.* **2004**, *75*, 807–821. [CrossRef]
60. Tunbridge, E.M.; Harrison, P.J.; Weinberger, D.R. Catechol-o-Methyltransferase, Cognition, and Psychosis: Val158Met and Beyond. *Biol. Psychiatry* **2006**, *60*, 141–151. [CrossRef]
61. Henquet, C.; Rosa, A.; Delespaul, P.; Papiol, S.; Fananás, L.; van Os, J.; Myin-Germeys, I. COMT ValMet moderation of cannabis-induced psychosis: A momentary assessment study of 'switching on' hallucinations in the flow of daily life. *Acta Psychiatr. Scand.* **2009**, *119*, 156–160. [CrossRef] [PubMed]
62. Pelayo-Terán, J.M.; Pérez-Iglesias, R.; Mata, I.; Carrasco-Marín, E.; Vázquez-Barquero, J.L.; Crespo-Facorro, B. Catechol-O-Methyltransferase (COMT) Val158Met variations and cannabis use in first-episode non-affective psychosis: Clinical-onset implications. *Psychiatry Res.* **2010**, *179*, 291–296. [CrossRef] [PubMed]
63. Costas, J.; Sanjuán, J.; Ramos-Ríos, R.; Paz, E.; Agra, S.; Tolosa, A.; Páramo, M.; Brenlla, J.; Arrojo, M. Interaction between COMT haplotypes and cannabis in schizophrenia: A case-only study in two samples from Spain. *Schizophr. Res.* **2011**, *127*, 22–27. [CrossRef] [PubMed]
64. Nieman, D.H.; Dragt, S.; van Duin, E.D.A.; Denneman, N.; Overbeek, J.M.; de Haan, L.; Rietdijk, J.; Ising, H.K.; Klaassen, R.M.C.; van Amelsvoort, T.; et al. COMT Val158Met genotype and cannabis use in people with an At Risk Mental State for psychosis: Exploring Gene x Environment interactions. *Schizophr. Res.* **2016**, *174*, 24–28. [CrossRef]
65. Gutiérrez, B.; Rivera, M.; Obel, L.; McKenney, K.; Martínez-Leal, R.; Molina, E.; Dolz, M.; Ochoa, S.; Usall, J.; Haro, J.M.; et al. Variability in the COMT gene and modification of the risk of schizophrenia conferred by cannabis consumption. *Rev. Psiquiatr. Salud Ment.* **2009**, *2*, 89–94. [CrossRef]
66. Zammit, S.; Owen, M.J.; Evans, J.; Heron, J.; Lewis, G. Cannabis, COMT and psychotic experiences. *Br. J. Psychiatry* **2011**, *199*, 380–385. [CrossRef]

67. Nawaz, R.; Siddiqui, S. Association of Single Nucleotide Polymorphisms in Catechol-OMethyltransferase and Serine-Threonine Protein Kinase Genes in the Pakistani Schizophrenic Population: A Study with Special Emphasis on Cannabis and Smokeless Tobacco. *CNS Neurol. Disord.-Drug Targets* **2015**, *14*, 1086–1095. [CrossRef]
68. Mané, A.; Bergé, D.; Penzol, M.J.; Parellada, M.; Bioque, M.; Lobo, A.; González-Pinto, A.; Corripio, I.; Cabrera, B.; Sánchez-Torres, A.M.; et al. Cannabis use, COMT, BDNF and age at first-episode psychosis. *Psychiatry Res.* **2017**, *250*, 38–43. [CrossRef]
69. Lodhi, R.J.; Wang, Y.; Rossolatos, D.; MacIntyre, G.; Bowker, A.; Crocker, C.; Ren, H.; Dimitrijevic, A.; Bugbee, D.A.; Loverock, A.; et al. Investigation of the COMT Val158Met variant association with age of onset of psychosis, adjusting for cannabis use. *Brain Behav.* **2017**, *7*, e00850. [CrossRef]
70. Husted, J.A.; Ahmed, R.; Chow, E.W.C.; Brzustowicz, L.M.; Bassett, A.S. Early environmental exposures influence schizophrenia expression even in the presence of strong genetic predisposition. *Schizophr. Res.* **2012**, *137*, 166–168. [CrossRef]
71. Colizzi, M.; Iyegbe, C.; Powell, J.; Ursini, G.; Porcelli, A.; Bonvino, A.; Taurisano, P.; Romano, R.; Masellis, R.; Blasi, G.; et al. Interaction between functional genetic variation of DRD2 and cannabis use on risk of psychosis. *Schizophr. Bull* **2015**, *41*, 1171–1182. [CrossRef] [PubMed]
72. Bioque, M.; Mas, S.; Costanzo, M.C.; Cabrera, B.; Lobo, A.; González-Pinto, A.; Rodriguez-Toscano, E.; Corripio, I.; Vieta, E.; Baeza, I.; et al. Gene-environment interaction between an endocannabinoid system genetic polymorphism and cannabis use in first episode of psychosis. *Eur. Neuropsychopharmacol.* **2019**, *29*, 786–794. [CrossRef] [PubMed]
73. Gage, S.H.; Jones, H.J.; Burgess, S.; Bowden, J.; Davey Smith, G.; Zammit, S.; Munafò, M.R. Assessing causality in associations between cannabis use and schizophrenia risk: A two-sample Mendelian randomization study. *Psychol Med.* **2017**, *47*, 971–980. [CrossRef] [PubMed]
74. Vaucher, J.; Keating, B.J.; Lasserre, A.M.; Gan, W.; Lyall, D.M.; Ward, J.; Smith, D.J.; Pell, J.P.; Sattar, N.; Paré, G.; et al. Cannabis use and risk of schizophrenia: A Mendelian randomization study. *Mol. Psychiatry* **2018**, *23*, 1287–1292. [CrossRef]
75. Pasman, J.A.; Verweij, K.J.H.; Gerring, Z.; Stringer, S.; Sanchez-Roige, S.; Treur, J.L.; Abdellaoui, A.; Nivard, M.G.; Baselmans, B.M.L.; Ong, J.S.; et al. GWAS of lifetime cannabis use reveals new risk loci, genetic overlap with psychiatric traits, and a causal effect of schizophrenia liability. *Nat. Neurosci.* **2018**, *21*, 1161–1170. [CrossRef]
76. Öngür, D.; Lin, L.; Cohen, B.M. Clinical characteristics influencing age at onset in psychotic disorders. *Compr. Psychiatry* **2009**, *50*, 13–19. [CrossRef]
77. Barrigón, M.L.; Gurpegui, M.; Ruiz-Veguilla, M.; Diaz, F.J.; Anguita, M.; Sarramea, F.; Cervilla, J. Temporal relationship of first-episode non-affective psychosis with cannabis use: A clinical verification of an epidemiological hypothesis. *J. Psychiatr. Res.* **2010**, *44*, 413–420. [CrossRef]
78. Sevy, S.; Robinson, D.G.; Napolitano, B.; Patel, R.C.; Gunduz-Bruce, H.; Miller, R.; McCormack, J.; Lorell, B.S.; Kane, J. Are cannabis use disorders associated with an earlier age at onset of psychosis? A study in first episode schizophrenia. *Schizophr. Res.* **2010**, *120*, 101–107. [CrossRef]
79. Schimmelmann, B.G.; Conus, P.; Cotton, S.M.; Kupferschmid, S.; Karow, A.; Schultze-Lutter, F.; McGorry, P.D.; Lambert, M. Cannabis use disorder and age at onset of psychosis-A study in first-episode patients. *Schizophr. Res.* **2011**, *129*, 52–56. [CrossRef]
80. Large, M.; Sharma, S.; Compton, M.T.; Slade, T.; Nielssen, O. Cannabis Use and Earlier Onset of Psychosis. *Arch. Gen. Psychiatry* **2011**, *68*, 555. [CrossRef]
81. De Hert, M.; Wampers, M.; Jendricko, T.; Franic, T.; Vidovic, D.; De Vriendt, N.; Sweers, K.; Peuskens, J.; van Winkel, R. Effects of cannabis use on age at onset in schizophrenia and bipolar disorder. *Schizophr. Res.* **2011**, *126*, 270–276. [CrossRef] [PubMed]
82. Dekker, N.; Meijer, J.; Koeter, M.; van den Brink, W.; van Beveren, N.; Kahn, R.S.; Linszen, D.H.; van Os, J.; Wiersma, D.; Bruggeman, R.; et al. Age at onset of non-affective psychosis in relation to cannabis use, other drug use and gender. *Psychol. Med.* **2012**, *42*, 1903–1911. [CrossRef] [PubMed]
83. Estrada, G.; Fatjó-Vilas, M.; Muñoz, M.J.; Pulido, G.; Miñano, M.J.; Toledo, E.; Illa, J.M.; Martín, M.; Miralles, M.L.; Miret, S.; et al. Cannabis use and age at onset of psychosis: Further evidence of interaction with COMT Val158Met polymorphism. *Acta Psychiatr. Scand.* **2011**, *123*, 485–492. [CrossRef] [PubMed]

84. Grech, A.; Camilleri, N.; Taylor, R. Cannabis use and age of admission to a psychiatric unit for first episode of psychosis. *Malta Med. J.* **2012**, *24*, 17–20.
85. Leeson, V.C.; Harrison, I.; Ron, M.A.; Barnes, T.R.E.; Joyce, E.M. The effect of cannabis use and cognitive reserve on age at onset and psychosis outcomes in first-episode schizophrenia. *Schizophr. Bull.* **2012**, *38*, 873–880. [CrossRef]
86. Galvez-Buccollini, J.A.; Proal, A.C.; Tomaselli, V.; Trachtenberg, M.; Coconcea, C.; Chun, J.; Manschreck, T.; Fleming, J.; Delisi, L.E. Association between age at onset of psychosis and age at onset of cannabis use in non-affective psychosis. *Schizophr. Res.* **2012**, *139*, 157–160. [CrossRef]
87. Allegri, F.; Belvederi Murri, M.; Paparelli, A.; Marcacci, T.; Braca, M.; Menchetti, M.; Michetti, R.; Berardi, D.; Tarricone, I. Current cannabis use and age of psychosis onset: A gender-mediated relationship? Results from an 8-year FEP incidence study in Bologna. *Psychiatry Res.* **2013**, *210*, 368–370. [CrossRef]
88. Stefanis, N.C.; Dragovic, M.; Power, B.D.; Jablensky, A.; Castle, D.; Morgan, V.A. Age at initiation of cannabis use predicts age at onset of psychosis: The 7-to 8-year trend. *Schizophr. Bull.* **2013**, *39*, 251–254. [CrossRef]
89. Tosato, S.; Lasalvia, A.; Bonetto, C.; Mazzoncini, R.; Cristofalo, D.; De Santi, K.; Bertani, M.; Bissoli, S.; Lazzarotto, L.; Marrella, G.; et al. The impact of cannabis use on age of onset and clinical characteristics in first-episode psychotic patients. Data from the Psychosis Incident Cohort Outcome Study (PICOS). *J. Psychiatr. Res.* **2013**, *47*, 438–444. [CrossRef]
90. Donoghue, K.; Doody, G.A.; Murray, R.M.; Jones, P.B.; Morgan, C.; Dazzan, P.; Hart, J.; Mazzoncini, R.; Maccabe, J.H. Cannabis use, gender and age of onset of schizophrenia: Data from the ÆSOP study. *Psychiatry Res.* **2014**, *215*, 528–532. [CrossRef]
91. O'Donoghue, B.; Lyne, J.; Madigan, K.; Lane, A.; Turner, N.; O'Callaghan, E.; Clarke, M. Environmental factors and the age at onset in first episode psychosis. *Schizophr. Res.* **2015**, *168*, 106–112.
92. Helle, S.; Ringen, P.A.; Melle, I.; Larsen, T.K.; Gjestad, R.; Johnsen, E.; Lagerberg, T.V.; Andreassen, O.A.; Kroken, R.A.; Joa, I.; et al. Cannabis use is associated with 3 years earlier onset of schizophrenia spectrum disorder in a naturalistic, multi-site sample (N = 1119). *Schizophr. Res.* **2016**, *170*, 217–221. [CrossRef] [PubMed]
93. Kantrowitz, J.T.; Nolan, K.A.; Sen, S.; Simen, A.A.; Lachman, H.M.; Bowers, M.B. Adolescent cannabis use, psychosis and catechol-O-methyltransferase genotype in african Americans and Caucasians. *Psychiatr. Q.* **2009**, *80*, 213–218. [CrossRef] [PubMed]
94. DeRosse, P.; Kaplan, A.; Burdick, K.E.; Lencz, T.; Malhotra, A.K. Cannabis use disorders in schizophrenia: Effects on cognition and symptoms. *Schizophr. Res.* **2010**, *120*, 95–100. [CrossRef]
95. Martin, A.K.; Robinson, G.; Reutens, D.; Mowry, B. Cannabis abuse and age at onset in schizophrenia patients with large, rare copy number variants. *Schizophr. Res.* **2014**, *155*, 21–25. [CrossRef]
96. Auther, A.M.; McLaughlin, D.; Carrión, R.E.; Nagachandran, P.; Correll, C.U.; Cornblatt, B.A. Prospective study of cannabis use in adolescents at clinical high risk for psychosis: Impact on conversion to psychosis and functional outcome. *Psychol. Med.* **2012**, *42*, 2485–2497. [CrossRef]
97. Valmaggia, L.R.; Day, F.L.; Jones, C.; Bissoli, S.; Pugh, C.; Hall, D.; Bhattacharyya, S.; Howes, O.; Stone, J.; Fusar-Poli, P.; et al. Cannabis use and transition to psychosis in people at ultra-high risk. *Psychol. Med.* **2014**, *44*, 2503–2512. [CrossRef]
98. Auther, A.M.; Cadenhead, K.S.; Carrión, R.E.; Addington, J.; Bearden, C.E.; Cannon, T.D.; McGlashan, T.H.; Perkins, D.O.; Seidman, L.; Tsuang, M.; et al. Alcohol confounds relationship between cannabis misuse and psychosis conversion in a high-risk sample. *Acta Psychiatr. Scand.* **2015**, *132*, 60–68. [CrossRef]
99. McHugh, M.J.; McGorry, P.D.; Yung, A.R.; Lin, A.; Wood, S.J.; Hartmann, J.A.; Nelson, B. Cannabis-induced attenuated psychotic symptoms: Implications for prognosis in young people at ultra-high risk for psychosis. *Psychol. Med.* **2017**, *47*, 616–626. [CrossRef]
100. Kraan, T.; Velthorst, E.; Koenders, L.; Zwaart, K.; Ising, H.K.; van den Berg, D.; de Haan, L.; van der Gaag, M. Cannabis use and transition to psychosis in individuals at ultra-high risk: Review and meta-analysis. *Psychol. Med.* **2016**, *46*, 673–681. [CrossRef]
101. Pries, L.K.; Lage-Castellanos, A.; Delespaul, P.; Kenis, G.; Luykx, J.J.; Lin, B.D.; Richards, A.L.; Akdede, B.; Binbay, T.; Altinyazar, V.; et al. Estimating Exposome Score for Schizophrenia Using Predictive Modeling Approach in Two Independent Samples: The Results From the EUGEI Study. *Schizophr. Bull.* **2019**, *45*, 960–965. [CrossRef] [PubMed]

102. Leweke, F.M.; Piomelli, D.; Pahlisch, F.; Muhl, D.; Gerth, C.W.; Hoyer, C.; Klosterkötter, J.; Hellmich, M.; Koethe, D. Cannabidiol enhances anandamide signaling and alleviates psychotic symptoms of schizophrenia. *Transl. Psychiatry* **2012**, 2, e94. [CrossRef] [PubMed]
103. McGuire, P.; Robson, P.; Cubala, W.J.; Vasile, D.; Morrison, P.D.; Barron, R.; Taylor, A.; Wright, S. Cannabidiol (CBD) as an Adjunctive Therapy in Schizophrenia: A Multicenter Randomized Controlled Trial. *Am. J. Psychiatry* **2018**, *175*, 225–231. [CrossRef] [PubMed]
104. Bhattacharyya, S.; Atakan, Z.; Martin-Santos, R.; Crippa, J.A.; Kambeitz, J.; Prata, D.; Williams, S.; Brammer, M.; Collier, D.A.; McGuire, P.K. Preliminary report of biological basis of sensitivity to the effects of cannabis on psychosis: AKT1 and DAT1 genotype modulates the effects of δ-9-tetrahydrocannabinol on midbrain and striatal function. *Mol. Psychiatry* **2012**, *17*, 1152–1155. [CrossRef]
105. Vaessen, T.S.J.; de Jong, L.; Schäfer, A.T.; Damen, T.; Uittenboogaard, A.; Krolinski, P.; Nwosu, C.V.; Pinckaers, F.M.E.; Rotee, I.L.M.; Smeets, A.P.W.; et al. The interaction between cannabis use and the Val158Met polymorphism of the COMT gene in psychosis: A transdiagnostic meta-analysis. *PLoS ONE* **2018**, *13*, e0192658. [CrossRef]
106. Radua, J.; Ramella-Cravaro, V.; Ioannidis, J.P.A.; Reichenberg, A.; Phiphopthatsanee, N.; Amir, T.; Yenn Thoo, H.; Oliver, D.; Davies, C.; Morgan, C.; et al. What causes psychosis? An umbrella review of risk and protective factors. *World Psychiatry* **2018**, *17*, 49–66. [CrossRef]
107. Assary, E.; Vincent, J.P.; Keers, R.; Pluess, M. Gene-environment interaction and psychiatric disorders: Review and future directions. *Semin. Cell Dev. Biol.* **2018**, *77*, 133–143. [CrossRef]
108. Schizophrenia Working Group of the Psychiatric Genomics Consortium. Biological insights from 108 schizophrenia-associated genetic loci. *Nature* **2014**, *511*, 421–427. [CrossRef]
109. Belsky, D.W.; Harden, K.P. Phenotypic Annotation: Using Polygenic Scores to Translate Discoveries From Genome-Wide Association Studies From the Top Down. *Curr. Dir. Psychol. Sci.* **2019**, *28*, 82–90. [CrossRef]
110. Baudin, G.; Godin, O.; Lajnef, M.; Aouizerate, B.; Berna, F.; Brunel, L.; Capdevielle, D.; Chereau, I.; Dorey, J.M.; Dubertret, C.; et al. Differential effects of childhood trauma and cannabis use disorders in patients suffering from schizophrenia. *Schizophr. Res.* **2016**, *175*, 161–167. [CrossRef]
111. Harley, M.; Kelleher, I.; Clarke, M.; Lynch, F.; Arseneault, L.; Connor, D.; Fitzpatrick, C.; Cannon, M. Cannabis use and childhood trauma interact additively to increase the risk of psychotic symptoms in adolescence. *Psychol. Med.* **2010**, *40*, 1627–1634. [CrossRef] [PubMed]
112. Vinkers, C.H.; Van Gastel, W.A.; Schubart, C.D.; Van Eijk, K.R.; Luykx, J.J.; Van Winkel, R.; Joëls, M.; Ophoff, R.A.; Boks, M.P.; Genetic Risk and OUtcome of Psychosis (GROUP) Investigators; et al. The effect of childhood maltreatment and cannabis use on adult psychotic symptoms is modified by the COMT Val158Met polymorphism. *Schizophr. Res.* **2013**, *150*, 303–311. [CrossRef] [PubMed]
113. Gurillo, P.; Jauhar, S.; Murray, R.M.; MacCabe, J.H. Does tobacco use cause psychosis? Systematic review and meta-analysis. *Lancet Psychiatry* **2015**, *2*, 718–725. [CrossRef]
114. Casajuana Kögel, C.; Balcells-Olivero, M.M.; López-Pelayo, H.; Miquel, L.; Teixidó, L.; Colom, J.; Nutt, D.J.; Rehm, J.; Gual, A. The Standard Joint Unit. *Drug Alcohol. Depend.* **2017**, *176*, 109–116. [CrossRef]
115. Freeman, T.P.; Lorenzetti, V. 'Standard THC units': A proposal to standardize dose across all cannabis products and methods of administration. *Addiction* **2019**, in press. [CrossRef]

Article

Do Adolescents Use Substances to Relieve Uncomfortable Sensations? A Preliminary Examination of Negative Reinforcement among Adolescent Cannabis and Alcohol Users

April C. May [1,2,*], Joanna Jacobus [1,2], Jennifer L. Stewart [3,4], Alan N. Simmons [2], Martin P. Paulus [3,4] and Susan F. Tapert [1,2]

1 San Diego State University/University of California, San Diego Joint Doctoral Program in Clinical Psychology, San Diego, CA 92120, USA; jjacobus@health.ucsd.edu (J.J.); stapert@health.ucsd.edu (S.F.T.)
2 Department of Psychiatry, University of California, San Diego, San Diego, CA 92093, USA; ansimmons@health.ucsd.edu
3 Laureate Institute for Brain Research, Tulsa, OK 74136, USA; JStewart@laureateinstitute.org (J.L.S.); mpaulus@laureateinstitute.org (M.P.P.)
4 Department of Community Medicine, University of Tulsa, Tulsa, OK 74104, USA
* Correspondence: acmay@ucsd.edu

Received: 7 March 2020; Accepted: 2 April 2020; Published: 5 April 2020

Abstract: Alcohol and cannabis use are highly prevalent among adolescents and associated with negative consequences. Understanding motivations behind substance use in youth is important for informing prevention and intervention efforts. The present study aims to examine negative reinforcement principles of substance use among adolescent cannabis and alcohol users by pairing a cue reactivity paradigm with an aversive interoceptive stimulus. Adolescents (ages 15–17), classified as controls (CTL; $n = 18$), cannabis and/or alcohol experimenters (CAN+ALC-EXP; $n = 16$), or individuals meeting clinical criteria for cannabis and/or alcohol use disorder (CAN+ALC-SUD; $n = 13$) underwent functional magnetic resonance imaging during which they experienced an aversive interoceptive probe delivered via breathing load while simultaneously performing a cue reactivity paradigm. Participants also provided self-report ratings of how their substance use is positively or negatively reinforced. While experiencing the breathing load, CAN+ALC-SUD exhibited greater ($p < 0.05$) deactivation in the right amygdala, the left inferior frontal gyrus, and the left parahippocampal gyrus than CAN+ALC-EXP and CTL, who did not differ. Across all substance users, activation during the breathing load within the left parahippocampal gyrus negatively correlated with cannabis and alcohol lifetime use episodes and the left inferior frontal gyrus activity negatively correlated with lifetime alcohol use episodes. CAN+ALC-SUD reported experiencing more positive and negative reinforcement of using their substance of choice than CAN+ALC-EXP; both user groups reported higher levels of positive than negative reinforcement. Adolescents with a cannabis/alcohol use disorder demonstrate an altered response to interoceptive perturbations. However, adolescent cannabis/alcohol use does not appear to be driven by negative reinforcement, as viewing substance images did not dampen this response. Based on self-report data, the experience of positive reinforcement may be stronger for adolescents. Future studies should examine whether positive reinforcement contributes to adolescent substance use.

Keywords: cannabis; alcohol; adolescents; fMRI; interoception; negative reinforcement; cue reactivity

1. Introduction

Increased risk-taking behavior is characteristic of adolescence, a critical time period marked by significant physical, cognitive, and behavioral development [1]. A common risky behavior initiated in adolescence is the use of illicit substances. Among 12th graders, approximately 44% report having used cannabis and approximately 59% report having used alcohol in their lifetime [2]. Adolescent substance use can also evolve into a substance use disorder (SUD). For example, in 2018, 2.1% of adolescents aged 12–17 met criteria for cannabis use disorder, while 1.6% met criteria for alcohol use disorder [3]. Substance use during adolescence also increases future risk of experiencing adverse consequences related to use; early adolescent cannabis use may contribute to low educational or occupational attainment, as well as increased use and development of a use disorder in adulthood [4]. Similarly, youth who initiate drinking before age 15 are at increased risk of developing an alcohol use disorder within their lifetime compared to youth who remain abstinent until age 21 [5,6]. Given the increased risks associated with adolescent substance use, it is important to improve our understanding of the motivations behind these behaviors in order to inform SUD prevention and intervention efforts.

Altered interoceptive-related neural processing has been implicated in SUD in combination with emotion dysregulation and decision-making deficits, resulting in suboptimal behavioral adjustments and the propensity to continue drug use despite negative consequences [7–12]. To date, examination of the brain mechanisms involved in interoception and negative reinforcement has focused on adult SUD and little research has examined these concepts among adolescent substance users [13–15].

Interoception is a biological and psychological process by which somatosensory information from inside and outside of the body is filtered and integrated within the brain to produce an overall representation of the bodily state [16]. Anterior cingulate cortex (ACC), thalamus, frontal regions, and insular cortex (IC) are components of brain circuitry essential for processing and integrating bodily afferents to generate an overall representation of the body [16–18]. Afferent signals pass through thalamocortical pathways to IC to be integrated with sensorimotor activity and emotional information delivered by ACC and frontal regions such as inferior frontal gyrus (IFG) [16]. This process results in complex interoceptive feeling states or emotional awareness [18] and may lead to a bodily prediction error if the experienced state differs from the expected state [19–22]. Body prediction errors motivate individuals to engage in goal-directed behavior (e.g., substance use) and either approach or avoid stimuli (e.g., substance-related stimuli) with the aim of reestablishing equilibrium [23].

Among non-substance-using individuals, frontocingulate regions, including IFG and ACC, are thought to act as a regulatory system of behavioral reactions in response to aversive stimuli [24,25]. However, among individuals with SUD this regulatory system appears altered. For example, IFG and ACC blood oxygen-level dependent (BOLD) signal reductions in response to negative interoceptive stimuli have been found to characterize young adults transitioning to stimulant use disorders [8,26] while adolescent substance users have also demonstrated an increased IFG BOLD signal in response to a negative interoceptive stimuli [27]. In general, differing patterns of ACC, IFG, and IC activation have distinguished substance users from healthy individuals [28]. These frontocingulate deficits may be linked to reduced motivation to engage in behavioral changes to reestablish equilibrium despite feeling or sensing consequences of aversive bodily stimulation [29]. In addition to interoceptive processing, poor emotion regulation, an inability to effectively reduce arousal and cope with negative emotions has been implicated in adolescent substance use and requires similar brain regions [30]. The IFG and amygdala comprise a brain circuit involved in determining the emotional significance of an external stimulus and signaling the physiological, behavioral, cognitive, and affective responses necessary to minimize the impact of unpleasant stimuli [25,30–33].

One conceptualization of SUD, based on negative reinforcement principles, posits that individuals use drugs in order to alleviate uncomfortable feelings in general (e.g., emotional dysregulation, uncomfortable interoceptive states) [34,35]. For example, dysfunctional interoceptive processing may result in substance users seeking out and consuming drugs in order to reduce uncomfortable interoceptive states. Neuroimaging research suggests that drug cues activate brain regions similar to those activated by aversive interoceptive stimuli; cannabis cues elicit activation in parahippocampal gyri and various frontal regions among non-treatment-seeking cannabis-using adolescents [36]. Adolescents who primarily use alcohol also demonstrate an exaggerated neural response within frontal regions including IFG, parahippocampus, amygdala, and posterior cingulate in response to cue images [37]. Accordingly, the present study pairs an aversive interoceptive stimulus with a cannabis and alcohol cue reactivity task during functional magnetic resonance imaging. This pairing is viewed as a proxy for negative reinforcement, allowing for the examination of whether the rewarding effects of substance images dampen the negative experience of the breathing load. Specifically, we posit that viewing rewarding drug-relevant cues will dampen the interoceptive BOLD response observed in adolescent substance users while experiencing an aversive interoceptive stimulus.

An inspiratory breathing load can be used as an aversive stimulus to induce a negative interoceptive state [38] and has previously been tested among young adult [8,39], adult [40], and adolescent substance users [27] as well as matched controls. While experiencing the breathing load, young adults with problem stimulant use show lower IFG, IC, and ACC activation compared to individuals who no longer use stimulants as well as non-using controls [8,39]. Similarly, adults with a significant history of methamphetamine use currently meeting criteria for a methamphetamine use disorder also show lower IC and ACC during the breathing load [40]. Despite these differences in brain activation, groups did not differ in their subjective ratings of the breathing load experience. Overall, the reduced activation seen in regions implicated in interoceptive processing is conceptualized as an overall diminished ability to regulate when one does not feel well, and that this inability contributes to continued substance use despite negative consequences. To date, only one study has utilized an inspiratory breathing load with adolescent substance users; these results revealed an overactivation in interoceptive regions. This inconsistent finding suggests that alterations in interoceptive processing may differ as a function of age, type of substance used, or amount of substance used.

The current study is the first to pair an aversive interoceptive stimulus with a cue reactivity paradigm to examine the role of negative reinforcement in substance use. In addition, the sample of the present study includes adolescents (ages 15–17) who report cannabis and alcohol use with and without use disorders. This will allow for the examination of negative reinforcement and interoceptive-related neural responses within diagnostically subthreshold adolescent substance users to investigate whether altered processing is simply a consequence of use or unique to adolescents experiencing functional impairments related to use (i.e., adolescents with use disorder diagnoses).

Participants included adolescents meeting criteria for either cannabis and/or alcohol use disorder, adolescents who use cannabis and alcohol but do not meet diagnostic criteria (experimenters) and healthy comparison participants. On the basis of prior work, it was hypothesized that substance users meeting diagnostic criteria compared to controls would show: (1) increased neural activation in response to the breathing load across all conditions of the cue task in brain regions involved in interoceptive processing, such as IC, ACC, and IFG, as well as regions implicated in emotion regulation, including amygdala and parahippocampal gyrus [27]; (2) increased striatal response while viewing substance images across all breathing load conditions, reflecting heightened reward responsivity to substance cues [41,42]; and (3) a blunted interoceptive neural response to the breathing load when paired with substance images (cannabis and alcohol images) suggesting exposure to a conditioned drug stimulus may help modulate reactions to internal and aversive states similar to negative reinforcement principals of drug use behavior. Additionally, adolescent substance users who did not meet criteria for SUD, referred to as "experimenters", were included to explore whether neural differences are more pronounced in adolescent substance users who endorse substance use-related functional impairment (i.e., adolescents meeting criteria for SUD) than those who do not. Therefore, it was hypothesized that experimenters would demonstrate a neural response more similar to controls than those meeting substance use disorder criteria, suggesting that impaired brain responses are a consequence of more severe use symptomatology.

2. Materials and Methods

2.1. Participants

Adolescent participants (n = 47, ages 15–17) were recruited through local high schools by flyers that advertised an adolescent neuroimaging research study consisting of a clinical interview and neuroimaging session. The University of California San Diego Human Research Protections Program approved the study protocol. Adolescent participants provided assent and informed consent was obtained from one parent or legal guardian prior to study enrollment. Participants were excluded if they endorsed any of the following: (1) lifetime Diagnostic and Statistical Manual (DSM-5) of Mental Disorders psychiatric disorder (other than substance use disorder, SUD); (2) current use of psychoactive medications; (3) history of major neurological or medical disorder; (4) head injuries or loss of consciousness > 5 min; (5) irremovable metal in body; (6) pregnancy; (7) non-correctable vision or hearing problems; (8) premature birth or prenatal alcohol/drug exposure; (9) left handedness; or (10) claustrophobia. Eligible participants received financial compensation for their participation.

The final sample consisted of 18 controls with very minimal histories of substance use (CTL; cannabis/alcohol maximum lifetime use episodes of 3 each, nicotine maximum lifetime use episodes of 10; 13M, 5F), 16 cannabis and alcohol experimenters (CAN+ALC-EXP; 12M, 4F), and 13 who met criteria for cannabis and/or alcohol use disorder (CAN+ALC-SUD; 9M, 4F). SUD group classification required a report of cannabis or alcohol use within the past three months, current endorsement of 2 or more DSM-5 SUD criteria for either cannabis or alcohol, and fewer than 15 lifetime uses of other drugs except for nicotine (see Table 1 for diagnostic details). On average, CAN+ALC-SUD participants reported 467 lifetime cannabis uses and 131 lifetime alcohol uses. CAN+ALC-EXP group classification required a report of no substance use history other than alcohol, cannabis, or nicotine, and no current or lifetime endorsement of DSM-5 SUD criteria. CAN+ALC-EXP reported significantly less cannabis ($t(12.48) = -5.31$, $p < 0.001$) and alcohol use ($t(12.28) = -3.12$, $p < 0.009$) than SUD but significantly more use of these substances than CTL (cannabis: $t(15) = -3.46$, $p = 0.003$; alcohol: $t(15.06) = -4.29$, $p = 0.001$) (see Table 1).

2.2. Clinical Interview

The clinical interview consisted of the Semi-Structured Assessment for Drug Dependence and Alcoholism (SSADDA; [43]) to assess for the presence of SUD and the Customary Drinking and Drug Use Record (CDDR) [44] to capture quantity of lifetime substance use, age of first use, and last substance use. Participants provided demographic information and a battery of self-report measures to assess characteristics related to SUD including the UPPS Impulsive Behavior Scale [45], the Multi-Dimensional Assessment of Interoceptive Awareness (MAIA) [46], and the Michigan Nicotine Reinforcement Questionnaire (MNRQ) [47]. The MNRQ was modified to assess negative reinforcement principles related to users' substance of choice rather than nicotine. Each participant was asked to indicate their drug of choice (cannabis, alcohol) and answer the MNRQ questions regarding their experiences with that drug rather than nicotine. The specific questions and scale were not altered.

2.3. Neuroimaging Procedures

Participants were asked to abstain from substance use for at least 72 h prior to their fMRI session as confirmed by combination of self-report, breathalyzer, and urine toxicology screens. A positive result for any substance other than cannabis excluded individuals from the study. Acute cannabis use is difficult to determine by examination of urinary metabolites and therefore use within the past 72 h is possible; however, all participants self-reported abstaining for the 72 h prior to the appointment and only 5 (4 CAN+ALC-SUD, 1 CAN+ALC-EXP) participants were positive for THC on the day of testing, which could reflect use from up to four weeks prior given the regularity of their use history.

The Cue Breathing fMRI paradigm paired a cue reactivity task with anticipation and experience of an unpleasant interoceptive stimulus, an inspiratory breathing load. Each participant received either a cannabis or alcohol version of the task, depending on their reported primary substance of choice. For the cue reactivity task, participants were presented with images of substances (cannabis or alcohol), comparison images consisting of closely matched objects resembling the substance images (e.g., dried leaves resembling cannabis, non-alcoholic beverages), or scrambled versions of the substance and comparison images where the object in the image was unidentifiable. CTL viewed the same version of the task (cannabis or alcohol) as an age-matched substance-using participant. While viewing each image, participants were asked to indicate whether they disliked, felt neutral, or liked the image. Participants provided ratings using the first three buttons of a four-button box and saw a red box appear on screen to confirm their selected answer.

Table 1. Characteristics of Substance Use.

CAN+ALC-SUDGroup Description	**% Meeting Diagnostic Criteria**	**Diagnostic Criteria Endorsed**		
		M(SD)	**Min**	**Max**
THC Use Disorder	92.31	3.42 (1.38)	2	6
Alcohol Use Disorder	61.54	2.63 (.74)	2	4

Substance Use	**CAN+ALC-SUD** Cannabis/Alcohol Substance Use Disorder	**CAN+ALC-EXP** Cannabis/Alcohol Experimenter	**CTL** Little to No Substance Use	**df**	*t*	*p*
Lifetime Cannabis Use	467.85 (288.05)	39.38 (45.15)	0.17 (0.514)	12.48	−5.31	<0.001
Days Since Last THC Use	18.69 (33.34)	71.69 (82.25)	46.11 (160.19)	20.63	2.35	0.029
Lifetime Alcohol Use	131.92 (131.55)	17.44 (15.87)	0.22 (0.73)	12.28	−3.12	0.009
Days Since Last Alcohol	16.46 (11.67)	45.38 (98.99)	22.22 (66.12)	27	1.04	0.306
Lifetime Alcohol Binge Episode	92.83 (71.90)	7.87 (7.97)	0.11 (0.47)	11.22	−4.07	0.002
Days Since Last Binge	24.70 (24.83)	90.93 (135. 77)	240 (−)	15.38	1.84	0.085
Lifetime Hallucinogen Use	2.69 (3.88)	0.13 (0.50)	–	12.32	−2.37	0.035
Days Since Last Hallucinogen	82.31 (93.58)	9.81 (39.25)	–	15.42	−2.61	0.019
Lifetime Sedative Use	0.77 (1.36)	–	–	12.00	−2.03	0.065
Days Since Last Sedative Use	179.15 (330.97)	–	–	12.00	−1.95	0.075
Lifetime Amphetamine Use	0.31 (1.11)	–	–	12.00	−1.00	0.337
Days Since Last Amphetamine Use	14.46 (52.14)	–	–	12	−1.00	0.337
Lifetime Rx Stimulant Use	1.92 (5.48)	0.06 (0.25)	–	12.04	−1.22	0.245
Days Since Last Rx Stimulant Use	148.23 (297.47)	17.94 (71.75)	–	13.14	−1.54	0.147
Lifetime Cocaine Use	0.92 (1.50)	–	–	12.00	−2.22	0.046
Days Since Last Cocaine Use	55.00 (91.33)	–	–	12.00	−2.17	0.051
Lifetime Ecstasy Use	14.85 (27.65)	–	–	12.00	−1.94	0.077
Days Since Last Ecstasy Use	293.62 (333.72)	–	–	12.00	−3.17	0.008
Lifetime Opiate Use	0.92 (2.75)	1.94 (7.49)	–	27	0.462	0.647
Days Since Last Opiate Use	139.31 (277.92)	26.56 (73.13)	–	13.35	−1.42	0.178
Lifetime Inhalant Use	2.38 (8.30)	–	–	12.00	−1.04	0.321
Days Since Last Inhalant Use	106.00 (259.42)	–	–	12.00	−1.47	0.166
Lifetime Nicotine Use	232.00 (409.19)	4.19 (6.73)	0.56 (2.36)	12.00	−2.01	0.068
Days Since Last Nicotine Use	92.69 (108.66)	130.69 (157.63)	21.94 (93.10)	26.39	0.766	0.451

Participants wore a nose clip and respired through a mouthpiece with a non-rebreathing valve (2600 series, Hans Rudolph). The breathing equipment was attached to the scanner head coil using Velcro straps to help hold the mouthpiece in position and eliminate the need for participants to contract their mouth muscles. The mouthpiece connected to a hose that allowed for an inspiratory resistance load of 40 cmH_2O/L/s to be attached. This breathing load consisted of a Plexiglas tube with a sintered bronze disk inside that partially limited airflow thereby producing a resistance load. A breathing load of 40 cm H_2O/L/s was selected based on previous research which has demonstrated that this load alters subjective symptoms without significantly affecting CO_2 or O_2 levels, and thereby does not impact the BOLD signal [48,49]. Prior to the scan, participants completed a training session during which they were introduced to the breathing equipment and practiced the task. Individuals experienced increasing levels of restriction up to the target load of 40 cm H_2O/L/s. The breathing load was described as feeling like "you are breathing through a straw" and participants were instructed to continue to breathe normally while experiencing the restriction. While in the scanner, participants experienced the breathing load at various times throughout the task for approximately 40 s at a time. Each block of images began with one null trial that lasted for 6 s. During this time, participants saw either a yellow or grey fixation screen. Yellow indicated there was a 1 in 4 (25%) chance of experiencing the breathing load during the next block of images. Alternatively, a grey fixation screen indicated there would be no chance of experiencing the breathing restriction during the upcoming block of images. Each null trial was followed by 6 pictures of the same type (substance, neutral, or scrambled) presented one at a time for 4 s each.

There was a total of 9 task conditions: anticipation neutral images, anticipation substance images, anticipation scrambled images, breathing load neutral images, breathing load substance images, breathing load scrambled images, neutral images only, substance images only, and scrambled images only. Trials during which neutral or scrambled images were presented without the anticipation or experience of the breathing load were combined into a baseline condition. This resulted in 5 conditions of interest: (1) *baseline:* neutral and scrambled images with no anticipation or breathing restriction; (2) *anticipation neutral images:* blocks of neutral images preceded by a yellow fixation screen during which the participant did not actually experience the breathing load; (3) *anticipation substance images:* blocks of substance images preceded by a yellow fixation screen during which the participant did not actually experience the breathing load; (4) *breathing load neutral images:* blocks of neutral images preceded by a yellow fixation screen during which the participant did experience the breathing load; (5) *breathing load substance images:* blocks of substance images preceded by a yellow fixation screen during which the participant did experience the breathing load (see Figure 1).

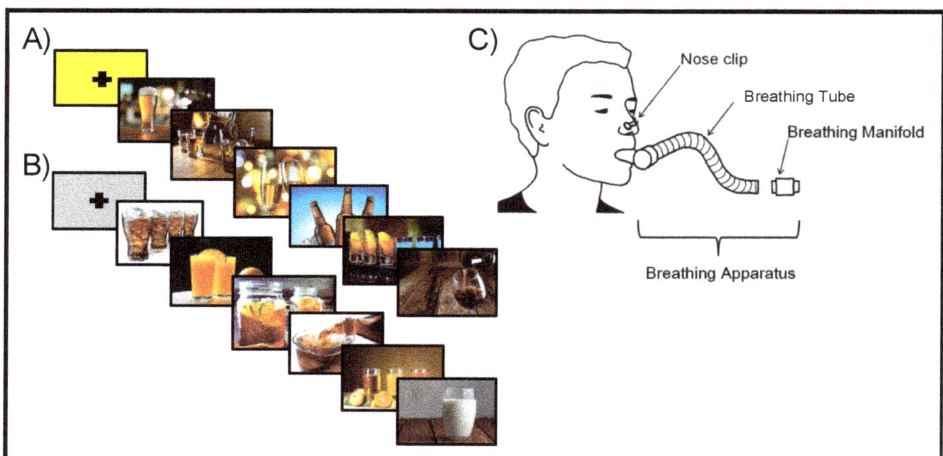

Figure 1. Depiction of the cue reactivity paradigm paired with interoceptive breathing load. (**A**) A yellow fixation screen is presented to the participant, indicating that there is a 1 in 4 chance they will experience the breathing load during the upcoming block of pictures. The fixation screen is immediately followed by 6 images—in this case, alcohol-related cue images. (**B**) A grey fixation screen is presented to the participant indicating that there is no chance they will experience the breathing load during the upcoming block of pictures. The fixation screen is immediately followed by 6 images—in this case, substance-matched comparison images. (**C**) Each participant wears the breathing apparatus while in the fMRI machine. They wear a nose clip to ensure they breathe through the tube only and a breathing manifold is attached at the end of the tube for periods of 40 s as indicated by the paired cue reactivity task.

Prior to the scan, participants underwent a training session to learn the task and become familiar with the breathing equipment. This ensured that participants would be able to complete the task within the scanner. Immediately after the scan, participants provided ratings of their in-scanner experience with the breathing load using a Visual Analog Scale (VAS). Participants rated the breathing load for pleasantness, unpleasantness, and intensity using a 10 cm scale ranging from 'not at all' to 'extremely'. After the scan, participants used the same VAS to rate their in-scanner experience of the breathing load.

2.4. Neuroimaging Data Acquisition

The cue reactivity paradigm was presented during one fMRI scan sensitive to blood oxygenation level-dependent (BOLD) contrast using a Signa EXCITE (GE Healthcare, Chicago, IL, USA) 3.0 Tesla scanner (T2*-weighted echo planar imaging (EPI) scans, TR = 2000 ms, TE = 30 ms, FOV = 24 cm (squared), 64 × 64 × 40 matrix, forty 3.0 mm axial slices with an in-plane resolution of 3.75 × 3.75 × 3 mm, flip angle = 90 degrees, 420 whole-brain acquisitions). For anatomical reference, a high-resolution T1-weighted image (spoiled gradient recalled [SPGR], TR = 8 ms, TE = 3 ms, slices = 172, FOV = 25 cm approximately 1 mm^3 voxels) was obtained.

2.5. Neuroimaging Data Analysis

2.5.1. Individual-Level Processing

All neuroimaging data was processed using the Analysis of Functional NeuroImages (AFNI) software package [50]. Following data acquisition, GE slices were reconstructed into AFNI BRIK format. Baseline volume for 3D registration was constructed using the largest temporal region containing the fewest voxel-wise outliers. Data was aligned to the baseline image using all other time points in dx, dy, dz, and roll, pitch, and yaw directions. The functional EPI underwent automatic coregistration to the high-resolution anatomical image and each alignment was manually inspected for each dataset. New outliers were generated for the volume-registered dataset based on whether a given time point greatly exceeded the mean number of voxel outliers for the time series. Six motion regressors (dx, dy, dz and roll, pitch, and yaw), a baseline and linear drift regressor, and nine task-related regressors (trials for anticipation neutral images, anticipation substance images, anticipation scrambled images, breathing load neutral images, breathing load substance images, breathing load scrambled images, neutral images only, substance images only, and scrambled images only) were convolved with a modified hemodynamic response function. The baseline condition, during which there was no cue or experience of the breathing load, served as the baseline for this analysis. A Gaussian Spatial Filter (6 mm full width-half maximum) was used to spatially blur data to account for anatomical differences. Automated transformations were applied to anatomical images and EPIs were subsequently transformed into Montreal neurological institute (MNI) space. Percent signal change (PSC) was determined by dividing each regressor of interest (anticipation neutral images, anticipation substance images, breathing load neutral images, breathing load substance images) by the baseline regressor and multiplying by 100.

2.5.2. Group-Level Analysis

A linear mixed-effects (LME) analysis (r-project.org) was performed to examine group differences in brain activation. Participants were treated as random effects, while group (CAN+ALC-SUD, CAN+ALC-EXP, CTL), interoceptive condition (no breathing load [anticipation], breathing load), and image type (neutral, substance) were treated as fixed effects. PSC from baseline (trials consisting of neutral and scrambled images and no chance or experience of the breathing load) was the dependent variable. The group main effect was examined to identify differences between CAN+ALC-SUD, CAN+ALC-EXP, and CTL across breathing load and cue image type conditions. The group by image type interaction was conducted to examine group differences while viewing substance images across all interoceptive conditions. The group by interoceptive condition interaction was examined to test hypotheses involving anticipation and receipt of the aversive interoceptive breathing load in CAN+ALC-SUD and CTL. The group by interoception by image type interaction was of interest because it allowed for examination of whether substance users show a blunted response to the aversive interoceptive stimuli when paired with the rewarding substance images. To guard against identifying false-positive areas of activation, a threshold adjustment method was applied using AFNI programs 3dFWHMx and 3dClustSim with the auto-correlation function (acf). The 3dClustSim identified a minimum cluster volume of 1280 μL (20 contiguous voxels) corresponding to a per-voxel p-value of 0.002 (bi-sided, NN = 3) to result in a voxel-wise probability of $p < 0.05$ (two-sided) corrected for multiple comparisons.

3. Results

3.1. Subject Characteristics

Groups did not differ in terms of demographics, including age ($F(2, 44) = 1.27$, $p = 0.290$), education ($F(2,43) = 0.956$, $p = 0.392$), racial ($\chi^2(8) = 9.043$, $p = 0.339$) and ethnic ($\chi^2(2) = 0.10$, $p = 0.953$) makeup, and gender distribution ($\chi^2(2) = 0.37$, $p = 0.830$); each group had more males than females. Moreover, there was no difference in subjective self-reported unpleasantness ($F(2,44) = 0.432$, $p = 0.652$) or intensity ($F(2,44) = 2.68$, $p = 0.08$) of the breathing load and the groups did not differ on self-reported interoceptive awareness and impulsivity. However, CAN+ALC-SUD compared to CAN+ALC-EXP reported higher levels of positive and negative reinforcement on the MNRQ. Additionally, both user groups reported higher levels of positive reinforcement than negative reinforcement on the MNRQ (see Table 2).

Table 2. Sample Characteristics.

	CAN+ALC-SUD Cannabis/Alcohol Substance Use Disorder	CAN+ALC-EXP Cannabis/Alcohol Experimenter	CTL Little to No Substance Use			
Demographics	*M*(*SD*)	*M*(*SD*)	*M*(*SD*)	df	*F*	*p*
Age (in years)	16.62 (0.51)	16.69 (0.70)	16.33 (0.77)	2,44	1.27	0.290
Education (in years)	10.46 (0.78)	10.47 (0.83)	10.11 (0.90)	2,43	0.956	0.392
WRAT 4 Verbal IQ	107.31 (14.29)	106.75 (12.37)	112.00 (13.82)	2,44	0.770	0.469
VAS Ratings	*M*(*SD*)	*M*(*SD*)	*M*(*SD*)	df	*F*	*p*
Unpleasant	5.69 (3.29)	4.63 (2.64)	5.34 (3.49)	2,44	0.432	0.652
Intensity	4.08 (3.47)	2.13 (2.77)	4.41 (2.89)	2,44	2.68	0.08
Questionnaires	*M*(*SD*)	*M*(*SD*)	*M*(*SD*)	df	*F/t*	*p*
MAIA						
Noticing	2.83 (1.52)	2.75 (1.03)	2.78 (1.19)	2,43	0.016	0.984
Not Distracting	2.14 (0.50)	2.37 (1.12)	2.52 (1.30)	2,43	0.443	0.645
Not Worrying	2.89 (1.43)	2.81 (1.42)	2.70 (1.05)	2,43	0.079	0.925
Attention Regulation	3.17 (.95)	3.45 (0.74)	3.14 (1.15)	2,43	0.494	0.613
Emotional Awareness	3.18 (1.43)	3.04 (1.31)	3.07 (.93)	2,43	0.054	0.948
Self-Regulation	3.10 (1.05)	3.23 (0.90)	3.01 (1.05)	2,43	0.207	0.814
Body Listening	1.36 (1.16)	1.96 (1.44)	1.79 (1.05)	2,43	0.844	0.437
Trusting	3.47 (1.40)	3.73 (1.08)	3.72 (0.92)	2,43	0.231	0.795
UPPS						
Lack of Premeditation	2.08 (0.39)	2.18 (0.49)	1.89 (0.42)	2,44	1.92	0.159
Urgency	2.30 (0.66)	2.17 (0.59)	2.06 (0.51)	2,44	0.672	0.516
Sensation Seeking	3.18 (0.29)	3.09 (0.48)	3.03 (0.44)	2,44	0.528	0.594
Lack of Perseverance	2.03 (0.59)	2.13 (0.58)	1.83 (0.34)	2,44	1.53	0.229
MNRQ						
Negative Reinforcement	2.85 (2.38)	0.875 (1.63)	–	20.52	2.55	0.019
Positive Reinforcement	11.38 (2.53)	7.25 (3.45)	–	27	3.59	0.001

3.2. Neuroimaging Results

No clusters met the thresholding requirement of 20 voxels for the main effect of group, the group by image type interaction, or the three-way group by interoceptive condition by cue image type interaction.

3.2.1. The Group by interoception interaction

Four brain regions survived thresholding: the right amygdala, the left IFG, the right posterior cingulate, and the left parahippocampal gyrus. (see Table 3). All interactions remained significant after controlling for lifetime nicotine use.

Table 3. fMRI results and between-group comparisons (SUD = CAN+ALC-SUD; EXP = CAN+ALC-EXP).

GROUP BY INTEROCEPTIVE CONDITION INTERACTION									
	R/L	Voxels	Volume	X	Y	Z	BA	Anticipation	Load
Amygdala	R	33	2112	28	−9	−30	28	SUD > EXP	EXP > SUD
Inferior Frontal Gyrus	L	28	1792	−13	24	−20	11	–	EXP = CTL > SUD
Posterior Cingulate	R	25	1600	13	−65	16	31	–	EXP = SUD > CTL
Parahippocampal Gyrus	L	21	1344	−24	−7	−19	35	CTL > EXP	CTL = EXP > SUD
MAIN EFFECT OF INTEROCEPTIVE CONDITION									
	R/L	**Voxels**	**Volume**	**X**	**Y**	**Z**	**BA**		**Condition Effect**
Cingulate Gyrus	R	3141	201024	8	−6	23	24		Load > Anticipation
Fusiform Gyrus	R	663	42432	40	−12	−24	20		Anticipation > Load
Superior Frontal Gyrus	R	334	21376	1	4	57	6		Load > Anticipation
Cingulate Gyrus	L	131	8384	−2	−25	37	31		Load > Anticipation
Cuneus	R	112	7168	18	−85	28	18		Load > Anticipation
Thalamus	R	64	4096	6	−18	4			Load > Anticipation
Declive	L	61	3904	−15	−63	−20			Load > Anticipation
Middle Frontal Gyrus	L	48	3072	−36	37	28	9		Anticipation > Load
Middle Occipital Gyrus	R	43	2752	34	−83	9	19		Load > Anticipation
Anterior Cingulate	L	39	2496	−6	31	15	24		Load > Anticipation
Precuneus	R	36	2304	5	−43	43	7		Load > Anticipation
Precentral Gyrus	R	29	1856	18	−26	64	4		Load > Anticipation
Precentral Gyrus	L	24	1536	−18	−29	63	4		Anticipation > Load
MAIN EFFECT OF CUE STIMULUS TYPE									
	R/L	**Voxels**	**Volume**	**X**	**Y**	**Z**	**BA**		**Stimulus Effect**
Medial Frontal Gyrus	R	43	2752	1	44	30	9		Substance > Comparison
Anterior Cingulate	L	23	1472	−1	46	8	32		Substance > Comparison

The right amygdala. A significant interaction within the right amygdala ($F(2,44) = 7.58$, $p = 0.001$, partial $\eta^2 = 0.256$) was examined. Here, groups significantly differed in activation for the anticipation only condition ($F(2, 44) = 4.28$, $p = 0.02$, partial $\eta^2 = 0.16$) with CAN+ALC-SUD showing significantly greater activation than CAN+ALC-EXP ($M = 0.16$, $SE = 0.06$, $p = 0.02$). Groups also significantly differed during the breathing load condition ($F(2, 44) = 4.59$, $p = 0.02$, partial $\eta^2 = 0.17$) with CAN+ALC-SUD showing lower activation than CAN+ALC-EXP ($M = 0.54$, $SE = 0.18$, $p = 0.004$). CTL did not significantly differ from either user group during either condition (see Figure 2).

The left inferior frontal gyrus. Within the left IFG ($F(2,44) = 5.66$, $p = 0.006$, partial $\eta^2 = 0.21$), groups did not differ during anticipation ($p = 0.28$) but did during the breathing load ($F(2,44) = 4.62$, $p = 0.015$, partial $\eta^2 = 0.17$). CAN+ALC-SUD exhibited lower activation than both CAN+ALC-EXP ($M = 0.27$, $SE = 0.11$, $p = 0.049$) and CTL ($M = 0.27$, $SE = 0.11$, $p = 0.049$), who did not differ from one another (see Figure 2).

The right posterior cingulate. An interaction within the right posterior cingulate ($F(2,44) = 4.11$, $p = 0.02$, partial $\eta^2 = 0.16$) was driven by a significant effect of condition for CTL only ($F(1,17) = 11.22$, $p = 0.004$, partial $\eta^2 = 0.39$) with greater deactivation while experiencing the breathing load; no simple main effect for group was seen in this region.

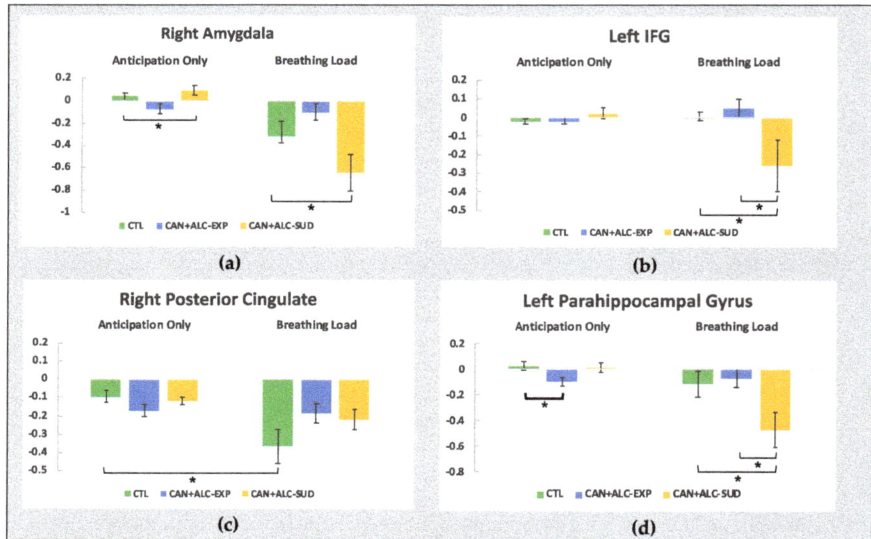

Figure 2. Neuroimaging results from the group by interoception condition interaction in (**a**) the right amygdala; (**b**) the left inferior frontal gyrus; (**c**) the right posterior cingulate; and (**d**) the left parahippocampal gyrus. * indicates significant differences.

The left parahippocampal gyrus. Within the left parahippocampal gyrus ($F(2,44) = 6.14$, $p = 0.004$, partial $\eta^2 = 0.22$), groups significantly differed during the anticipation condition ($F(2,44) = 3.98$, $p = 0.02$, partial $\eta^2 = 0.15$) and during the breathing load trials ($F(2,44) = 4.23$, $p = 0.02$, partial $\eta^2 = 0.16$). Specifically, during anticipation only trials, CTL exhibited significantly greater activation than CAN+ALC-EXP ($M = 0.12$, $SE = 0.05$, $p = 0.03$), and CAN+ALC-SUD did not differ from either group. For the breathing load, CAN+ALC-SUD showed significantly lower activation than both CTL ($M = 0.35$, $SE = 0.15$, $p = 0.05$) and CAN+ALC-EXP ($M = 0.40$, $SE = 0.15$, $p = 0.03$; see Figure 2).

3.2.2. Follow-Up Correlations

Follow-up correlations were conducted within CAN+ALC-SUD and CAN+ALC-EXP between activation in significant regions and lifetime episodes of cannabis and alcohol use. Within the left IFG, activation during the breathing load condition negatively correlated with lifetime episodes of alcohol use ($r = -0.546$, $p = 0.002$, $R^2 = 0.298$). Within CAN+ALC-SUD and CAN+ALC-EXP, PHG activation during the breathing load condition negatively correlated with lifetime episodes of cannabis ($r = -0.570$, $p = 0.001$, $R^2 = 0.325$) and alcohol use ($r = -0.473$, $p = 0.009$, $R^2 = 0.224$; see Figure 3).

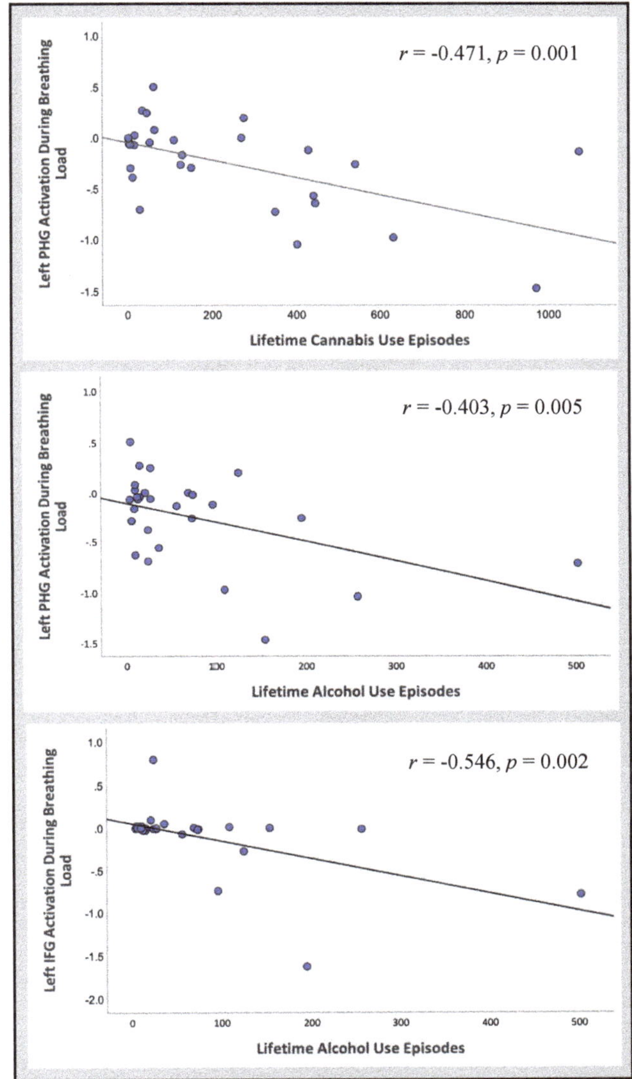

Figure 3. Follow-up correlations between activation in significant regions of interest and reported lifetime episodes of cannabis and alcohol use across all substance users.

4. Discussion

The present investigation aimed to examine the role of negative reinforcement in adolescent substance use by pairing a cue reactivity paradigm with an aversive interoceptive probe. It was hypothesized that viewing rewarding substance images would dampen the exaggerated interoceptive response to an aversive probe that has previously been observed in adolescent substance users [27]. Specifically, CAN+ALC-SUD compared to CTL was hypothesized to show: (1) heightened neural activation during the breathing load experience in brain regions involved in interoceptive processing and emotion regulation; (2) heightened neural reward responsivity to substance images; and (3) a decreased interoceptive neural response to the breathing load when paired with substance images.

It was also hypothesized that, overall, CAN+ALC-EXP would demonstrate a neural response more similar to CTL than CAN+ALC-SUD.

The hypotheses were partially supported. In relation to hypothesis one, a consistent pattern was observed within the left IFG and the left parahippocampal gyrus, wherein CAN+ALC-SUD exhibited a differential BOLD response to the breathing load compared to CAN+ALC-EXP and CTL (Figure 3; Table 3). Based on previous work demonstrating that adolescent SUD showed an increased response to the breathing load [27], it was hypothesized that CAN+ALC-SUD in the present investigation would also exhibit an increased BOLD response. However, compared to CAN+ALC-EXP and CTL, CAN+ALC-SUD showed greater deactivation during the breathing load. Although this result is inconsistent with previous findings among adolescent substance users [27], it is consistent with previous findings among young adults transitioning from recreational to problematic substance use [39]. A similar pattern was observed in the right amygdala, with CAN+ALC-SUD demonstrating greater deactivation than CAN+ALC-EXP. However, CTL did not differ from either group. Hypothesis two was not supported, as CAN+ALC-SUD did not show a differential reward response to substance images compared to CTL. In line with the lack of an exaggerated reward response to the substance images within CAN+ALC-SUD, viewing these cues did not attenuate the exaggerated interoceptive response exhibited by CAN+ALC-SUD (hypothesis 3). Lastly, it was hypothesized that CAN+ALC-EXP would demonstrate brain responses more similar to CTL than CAN+ALC-SUD; this was partially supported. During the anticipation only condition, CAN+ALC-EXP showed an inconsistent pattern. However, during the breathing load condition, CAN+ALC-EXP did not differ from CTL in the right amygdala, the left IFG, or the left parahippocampal gyrus.

The overall findings suggest that CAN+ALC-SUD experience the aversive breathing load differently than CTL and CAN+ALC-EXP in brain regions implicated in interoception and emotion regulation. However, this observation is in the opposite direction of previous findings. Adolescent SUD has previously shown exaggerated activation rather than deactivation in interoceptive regions when experiencing an aversive breathing load [27]. Additionally, viewing images of alcohol and cannabis did not appear to dampen the blunted interoceptive response seen among CAN+ALC-SUD. This finding would suggest that substance use may not be negatively reinforced by dampening uncomfortable sensations. The pattern of use demonstrated by adolescent substance users (non-treatment-seeking users meeting diagnostic criteria) may not be substantial enough to invoke withdrawal-related symptoms compared to adults who have used heavily for years and/or treatment seekers. Therefore, using in order to relieve uncomfortable sensations may be less common among adolescent users or individuals with less significant use patterns. Future studies should collect subjective ratings of how 'unpleasant' and 'aversive' participants found the breathing load to be while viewing substance and neutral images separately, as this would provide a clearer understanding of whether or not viewing substance images can contribute to an overall reduction in the aversiveness of the breathing load. Lastly, adolescents with SUD showed amygdala deactivation while experiencing the breathing load but increased activation when anticipating the upcoming load. Previous research has demonstrated that cannabis users exhibit deactivation in the amygdala while viewing emotional images, indicative of altered emotion regulation. This may suggest that the observed group differences in the present study are due to differences in emotion regulation. Although, emotion regulation was not directly assessed in this study, this is a potential avenue for future research.

Interestingly, there were also no significant findings within the insular cortex despite its central role in interoception. This contradicts previous research demonstrating that adolescents meeting criteria for SUD exhibit an increased insular response to the breathing load [27]. It is possible that this lack of insular cortex findings is due to the more stringent thresholding methods employed in the present investigation based on current methodological recommendations for the analysis of fMRI data, as insular activation was present at lower thresholds [51]. Overall, this could suggest that experiencing the breathing load within the context of an experimental manipulation may not be significant enough to elicit a strong insular response among adolescents. Future research should examine whether there is an age-related difference in response to aversive interoceptive perturbations.

The lack of evidence demonstrating any negative reinforcement-related neural response may also be because CAN+ALC-SUD did not find the images rewarding enough, given that an exaggerated reward response was not observed. Altered reward responsivity to substance cues is an established finding among adult substance users [52]. It is possible, given that adolescents with CAN+ALC-SUD typically have significantly less use history than adults with CAN+ALC-SUD, that adolescent reward networks have not yet been altered to show an exaggerated response to substance images. This would suggest that altered reward responsivity is not a predisposition among CAN+ALC-SUD but rather a consequence of use. However, an exaggerated neural response to alcohol and cannabis images in limbic regions has been observed among alcohol-using adolescents and young adults [36,37,53]. A possible reason for our discrepant finding could be differences in characteristics defining each sample. For example, participants in the present study used both alcohol and cannabis. Reward circuitry among alcohol and cannabis users may differ from individuals who only consume alcohol and/or cannabis like those in the previously mentioned investigation [36,37]. Future examination of reward circuitry in single- and multi-substance users with a larger sample could help to elucidate this question.

Clinically, our findings suggest that interventions aiming to improve coping through emotion regulation may not be the most effective for adolescent substance users given the lack of evidence that substance use is driven by negative reinforcement. Alternatively, adolescent substance use may be driven more by positive reinforcement; CAN+ALC-SUD self-reported significantly more motivations for use related to positive, as opposed to negative, reinforcement than CAN+ALC-EXP (Table 1). This aligns with the neurobiological imbalance model, which posits that the development of cognitive control regions is more protracted from childhood to young adulthood, while reward regions follow a curvilinear path of development, with a peak in reward responsivity during adolescence [1,54]. This heightened reward response during adolescence can be seen in reward-processing brain regions (i.e., striatum, insula, anterior cingulate cortex) [55–58] when anticipating and receiving various types of rewards [59,60]. Behaviorally, this imbalance may contribute to an increase in reward-seeking behaviors, including drug and alcohol experimentation [59] and increased susceptibility to the motivational properties of these substances. This may suggest that interventions aimed at helping adolescents learn alternative ways of experiencing reward may be more effective than those aimed at reducing uncomfortable sensations [61].

Although adolescent substance users report negative reinforcement of substance use, this was not observed using a functional imaging paradigm. As reported above, groups also did not differ in their neural responsivity to the substance images, but this finding may be due to a limitation of study design. The substance images used in the cue reactivity paradigm may not be potent or personally relevant enough to elicit a sufficient neural response to overcome the undesirable impact of the breathing load trials [61–63]. In daily life, adolescents may experience uncomfortable interoceptive signals on par with the breathing load experienced within the scanner while the rewarding effects of actual substance use may not be comparable to viewing images. Experimentally administering alcohol and drugs in conjunction with fMRI is an increasingly popular research method that may be more powerful for detecting neural changes related to negative reinforcement [64,65]. Alternatively, creating personalized cue reactivity paradigms using substance-related images from adolescents' social media accounts may be an alternative method of increasing the valence of the substance cues. Future researchers investigating negative reinforcement principles within adolescent substance users should consider these methods to determine whether a more robust substance cue can elicit neural differences.

An additional limitation of the present study may be the categorization of adolescents based on meeting criteria for CAN+ALC-SUD. The observed correlations between substance use and neural response suggest that future examinations of adolescent substance users may be improved using a dimensional, rather than categorical, approach. Although significant differences in BOLD response to interoceptive stimulation have been observed among adult substance users with and without

CAN+ALC-SUD [14,66,67], amount of substance use may be a more differentiating factor than reported CAN+ALC-SUD criteria in young users with comparatively little substance use experience. The small sample size of 13 CAN+ALC-SUD, 16 CAN+ALC-EXP, and 18 CTL also limits the conclusions that can be drawn from the current study and the ability to look at substance-use groups individually (e.g., cannabis vs. alcohol) although comorbid cannabis and alcohol use is common among adolescents [68]. Inclusion of more substance-using adolescents in future studies could help better differentiate between youth who experiment with drugs and alcohol and those who experience more negative consequences related to their use. Lastly, CAN+ALC-SUD and CAN+ALC-EXP significantly differed in the amount of time reported since their last cannabis use. Given that cannabis metabolites can remain in the body for up to three weeks after regular use, it is possible that the differences observed between groups could be due to residual effects in the CAN+ALC-SUD group. Therefore, it is possible that the reported findings are more reflective of the effects of current use and that these differences may resolve with continued abstinence, highlighting another potential avenue for future research.

Despite these limitations, the present study contributes preliminary findings to our overall understanding of substance use in adolescence. The findings further support the hypothesis that interoceptive processing may be altered in substance users. Further, the results suggest that adolescents may not seek substances to reduce negative or uncomfortable sensations, rather use may be driven more by increased sensation-seeking and reward responsivity in adolescence. Examining positive reinforcement in adolescent substance use is an important avenue for future research.

Author Contributions: Conceptualization, A.C.M., J.J., J.L.S., M.P.P. and S.F.T.; data curation, A.C.M. and J.L.S. formal analysis, A.C.M., J.L.S. and A.N.S.; funding acquisition, A.C.M., J.L.S., M.P.P. and S.F.T.; investigation, A.C.M. and J.L.S.; methodology, A.C.M., J.J., J.L.S., A.N.S., M.P.P. and S.F.T.; project administration, A.C.M. and J.L.S.; resources, M.P.P. and S.F.T.; software, A.N.S. and M.P.P.; supervision, J.J., J.L.S., A.N.S., M.P.P. and S.F.T.; validation, A.N.S., M.P.P. and S.F.T.; visualization, A.C.M., J.J. and J.L.S.; writing—original draft, A.C.M.; writing—review and editing, J.J., J.L.S., A.N.S., M.P.P. and S.F.T. All authors have read and agreed to the published version of the manuscript.

Funding: This research has been supported in part by NIDA 5P20DA027843, NIH/NIDA U01DA041089, NIH/NIAAA T32AA013525, NIAAA F31AA027169, NIDA R21DA047953, NIGMS 1P20GM121312, California Tobacco-Related Disease Research Grants Program Office of the University of California Grant 580264, and The William K. Warren Foundation. Financial Disclosures: Paulus is an advisor to Spring Care, Inc., a behavioral health startup, he has received royalties for an article about methamphetamine in UpToDate.

Conflicts of Interest: The authors declare no conflict of interest.

References

1. Casey, B.J.; Jones, R.M. Neurobiology of the adolescent brain and behavior: Implications for substance use disorders. *J. Am. Acad. Child Adolesc. Psychiatry* **2010**, *49*, 1189–1201. [CrossRef] [PubMed]
2. Johnston, L.D.; Miech, R.A.; O'Malley, P.M.; Bachman, J.G.; Schulenberg, J.E.; Patrick, M.E. *Monitoring the Future National Survey Results on Drug Use 1975–2019: Overview, Key Findings on Adolescent Drug Use*; Education Resources Information Center: Washington, DC, USA, 2020.
3. Substance Abuse and Mental Health Services Administration. *Key Substance Use and Mental Health Indicators in the United States: Results from the 2018 National Survey on Drug Use and Health, HHS*; HHS Publ. No. PEP19-5068, NSDUH Ser. H-54; Center for Behavioral Health Statistics and Quality, Substance Abuse and Mental Health Services Administration: Rockville, MD, USA, 2019. [CrossRef]
4. Hasin, D.S. US Epidemiology of Cannabis Use and Associated Problems. *Neuropsychopharmacology* **2018**, *43*, 195–212. [CrossRef] [PubMed]
5. Grant, B.F.; Dawson, D.A. Age at onset of alcohol use and its association with DSM-IV alcohol abuse and dependence: Results from the national longitudinal alcohol epidemiologic survey. *J. Subst. Abuse* **1997**, *9*, 103–110. [CrossRef]
6. Substance Abuse and Mental Health Services Administration. *Results from the 2013 National Survey on Drug Use and Health: Summary of National Findings*; NSDUH Series H-48; HHS Publication No. (SMA) 14-4863; Center for Behavioral Health Statistics and Quality, Substance Abuse and Mental Health Services Administration: Rockville, MD, USA, 2014.

7. London, E.D.; Ernst, M.; Grant, S.; Bonson, K.; Weinstein, A. Orbitofrontal cortex and human drug abuse: Functional imaging. *Cereb. Cortex* **2000**, *10*, 334–342. [CrossRef] [PubMed]
8. Stewart, J.L.; Juavinett, A.L.; May, A.C.; Davenport, P.W.; Paulus, M.P. Do you feel alright? Attenuated neural processing of aversive interoceptive stimuli in current stimulant users. *Psychophysiology* **2015**, *52*, 249–262. [CrossRef] [PubMed]
9. Bechara, A.; Tranel, D.; Damasio, H. Characterization of the decision-making deficit of patients with ventromedial prefrontal cortex lesions. *Brain* **2000**, *123*. [CrossRef]
10. Nestor, L.J.; Ghahremani, D.G.; Monterosso, J.; London, E.D. Prefrontal hypoactivation during cognitive control in early abstinent methamphetamine-dependent subjects. *Psychiatry Res. Neuroimaging* **2011**, *194*. [CrossRef]
11. Paulus, M.P.; Feinstein, J.S.; Tapert, S.F.; Liu, T.T. Trend detection via temporal difference model predicts inferior prefrontal cortex activation during acquisition of advantageous action selection. *Neuroimage* **2004**. [CrossRef]
12. Verdejo-Garcia, A.; Chong, T.T.J.; Stout, J.C.; Yücel, M.; London, E.D. Stages of dysfunctional decision-making in addiction. *Pharmacol. Biochem. Behav.* **2018**. [CrossRef]
13. Sönmez, M.B.; Kahyacı Kılıç, E.; Ateş Çöl, I.; Görgülü, Y.; Köse Çınar, R. Decreased interoceptive awareness in patients with substance use disorders. *J. Subst. Use* **2017**, *22*. [CrossRef]
14. May, A.C.; Stewart, J.L.; Migliorini, R.; Tapert, S.F.; Paulus, M.P. Methamphetamine Dependent Individuals Show Attenuated Brain Response to Pleasant Interoceptive Stimuli. *Drug Alcohol Depend.* **2013**, *131*, 238–246. [CrossRef] [PubMed]
15. Stewart, S.H.; Zvolensky, M.J.; Eifert, G.H. Negative-reinforcement drinking motives mediate the relation between anxiety sensitivity and increased drinking behavior. *Pers. Individ. Differ.* **2001**. [CrossRef]
16. Craig, A.D. How do you feel? Interoception: The sense of the physiological condition of the body. *Nat. Rev. Neurosci.* **2002**. [CrossRef] [PubMed]
17. Pollatos, O.; Gramann, K.; Schandry, R. Neural systems connecting interoceptive awareness and feelings. *Hum. Brain Mapp.* **2007**. [CrossRef] [PubMed]
18. Craig, A.D. Interoception: The sense of the physiological condition of the body. *Curr. Opin. Neurobiol.* **2003**, *13*, 500–505. [CrossRef]
19. Barrett, L.F.; Simmons, W.K. Interoceptive predictions in the brain. *Nat. Rev. Neurosci.* **2015**. [CrossRef]
20. Paulus, M.P.; Tapert, S.F.; Schulteis, G. The role of interoception and alliesthesia in addiction. *Pharmacol. Biochem. Behav.* **2009**. [CrossRef]
21. Verdejo-Garcia, A.; Clark, L.; Dunn, B.D. The role of interoception in addiction: A critical review. *Neurosci. Biobehav. Rev.* **2012**. [CrossRef]
22. Naqvi, N.H.; Bechara, A. The Insula: A Critical Neural Substrate for Drug Seeking under Conflict and Risk. *Wiley Handb. Cogn.* **2015**. [CrossRef]
23. Paulus, M.P.; Stewart, J.L. Interoception and drug addiction. *Neuropharmacology* **2014**, *76*, 342–350. [CrossRef]
24. Critchley, H.D.; Wiens, S.; Rotshtein, P.; Öhman, A.; Dolan, R.J. Neural systems supporting interoceptive awareness. *Nat. Neurosci.* **2004**. [CrossRef] [PubMed]
25. Etkin, A.; Buechel, C.; Gross, J.J. The neural bases of emotion regulation. *Nat. Rev. Neurosci.* **2015**. [CrossRef] [PubMed]
26. Stewart, J.L.; Flagan, T.M.; May, A.C.; Reske, M.; Simmons, A.N.; Paulus, M.P. Young adults at risk for stimulant dependence show reward dysfunction during reinforcement-based decision making. *Biol. Psychiatry* **2013**, *73*, 235–241. [CrossRef] [PubMed]
27. Berk, L.; Stewart, J.L.; May, A.C.; Wiers, R.W.; Davenport, P.W.; Paulus, M.P.; Tapert, S.F. Under pressure: Adolescent substance users show exaggerated neural processing of aversive interoceptive stimuli. *Addiction* **2015**, *110*, 2025–2036. [CrossRef]
28. Stewart, J.L.; Butt, M.; May, A.C.; Tapert, S.F.; Paulus, M.P. Insular and cingulate attenuation during decision making is associated with future transition to stimulant use disorder. *Addiction* **2017**, *112*, 1567–1577. [CrossRef] [PubMed]
29. Verdejo-García, A.; Bechara, A. A somatic marker theory of addiction. *Neuropharmacology* **2009**. [CrossRef] [PubMed]

30. Wills, T.A.; Pokhrel, P.; Morehouse, E.; Fenster, B. Behavioral and Emotional Regulation and Adolescent Substance Use Problems: A Test of Moderation Effects in a Dual-Process Model. *Psychol. Addict. Behav.* **2011**, *25*, 279–292. [CrossRef] [PubMed]
31. Payer, D.; Lieberman, M.; London, E. Neural correlates of affect processing and aggression in methamphetamine dependence. *Arch. Gen. Psychiatry* **2011**, *68*, 271–282. [CrossRef]
32. Veinante, P.; Yalcin, I.; Barrot, M. The amygdala between sensation and affect: A role in pain. *J. Mol. Psychiatry* **2013**. [CrossRef]
33. Morawetz, C.; Bode, S.; Derntl, B.; Heekeren, H.R. The effect of strategies, goals and stimulus material on the neural mechanisms of emotion regulation: A meta-analysis of fMRI studies. *Neurosci. Biobehav. Rev.* **2017**. [CrossRef]
34. Conrod, P.; Nikolaou, K. Annual Research Review: On the developmental neuropsychology of substance use disorders. *J. Child Psychol. Psychiatry Allied Discip.* **2016**. [CrossRef] [PubMed]
35. Wolitzky-Taylor, K.; McBeth, J.; Guillot, C.R.; Stone, M.D.; Kirkpatrick, M.G.; Zvolensky, M.J.; Buckner, J.D.; Leventhal, A.M. Transdiagnostic processes linking anxiety symptoms and substance use problems among adolescents. *J. Addict. Dis.* **2016**. [CrossRef] [PubMed]
36. Karoly, H.C.; Schacht, J.P.; Meredith, L.R.; Jacobus, J.; Tapert, S.F.; Gray, K.M.; Squeglia, L.M. Investigating a novel fMRI cannabis cue reactivity task in youth. *Addict. Behav.* **2019**, *89*, 20–28. [CrossRef] [PubMed]
37. Tapert, S.F.; Cheung, E.H.; Brown, G.G.; Frank, L.R.; Paulus, M.P.; Schweinsburg, A.D.; Meloy, M.J.; Brown, S.A. Neural Response to Alcohol Stimuli in Adolescents with Alcohol Use Disorder. *Arch. Gen. Psychiatry* **2003**, *60*, 727. [CrossRef] [PubMed]
38. Lopata, M.; La Fata, J.; Evanich, M.J.; Lourenço, R.V. Effects of flow-resistive loading on mouth occlusion pressure during CO_2 rebreathing. *Am. Rev. Respir. Dis.* **1977**. [CrossRef] [PubMed]
39. Stewart, J.L.; Parnass, J.M.; May, A.C.; Davenport, P.W.; Paulus, M.P. Altered frontocingulate activation during aversive interoceptive processing in young adults transitioning to problem stimulant use. *Front. Syst. Neurosci.* **2013**. [CrossRef] [PubMed]
40. Stewart, J.L.; May, A.C.; Poppa, T.; Davenport, P.W.; Tapert, S.F.; Paulus, M.P. You are the danger: Attenuated insula response in methamphetamine users during aversive interoceptive decision-making. *Drug Alcohol Depend.* **2014**, *142*, 110–119. [CrossRef]
41. Schacht, J.P.; Anton, R.F.; Myrick, H. Functional neuroimaging studies of alcohol cue reactivity: A quantitative meta-analysis and systematic review. *Addict. Biol.* **2013**. [CrossRef]
42. Zhou, X.; Zimmermann, K.; Xin, F.; Zhao, W.; Derckx, R.T.; Sassmannshausen, A.; Scheele, D.; Hurlemann, R.; Weber, B.; Kendrick, K.M.; et al. Cue Reactivity in the Ventral Striatum Characterizes Heavy Cannabis Use, Whereas Reactivity in the Dorsal Striatum Mediates Dependent Use. *Biol. Psychiatry Cogn. Neurosci. Neuroimaging* **2019**. [CrossRef]
43. Pierucci-Lagha, A.; Gelernter, J.; Feinn, R.; Cubells, J.F.; Pearson, D.; Pollastri, A.; Farrer, L.; Kranzler, H.R. Diagnostic reliability of the semi-structured assessment for drug dependence and alcoholism (SSADDA). *Drug Alcohol Depend.* **2005**. [CrossRef]
44. Brown, S.A.; Myers, M.G.; Lippke, L.; Tapert, S.F.; Stewart, D.G.; Vik, P.W. Psychometric evaluation of the Customary Drinking and Drug Use Record (CDDR): A measure of adolescent alcohol and drug involvement. *J. Stud. Alcohol* **1998**, *59*, 427–438. [CrossRef] [PubMed]
45. Whiteside, S.P.; Lynam, D.R. The Five Factor Model and impulsivity: Using a structural model of personality to understand impulsivity. *Pers. Individ. Differ.* **2001**, *30*, 669–689. [CrossRef]
46. Mehling, W.E.; Price, C.; Daubenmier, J.J.; Acree, M.; Bartmess, E.; Stewart, A. The Multidimensional Assessment of Interoceptive Awareness (MAIA). *PLoS ONE* **2012**. [CrossRef] [PubMed]
47. Pomerleau, O.F.; Fagerström, K.O.; Marks, J.L.; Tate, J.C.; Pomerleau, C.S. Development and validation of a self-rating scale for positive- and negative-reinforcement smoking: The Michigan nicotine reinforcement questionnaire. *Nicotine Tob. Res.* **2003**. [CrossRef] [PubMed]
48. Paulus, M.P.; Flagan, T.; Simmons, A.N.; Gillis, K.; Kotturi, S.; Thom, N.; Johnson, D.C.; Van Orden, K.F.; Davenport, P.W.; Swain, J.L. Subjecting elite athletes to inspiratory breathing load reveals behavioral and neural signatures of optimal performers in extreme environments. *PLoS ONE* **2012**, *7*, e29394. [CrossRef]
49. Davenport, P.W.; Vovk, A. Cortical and subcortical central neural pathways in respiratory sensations. *Respir. Physiol. Neurobiol.* **2009**. [CrossRef]

50. Cox, R.W. AFNI: Software for analysis and visualization of functional magnetic resonance neuroimages. *Comput. Biomed. Res.* **1996**. [CrossRef]
51. Eklund, A.; Nichols, T.E.; Knutsson, H. Cluster failure: Why fMRI inferences for spatial extent have inflated false-positive rates. *Proc. Natl. Acad. Sci. USA* **2016**. [CrossRef]
52. Luijten, M.; Schellekens, A.F.; Kühn, S.; Machielse, M.W.J.; Sescousse, G. Disruption of Reward Processing in Addiction. *JAMA Psychiatry* **2017**. [CrossRef]
53. Dager, A.D.; Anderson, B.M.; Rosen, R.; Khadka, S.; Sawyer, B.; Jiantonio-Kelly, R.E.; Austad, C.S.; Raskin, S.A.; Tennen, H.; Wood, R.M.; et al. Functional magnetic resonance imaging (fMRI) response to alcohol pictures predicts subsequent transition to heavy drinking in college students. *Addiction* **2014**. [CrossRef]
54. Somerville, L.H.; Jones, R.M.; Casey, B.J. A time of change: Behavioral and neural correlates of adolescent sensitivity to appetitive and aversive environmental cues. *Brain Cogn.* **2010**. [CrossRef] [PubMed]
55. Delgado, M.R.; Locke, H.M.; Stenger, V.A.; Fiez, J.A. Dorsal striatum responses to reward and punishment: Effects of valence and magnitude manipulations. *Cogn. Affect. Behav. Neurosci.* **2003**, *3*, 27–38. [CrossRef] [PubMed]
56. Ernst, M.; Nelson, E.E.; Jazbec, S.; McClure, E.B.; Monk, C.S.; Leibenluft, E.; Blair, J.; Pine, D.S. Amygdala and nucleus accumbens in responses to receipt and omission of gains in adults and adolescents. *Neuroimage* **2005**, *25*, 1279–1291. [CrossRef] [PubMed]
57. Elliott, R.; Newman, J.L.; Longe, O.A.; Deakin, J.F.W. Differential response patterns in the striatum and orbitofrontal cortex to financial reward in humans: A parametric functional magnetic resonance imaging study. *J. Neurosci.* **2003**, *23*, 303–307. [CrossRef]
58. Seymour, B.; Daw, N.; Dayan, P.; Singer, T.; Dolan, R. Differential Encoding of Losses and Gains in the Human Striatum. *J. Neurosci.* **2007**, *27*, 4826–4831. [CrossRef]
59. Hardin, M.G.; Ernst, M. Functional brain imaging of development-related risk and vulnerability for substance use in adolescents. *Annu. Rev. Clin. Psychol.* **2016**, *11*, 361–377. [CrossRef]
60. May, A.C.; Stewart, J.L.; Tapert, S.F.; Paulus, M.P. The effect of age on neural processing of pleasant soft touch stimuli. *Front. Behav. Neurosci.* **2014**, *8*, 52. [CrossRef]
61. Aguinaldo, L.D.; Squeglia, L.M.; Gray, K.M.; Coronado, C.; Lees, B.; Tomko, R.L.; Jacobus, J. Behavioral Treatments for Adolescent Cannabis Use Disorder: A Rationale for Cognitive Retraining. *Curr. Addict. Rep.* **2019**. [CrossRef]
62. Fatseas, M.; Serre, F.; Alexandre, J.M.; Debrabant, R.; Auriacombe, M.; Swendsen, J. Craving and substance use among patients with alcohol, tobacco, cannabis or heroin addiction: A comparison of substance- and person-specific cues. *Addiction* **2015**. [CrossRef]
63. Karoly, H.C.; Schacht, J.P.; Jacobus, J.; Meredith, L.R.; Taylor, C.T.; Tapert, S.F.; Gray, K.M.; Squeglia, L.M. Preliminary evidence that computerized approach avoidance training is not associated with changes in fMRI cannabis cue reactivity in non-treatment-seeking adolescent cannabis users. *Drug Alcohol Depend.* **2019**. [CrossRef]
64. Filbey, F.M.; Claus, E.; Audette, A.R.; Niculescu, M.; Banich, M.T.; Tanabe, J.; Du, Y.P.; Hutchison, K.E. Exposure to the taste of alcohol elicits activation of the mesocorticolimbic neurocircuitry. *Neuropsychopharmacology* **2008**. [CrossRef] [PubMed]
65. Strang, N.M.; Claus, E.D.; Ramchandani, V.A.; Graff-Guerrero, A.; Boileau, I.; Hendershot, C.S. Dose-dependent effects of intravenous alcohol administration on cerebral blood flow in young adults. *Psychopharmacology* **2015**. [CrossRef] [PubMed]
66. Bjork, J.M.; Smith, A.R.; Hommer, D.W. Striatal sensitivity to reward deliveries and omissions in substance dependent patients. *Neuroimage* **2008**. [CrossRef] [PubMed]
67. Gowin, J.L.; Stewart, J.L.; May, A.C.; Ball, T.M.; Wittmann, M.; Tapert, S.F.; Paulus, M.P. Altered cingulate and insular cortex activation during risk-taking in methamphetamine dependence: Losses lose impact. *Addiction* **2014**, *109*, 237–247. [CrossRef] [PubMed]
68. Mason, W.A.; Chmelka, M.B.; Howard, B.K.; Thompson, R.W. Comorbid alcohol and cannabis use disorders among high-risk youth at intake into residential care. *J. Adolesc. Health* **2013**. [CrossRef] [PubMed]

Article

Marijuana Use among African American Older Adults in Economically Challenged Areas of South Los Angeles

Sharon Cobb [1], Mohsen Bazargan [2,3,4], James Smith [2], Homero E. del Pino [5,6], Kimberly Dorrah [3] and Shervin Assari [2,*]

1 School of Nursing, Charles R Drew University of Medicine and Science, Los Angeles, CA 90059, USA
2 Department of Family Medicine, College of Medicine, Charles R Drew University of Medicine and Science, Los Angeles, CA 90059, USA
3 Department of Public Health, Charles R Drew University of Medicine and Science, Los Angeles, CA 90059, USA
4 Department of Family Medicine, University of California, Los Angeles (UCLA), Los Angeles, CA 90095, USA
5 Department of Psychiatry and Human Behavior, College of Medicine, Charles R Drew University of Medicine and Science, Los Angeles, CA 90059, USA
6 Department of Psychiatry and Biobehavioral Sciences, University of California, Los Angeles (UCLA), Los Angeles, CA 90095, USA
* Correspondence: shervinassari@cdrewu.edu

Received: 1 July 2019; Accepted: 15 July 2019; Published: 16 July 2019

Abstract: Purpose: This study explored demographic, social, behavioral, and health factors associated with current marijuana use (MU) among African American older adults who were residing in economically challenged areas of south Los Angeles. **Methods:** This community-based study recruited a consecutive sample of African American older adults (n = 340), age ≥ 55 years, residing in economically challenged areas of South Los Angeles. Interviews were conducted to collect data. Demographics (age and gender), socioeconomic status (educational attainment, income, and financial strain), marital status, living alone, health behaviors (alcohol drinking and cigarette smoking), health status (number of chronic medical conditions, body mass index, depression, and chronic pain), and current MU were collected. Logistic regression was used to analyze the data. **Results:** Thirty (9.1%) participants reported current MU. Age, educational attainment, chronic medical conditions, and obesity were negatively associated with current MU. Gender, income, financial strain, living alone, marital status, smoking cigarettes, drinking alcohol, depression, and pain did not correlate with MU. **Conclusion:** Current MU is more common in younger, healthier, less obese, less educated African American older adults. It does not seem that African American older adults use marijuana for the self-medication of chronic disease, pain, or depression. For African American older adults, MU also does not co-occur with cigarette smoking and alcohol drinking. These results may help clinicians who provide services for older African Americans in economically challenged urban areas.

Keywords: African American; black; older adult; marijuana use

---

1. Background

Very little epidemiological information exists on marijuana use (MU) among African American older adults, specifically those who live in economically challenged urban settings [1]. This information is essential for the design and implementation of cessation and treatment programs for African American older adults in such settings [2]. The patterns and predictors of MU in the African American community differ from those of other communities [3,4]. Economically challenged African American communities may have high availability of marijuana, marijuana initiation, and use, combined with poor access to cessation programs. Low access to MU cessation programs may operate as a vulnerability factor for this population, which increases the risk of substance use problems [5–10]. Telescoping effect is a phenomenon that describes the more rapid transition of marginalized African Americans from substance use to poor outcomes. Such a phenomenon explains why African American communities are at an increased risk of risky trajectories of substance use [5–10]. As a result, although African Americans show a lower prevalence of substance use, they are more likely to develop undesired substance use outcomes [5–10].

Demographic factors [11] such as age, period, cohort effects, and gender all influence MU [12–18]. Among adults, age increases the risk of lifetime substance use; however, current use is more common at younger ages and in more recent cohorts [12]. Gender is a main determinant of substance use—across studies, males are more likely to use tobacco, alcohol, and marijuana [19].

Low socioeconomic status (SES) reduces population health [20] and increases the risk of use of a wide range of substances such as tobacco [21,22], alcohol [23], and marijuana [24,25]. Low SES is also a major driver of racial and ethnic health disparities that explain the worse behavioral and health outcomes in the African American community, relative to Whites [26]. SES indicators, such as educational attainment, financial strain, and income, impact health [20] and health behaviors such as MU [24,25]. Some recent literature, however, suggests that the health effects of SES are smaller for African Americans than Whites [20]. For example, as a result of practices and preferences of the labor market which marginalizes highly educated African Americans, highly educated African Americans are less likely to secure high paying jobs than Whites. This phenomenon reduces the central role of education level on health and behaviors of African Americans [27]. As a result, education attainment may generate less health for African Americans than Whites [21,27], also known as minorities diminished returns (MDRs) [21,27,28]. As a result, education attainment may have smaller effects for African American individuals [21,27] than for Whites. While education attainment, income [22], and financial strain [29] shape health and health behaviors of populations, there is a need to test how these SES indicators impact the MU of African American older adults.

Use of tobacco and alcohol may be associated with MU [30]. This is in part because the use of various substances has shared risk factors [31]. One mechanism for the association between tobacco, alcohol, and marijuana use is that some of these substances may operate as a gateway to the other substances [3]. For example, individuals who currently use alcohol or tobacco are more likely to use marijuana in the future [30].

Health problems may also covary with MU. First, there is large body of research suggesting a negative association between obesity and MU [32]. This literature has proposed multiple mechanisms, including that drugs and food may be two non-overlapping reward pathways [32]. Thus, people may have great interest in either food or drugs, but not both, to cope with stress [33]. Second, people may also turn to MU to cope with pain [13] or depression [14]. In this case, we expect higher risk of MU in individuals who have high levels of depressive symptoms [14] or pain [13].

Most of the literature on MU among African Americans is on younger age groups [11,14]. As a result, we have limited knowledge about how demographics, socioeconomics, health behaviors, and health correlate with MU among African American older adults [15]. There is also a need to expand the existing literature on demographic, social, behavioral, and health determinants of MU in African American older adults who reside in economically challenged urban areas. This is particularly important given the transition in the patterns of MU following the legalization of marijuana. Similarly, medical marijuana may be used by individuals, particularly older adults, and this effect may depend on whether recreational or medical marijuana is legal or not [13].

Aims

The current study explored demographic, social, behavioral, and health determinants of current MU in economically challenged African American older adults. We hypothesized that MU is more common in people who are younger, fit, healthy, male, have low educational attainment, have high levels of financial strain, are those who smoke and drink, and have depression and pain.

2. Methods

2.1. Design and Setting

A survey was performed in economically challenged areas of south Los Angeles between 2015 and 2018 [16–18,34]. The survey included a structured face-to-face interview which collected data on demographic factors, SES, health behaviors, health status, and MU. Participants were living in the Service Planning Area 6 (SPA 6), Los Angeles County, California. SPA 6 is one of the most economically challenged urban areas, with 58% of adults having income levels less than 200% of the federal poverty line (FPL) and 36% of the population being uninsured [35,36]. 49% of older adults residing in SPA 6 are African Americans. Between 2013 and 2015, the percentage of homeless AA individuals in SPA 6 rose from 39% to 70%.

2.2. Participants and Sampling

A non-random sample of African American older adults was recruited for this study. Participants were sampled from predominantly African American housing units and senior centers that were located in south Los Angeles. Participants were 340 African Americans. Individuals were eligible if they were (a) African American/Black, (b) non-institutionalized, (c) aged 55 years or older, and (d) lived in south Los Angeles (LA). Exclusion criteria were (a) enrollment in skilled nursing facilities, (b) current enrollment in a clinical trial (because the intervention can interfere with MU and other health behaviors), and (c) severe cognitive deficit (not being able to consent and conduct the interview).

2.3. Institutional Review Board (IRB)

The Charles R. Drew University of Medicine and Science (CDU) IRB approved the study protocol. All respondents signed a written informed consent.

2.4. Study Measures

2.4.1. Independent Variables

Socio-economic status (SES): Three SES indicators were included. Educational attainment, financial strain, and income. Education attainment was conceptualized as years of schooling. This variable was treated as an interval variable. Financial strain was measured using three items borrowed from the Pearlin's list of financial difficulties that are commonly experienced by low SES people [37]. These items cover not having enough money for essential needs such as food, clothes, rent/mortgage, and utility bills. Responses were on a Likert scale ranging from 1 'never' to 5 'always'. We calculated a total score a score that reflected overall financial difficulties. (Cronbach's alpha = 0.923). Household monthly income was a continuous measure (in USD $1000).

Demographic Characteristics: Age (years) and gender (male, female) were measured. Age was a continuous variable. Gender was a dichotomous measure.

Living Arrangements: A dichotomous variable reflected participants' living arrangement. Participants' living arrangement was measured using a single item. The variables were 1 (living alone) and 0 (there are any other members accompanying them) [38].

Marital Status: Participants' family type (marital status) was measured using a single item self-report. This variable was treated as a dichotomous variable: married = 1, non-married = 0.

Obesity: Obesity was measured by measurement of weight and height. Height and weight were measured in inches and pounds, respectively. Then, height and weight were converted to meters and kilograms. Body mass index (BMI) was then defined as weight (kilograms) divided by height (meters) squared.

Number of Chronic Medical Conditions (CMC): Individuals were asked if they were ever told that they had the following chronic medical conditions: hypertension (HTN), heart disease, stroke, cancer, diabetes (MD), thyroid disease, chronic obstructive pulmonary disease (COPD), asthma, osteoarthritis, gastrointestinal (GI) disease, rheumatoid arthritis (RA), and lipid disorder. Self-reported CMCs were valid and reliable [39]. We calculated the total number of CMCs as reported by the individual.

Depression: This study measured depression using the 15-item Geriatric Depression Scale (Short Form) (GDS-SF) [40]. Results range from 1 to 15, with a higher score indicative of severe depressive symptoms. The GDS-SF has shown very good reliability and validity, and has been commonly used in community and clinical settings [40].

Pain: Pain intensity was measured by the McGill Pain Questionnaire (Short Form 2) (MPQ-SF-2) [41]. This scale has 22 pain items asking about the experience of various types of pain in the past week. Each item was on an 11-point rating scale ranging from 0 (none) to 10 (worst possible). A total pain score was calculated. A higher score reflected more intense chronic pain [41].

Tobacco Use: Participants were asked whether they smoke cigarettes. The exact question was: "How would you describe your cigarette smoking habits?" Response items included never smoked, previously smoked, and current smoker. We defined a dichotomous variable as current smoker versus other statuses.

Drinking Alcohol: Participants' alcohol use was asked using this question: "Do you drink alcohol?" The response items included *yes* and *no*. Drinking alcohol was a dichotomous variable.

2.4.2. Outcome Variables

Current Marijuana Use (MU): Two items were used to measure current use of marijuana: (1) *"Are you taking marijuana or any related products for pain?"*, and (2) *"In the past year have you been treated by Compassion provider for marijuana related products?"* [42].

2.5. Statistical Analysis

We used SPSS 23.0 (IBM, New York, NY, USA) for data analysis. To describe the sample, we reported frequencies (*n*) and relative frequencies (%) of the categorical variables. We calculated the average number of CMCs for the analysis. Means and standard deviations (SD) were reported for continuous measures. We used the non-parametric Spearman correlation test (zero order correlation) to estimate the bivariate correlations between the study variables. We applied logistic regression models with MU as the outcome (the dependent variable) and demographics, SES, health behaviors, and health as the predictors (independent variables). As almost all participants had some type of health care coverage, we did not include health insurance to our logistic regression models. We reported the odds ratio (OR), and their associated standard error (SE), 95% confidence intervals (95% CI), and *p* values from our logistic regression models.

3. Results

3.1. Descriptive Statistics

Table 1 describes the study variables. All participants were at least 55 years old. Participants had an average age of 69.6 (SD = 9.3) years old. From all our participants, 63.2% were female. From our participants, 9.1% ($n = 30$) reported MU.

Table 1. Descriptive statistics (n = 340).

Characteristics		
	Mean	**SD**
Age (years)	69.60	9.33
Educational attainment (years)	12.73	2.13
Financial strain	12.10	6.15
Monthly household income (USD $1000)	2.69	1.10
Chronic medical conditions (CMC)	4.35	1.83
Depressive symptoms	3.36	3.04
Chronic pain	2.51	2.44
	n	%
Gender		
Men	125	36.8
Women	215	63.2
Cigarette smoking (current)		
No	107	31.7
Yes	231	68.3
Alcohol drinking		
No	190	56.0
Yes	149	44.0
Obesity		
No	188	55.3
Yes	151	44.4
Current marijuana use		
No	309	90.9
Yes	31	9.1

3.2. *Bivariate Analysis*

Table 2 shows a summary of the bivariate correlations between the study variables, using a non-parametric correlation Spearman test. Age, education attainment, obesity, and number of CMCs were negatively correlated with current MU, however, other variables, such as financial strain, depression, and pain, were not associated with MU. Smoking cigarettes and drinking alcohol also did not correlate with current MU.

Table 2. Bivariate correlations.

	1	2	3	4	5	6	7	8	9	10	11	12
1 Gender (female)	1	0.12 *	0.10	−0.14 *	−0.09	0.13 *	−0.17 **	0.15 **	0.23 **	−0.05	0.06	−0.12 *
2 Age		1	−0.19 **	−0.29 **	0.04	0.43 **	−0.34 **	0.07	−0.07	−0.24 **	−0.22 **	−0.11 *
3 Educational attainment (years)			1	−0.09	0.22 **	−0.06	0.06	−0.10	0.01	−0.15 **	−0.03	−0.07
4 Financial strain				1	−0.15 **	−0.16 **	0.14 *	0.13 *	0.04	0.44 **	0.37 **	0.03
5 Monthly household income (USD $1000)					1	0.07	−0.05	−0.16 **	0.10	−0.13 *	−0.18 **	0.02
6 Smoking						1	−0.33 **	0.04	0.19 **	−0.16 **	−0.16 **	−0.07
7 Drinking							1	0.04	−0.05	0.05	0.17 **	0.07
8 Chronic medical conditions (CMCs)								1	0.12 *	0.25 **	0.42 **	−0.17 **
9 Obesity									1	0.09	0.09	−0.12 *
10 Depression										1	0.44 **	−0.04
11 Chronic pain											1	−0.02
12 Current marijuana use												1

*, $p < 0.05$; **, $p < 0.01$.

3.3. Multivariable Analysis

Table 3 shows the results of a logistic regression model with current MU as the outcome. According to this model, age, educational attainment, chronic medical conditions, and obesity were negatively associated with current MU. Gender, monthly household income, financial strain, living alone, marital status, smoking cigarettes, drinking alcohol, depression, and pain did not predict current MU (Table 3).

Table 3. Summary of multivariable logistic regression models with marijuana use (MU) as the outcomes. OR = odds ratio, CI = confidence intervals.

	OR	95% CI	p
Gender (female)	0.73	0.32–1.67	0.45
Age	0.94	0.89–1.00	0.04
Educational attainment (years)	0.81	0.68–0.96	0.02
Financial strain	1.01	0.94–1.08	0.78
Monthly household income (USD $1000)	1.10	0.77–1.58	0.60
Smoking	1.27	0.49–3.28	0.62
Drinking	1.22	0.51–2.90	0.65
Chronic medical conditions (CMC)	0.69	0.52–0.91	0.01
Obesity	0.42	0.17–1.04	0.05
Depression	0.92	0.77–1.09	0.32
Chronic pain	1.13	0.92–1.39	0.24
Constant	276.52		0.03

4. Discussion

Age, educational attainment, obesity, and CMCs were associated with current MU in our sample of African American older adults in economically challenged areas of south Los Angeles. While lower age was correlated with current MU, gender was not linked to the same behavior. Educational attainment, but not income or financial strain, was associated with MU. Neither cigarette nor alcohol use was associated with current MU in this population. However, current MU was associated with a lower risk of obesity, and individuals who reported MU were less likely to have CMCs. Our findings were in line with the research that shows a negative association between BMI and current MU [43], although not all studies have shown such negative associations [44].

Prevalence of MU was very low in this study, which is slightly higher than the national prevalence rate of 2.9% among older adults [15]. Currently, white women are identified as the most prevalent users of marijuana among older adults [45]. However, the results should also be interpreted in the full context of marijuana use in the whole age range of the African American population. MU use is higher among African American young adults compared to age-matched Whites, as they are more likely to use marijuana before tobacco [46]. However, this may not translate to older African Americans, who may be less knowledgeable about MU-associated health risks. In addition, MU is heavily stigmatized, and individuals may be reluctant to report usage, which may be a contributing factor to the reported low rate of MU among this population [47]. MU has been heavily criminalized for decades within the African American population, with high rates of marijuana-related arrests and negative interactions with the criminal justice system for use and possession [48]. Even though there is a national wave of legalizing medical and/or recreational marijuana in multiple states—including California—these African American older adults may still not want to report MU because of fear and stigma. The results should also be interpreted with the knowledge regarding the types of consumption methods of MU, which may include marijuana in cigarillos wraps, commonly known as "blunts", that are prevalent in stores in economically challenged areas and physically more harmful. This presents a striking difference, as Whites are more likely to use other methods of marijuana consumption, such as edibles, which are more costly and harder to access for this sample population [49]. Finally, we cannot rule out the likelihood of measurement bias, and should use multiple items to measure MU. Thus, there is a need for more research on this topic.

We did not find any linkage between MU and depression or pain, suggesting that African American older adults in economically challenged areas of Los Angeles do not use marijuana to self-medicate their depression and pain. Although this is a plausible explanation for our findings, other explanations should also be considered. Among African Americans, early-starting MU is linked to depressive symptoms and/or depressive disorders with underlying adverse child experiences [50,51]. Yet, this relationship has been primarily explored among adolescent and young African American adults. As there was a low rate of depression and pain among African American older adults in this sample, use of marijuana for depression and pain cannot be accurately assessed in this study. More research is needed on the motivations behind MU in African American older adults.

The profile of the typical marijuana user is younger, less educated, healthier (less CMCs), and more physically fit (less obese). MU among African American older adults does not co-occur with smoking cigarettes and drinking alcohol, which differs from the typical pattern of an older adult marijuana user [2]. Instead, MU may be a sporadic or inconsistent behavior, rather than consistent with other health risk behaviors, particularly substance use. Still, understanding the profile of marijuana users in African American older adults may help the delivery of health services and increase health education [52].

While educational attainment was associated with MU, we did not find protective effects of income and low financial strain on MU. The lack of association between MU and other SES indicators could be due to MDRs; hence, income may not be particularly protective against risk behaviors such as substance use [21,27,28].

More efforts should be geared toward the health education of African American older adults, who may not be aware of the health implications of marijuana. There is an increase in the number of policies, both at the local and state levels, addressing medical and recreational MU. There is a need for health systems, including providers and various community organizations, to provide knowledge about marijuana to the African American community. Our findings showed that adults likely to use marijuana are younger and have fewer CMCs. Yet, chronic MU may lead to future health issues as individuals age. Further research should focus on the health profile of African American older adults with MU.

Limitations

The study is not without methodological and conceptual limitations. First, its cross-sectional design limits any causal inferences. Second, the non-random sample limits the generalizability of the results to the broader African American community. Third, the study used a simplistic measure of current MU. As a result, this study was unable to assess either past or lifetime use of marijuana. Frequency of MU, types of MU consumption, and access and availability of MU were also not obtained. In addition, we cannot rule out the possibility of measurement bias, as we relied on self-reported marijuana use. Still, because of the stigma and sensitivity of marijuana, especially within this population, self-reported MU may be the best method to obtain information. In addition, data collection occurred prior to California legalizing recreational marijuana. Therefore, marijuana users within this sample may have not reported use because it may have been potentially illegal. These are important factors to consider when examining the relatively low prevalence of marijuana use in this population. Another limitation of the study is the lack of epidemic data from younger African Americans for comparison. If such data were available, the analysis and arguments could be significantly improved. Yet, there is a need to consider generational patterns when examining the two groups, as attitudes toward MU have become more positive among younger populations, which may be linked to higher prevalence among younger groups. Fourth, the study only included older African Americans who were residing in low income inner cities. Even though this decreased generalizability, few studies have focused on substance use among African American older adults residing in underserved areas, especially with MU. This is an important study to assess the prevalence and significant factors to start to understand the profile of African American older adults with MU. Fifth, our measure of CMCs was not exclusive. Other conditions, such as respiratory diseases, neurological diseases, and other diseases impacted by chronic MU could be included in future studies. The findings thus may differ for any other groups, even African Americans who have higher SES or those who are biracial or multi-racial.

The results reported here should be regarded as preliminary. More research is needed with more detailed information on MU. Despite these limitations, this study contributes to our knowledge of MU use in a population (older African American adults) about which little is currently known.

5. Conclusions

Educational attainment seems to protect African American older adults against current MU. At the same time, current MU is more common in younger, healthier, and more physically fit older African American adults. For African American older adults in poor urban areas, current MU may not co-occur with the use of tobacco and alcohol. Further studies should focus on assessing various aspects of MU, such as lifetime use, accessibility, and consumption of marijuana-related products, such edibles, among these adults. We did not find strong evidence suggesting that African American older adults turn to MU for self-medication of pain and depression.

Author Contributions: M.B.: Conceptualization of the study, study design, funding acquisition, overseeing the study, data analysis, and revision of the paper. S.A. and S.C. conceptualization, data analysis, and contribution to the first draft and revision. Other authors including H.E.d.P., K.D. and J.S., contributed to the first draft and revision. K.D. also conducted an extensive literature review on the topic. J.S. also conducted the study and gathered the data. All authors approved the final draft.

Funding: This study was supported by the Center for Medicare and Medicaid Services (CMS), grant1H0CMS331621 to Charles R. Drew University of Medicine and Science (PI: Bazargan). Additionally, Bazargan is supported by the NIH under Awards "54MD008149", R25 MD007610, 2U54MD007598 (PI: Vadgama), and U54 TR001627 (PIs: Dubinett, and Jenders). Assari is also supported by the National Cancer Institute (NCI), grant CA201415-02 (Co-PI = Mistry).

Conflicts of Interest: The authors declare no conflict of interest.

References

1. Lloyd, S.L.; Striley, C.W. Marijuana Use Among Adults 50 Years or Older in the 21st Century. *Gerontol. Geriatr. Med.* **2018**, *4*, 2333721418781668. [CrossRef] [PubMed]
2. Dinitto, D.M.; Choi, N.G. Marijuana use among older adults in the U.S.A.: User characteristics, patterns of use, and implications for intervention. *Int. Psychogeriatr.* **2011**, *23*, 732–741. [CrossRef] [PubMed]
3. Chung, T.; Kim, K.H.; Hipwell, A.E.; Stepp, S.D. White and Black adolescent females differ in profiles and longitudinal patterns of alcohol, cigarette, and marijuana use. *Psychol. Addict. Behav.* **2013**, *27*, 1110–1121. [CrossRef] [PubMed]
4. Smiley-McDonald, H.M.; Moore, K.N.; Heller, D.C.; Ropero-Miller, J.D.; McIntire, G.L.; Wallace, F.N. Patterns of Marijuana Use in a 6-Month Pain Management Sample in the United States. *Subst. Abus. Res. Treat.* **2017**, *11*, 1178221817724783. [CrossRef] [PubMed]
5. Assari, S. Separate and Combined Effects of Anxiety, Depression and Problem Drinking on Subjective Health among Black, Hispanic and Non-Hispanic White Men. *Int. J. Prev. Med.* **2014**, *5*, 269–279. [PubMed]
6. Avalos, L.A.; Mulia, N. Formal and informal substance use treatment utilization and alcohol abstinence over seven years: Is the relationship different for blacks and whites? *Drug Alcohol Depend.* **2012**, *121*, 73–80. [CrossRef] [PubMed]
7. Caetano, R.; Baruah, J.; Ramisetty-Mikler, S.; Ebama, M.S. Sociodemographic Predictors of Pattern and Volume of Alcohol Consumption across Hispanics, Blacks, and Whites: 10-year trend (1992–2002). *Alcohol. Clin. Exp. Res.* **2010**, *34*, 1782–1792. [CrossRef]
8. Mulia, N.; Ye, Y.; Greenfield, T.K.; Zemore, S.E. Disparities in Alcohol-related Problems among White, Black and Hispanic Americans. *Alcohol. Clin. Exp. Res.* **2009**, *33*, 654–662. [CrossRef]
9. Mulia, N.; Ye, Y.; Zemore, S.E.; Greenfield, T.K. Social Disadvantage, Stress, and Alcohol Use Among Black, Hispanic, and White Americans: Findings From the 2005 U.S. National Alcohol Survey. *J. Stud. Alcohol Drugs* **2008**, *69*, 824–833. [CrossRef]

10. Nyaronga, D.; Greenfield, T.K.; McDaniel, P.A. Drinking context and drinking problems among black, white, and Hispanic men and women in the 1984, 1995, and 2005 U.S. National Alcohol Surveys. *J. Stud. Alcohol Drugs* **2009**, *70*, 16–26. [CrossRef]
11. Assari, S.; Mistry, R.; Lee, D.B.; Caldwell, C.H.; Zimmerman, M.A. Perceived Racial Discrimination and Marijuana Use a Decade Later; Gender Differences Among Black Youth. *Front. Pediatr.* **2019**, *7*, 78. [CrossRef] [PubMed]
12. Kerr, W.C.; Greenfield, T.K.; Bond, J.; Ye, Y.; Rehm, J. Age-period-cohort influences on trends in past year marijuana use in the US from the 1984, 1990, 1995 and 2000 National Alcohol Surveys. *Drug Alcohol Depend.* **2007**, *86*, 132–138. [CrossRef] [PubMed]
13. Caulley, L.; Caplan, B.; Ross, E. Medical Marijuana for Chronic Pain. *New Engl. J. Med.* **2018**, *379*, 1575–1577. [CrossRef] [PubMed]
14. Assari, S.; Mistry, R.; Caldwell, C.H.; Zimmerman, M.A. Marijuana Use and Depressive Symptoms; Gender Differences in African American Adolescents. *Front. Psychol.* **2018**, *9*, 2135. [CrossRef] [PubMed]
15. Han, B.H.; Palamar, J.J. Marijuana use by middle-aged and older adults in the United States, 2015–2016. *Drug Alcohol Depend.* **2018**, *191*, 374–381. [CrossRef] [PubMed]
16. Bazargan, M.; Smith, J.; Movassaghi, M.; Martins, D.; Yazdanshenas, H.; Mortazavi, S.S.; Orum, G. Polypharmacy among Underserved Older African American Adults. *J. Aging Res.* **2017**, *2017*, 1–8. [CrossRef] [PubMed]
17. Bazargan, M.; Smith, J.; Yazdanshenas, H.; Movassaghi, M.; Martins, D.; Orum, G. Non-adherence to medication regimens among older African-American adults. *BMC Geriatr.* **2017**, *17*, 163. [CrossRef] [PubMed]
18. Bazargan, M.; Smith, J.L.; King, E.O. Potentially inappropriate medication use among hypertensive older African-American adults. *BMC Geriatr.* **2018**, *18*, 238. [CrossRef] [PubMed]
19. Zilberman, M.; Tavares, H.; El-Guebaly, N. Gender similarities and differences: The prevalence and course of alcohol- and other substance-related disorders. *J. Addict. Dis.* **2003**, *22*, 61–74. [CrossRef]
20. Link, B.G.; Phelan, J. The social shaping of health and smoking. *Drug Alcohol Depend.* **2009**, *104*, 6–10. [CrossRef]
21. Assari, S.; Mistry, R. Educational Attainment and Smoking Status in a National Sample of American Adults; Evidence for the Blacks' Diminished Return. *Int. J. Environ. Res. Public Heal.* **2018**, *15*, 763. [CrossRef] [PubMed]
22. Kubzansky, L.D.; Berkman, L.F.; Glass, T.A.; Seeman, T.E. Is Educational Attainment Associated with Shared Determinants of Health in the Elderly? Findings from the MacArthur Studies of Successful Aging. *Psychosom. Med.* **1998**, *60*, 578–585. [CrossRef] [PubMed]
23. Assari, S.; Farokhnia, M.; Mistry, R. Education Attainment and Alcohol Binge Drinking: Diminished Returns of Hispanics in Los Angeles. *Behav. Sci.* **2019**, *9*, 9. [CrossRef] [PubMed]
24. Miech, R.; Chilcoat, H. The Formation of a Socioeconomic Disparity: A Case Study of Cocaine and Marijuana Use in the 1990s. *Am. J. Prev. Med.* **2007**, *32*, S171–S176. [CrossRef]
25. Williams, C.T.; Juon, H.-S.; Ensminger, M.E. Marijuana and cocaine use among female African-American welfare recipients. *Drug Alcohol Depend.* **2004**, *75*, 185–191. [CrossRef] [PubMed]
26. Levin, S.; Mayer-Davis, E.J.; Ainsworth, B.E.; Addy, C.L.; Wheeler, F.C. Racial/Ethnic Health Disparities in South Carolina and the Role of Rural Locality and Educational Attainment. *South. Med J.* **2001**, *94*, 711–718. [CrossRef]
27. Assari, S.; Lankarani, M.M. Workplace Racial Composition Explains High Perceived Discrimination of High Socioeconomic Status African American Men. *Brain Sci.* **2018**, *8*, 139. [CrossRef]
28. Assari, S.; Caldwell, C.H.; Zimmerman, M.A. Family Structure and Subsequent Anxiety Symptoms; Minorities' Diminished Return. *Brain Sci.* **2018**, *8*, 97. [CrossRef]
29. Andrade, F.; Kramer, K.; Monk, J.; Greenlee, A.; Mendenhall, R. Financial stress and depressive symptoms: The impact of an intervention of the Chicago Earned Income Tax Periodic Payment. *Public Heal.* **2017**, *153*, 99–102. [CrossRef]
30. Roche, D.; Bujarski, S.; Green, R.; Hartwell, E.; Leventhal, A.; Ray, L. Alcohol, tobacco, and marijuana consumption is associated with increased odds of same-day substance co- and tri-use. *Drug Alcohol Depend.* **2019**, *200*, 40–49. [CrossRef]

31. Oser, C.B.; Harp, K.; Pullen, E.; Bunting, A.M.; Stevens-Watkins, D.; Staton, M. African–American Women's Tobacco and Marijuana Use: The Effects of Social Context and Substance Use Perceptions. *Subst. Use Misuse* **2019**, *54*, 873–884. [CrossRef] [PubMed]
32. Dileone, R.J.; Taylor, J.R.; Picciotto, M.R. The drive to eat: comparisons and distinctions between mechanisms of food reward and drug addiction. *Nat. Neurosci.* **2012**, *15*, 1330–1335. [CrossRef] [PubMed]
33. Blum, K.; Liu, Y.; Shriner, R.; Gold, M.S. Reward Circuitry Dopaminergic Activation Regulates Food and Drug Craving Behavior. *Curr. Pharm. Des.* **2011**, *17*, 1158–1167. [CrossRef] [PubMed]
34. Bazargan, M.; Yazdanshenas, H.; Gordon, D.; Orum, G. Pain in Community-Dwelling Elderly African Americans. *J. Aging Health* **2016**, *28*, 403–425. [CrossRef]
35. Key Indicators of Health by Service Planning Area. Available online: https://www.google.com/url?sa=t&rct=j&q=&esrc=s&source=web&cd=1&ved=2ahUKEwjmgtTY37bjAhWVA4gKHQFUCFQQFjAAegQIABAC&url=http%3A%2F%2Fpublichealth.lacounty.gov%2Fha%2Fdocs%2F2015LACHS%2FKeyIndicator%2FPH-KIH_2017-sec%2520UPDATED.pdf&usg=AOvVaw3BBiFUkxHMxKgw3srfrCe8 (accessed on 5 May 2019).
36. Community Health Assessment 2015. Available online: https://www.google.com/url?sa=t&rct=j&q=&esrc=s&source=web&cd=9&ved=2ahUKEwjOoLPH4LbjAhWGfXAKHSqbD6MQFjAIegQIBBAC&url=https%3A%2F%2Fwww.thinkhealthla.org%2Fcontent%2Fsites%2Flosangeles%2FCommunityHealthAssesmentJune2015Revised_Logo_121916.pdf&usg=AOvVaw15vlzXCsSr9Iu_LCesXgJE (accessed on 5 May 2019).
37. Laaksonen, E.; Lallukka, T.; Lahelma, E.; Ferrie, J.E.; Rahkonen, O.; Head, J.; Marmot, M.G.; Martikainen, P. Economic difficulties and physical functioning in Finnish and British employees: contribution of social and behavioural factors. *Eur. J. Public Health* **2011**, *21*, 456–462. [CrossRef] [PubMed]
38. Theeke, L.A. Sociodemographic and Health-Related Risks for Loneliness and Outcome Differences by Loneliness Status in a Sample of U.S. Older Adults. *Res. Gerontol. Nurs.* **2010**, *3*, 113–125. [CrossRef] [PubMed]
39. Fowles, J.B.; Fowler, E.J.; Craft, C. Validation of Claims Diagnoses and Self-Reported Conditions Compared with Medical Records for Selected Chronic Diseases. *J. Ambul. Care Manag.* **1998**, *21*, 24–34. [CrossRef]
40. Bass, D.S.; Attix, D.K.; Phillips-Bute, B.; Monk, T.G. An Efficient Screening Tool for Preoperative Depression: The Geriatric Depression Scale-Short Form. *Anesthesia Analg.* **2008**, *106*, 805–809. [CrossRef] [PubMed]
41. Melzack, R. The short-form McGill pain questionnaire. *Pain* **1987**, *30*, 191–197. [CrossRef]
42. Shelton, K.; Lifford, K.; Fowler, T.; Rice, F.; Neale, M.; Harold, G.; Thapar, A.; van den Bree, M. The association between conduct problems and the initiation and progression of marijuana use during adolescence: A genetic analysis across time. *Behav. Genet.* **2007**, *37*, 314–325. [CrossRef] [PubMed]
43. Pickering, R.P.; Grant, B.F.; Chou, S.P.; Compton, W.M. Are overweight, obesity, and extreme obesity associated with psychopathology? Results from the national epidemiologic survey on alcohol and related conditions. *J. Clin. Psychiatry* **2007**, *68*, 998–1009. [CrossRef] [PubMed]
44. Zimlichman, E.; Kochba, I.; Mimouni, F.B.; Shochat, T.; Grotto, I.; Kreiss, Y.; Mandel, D. Smoking habits and obesity in young adults. *Addiction* **2005**, *100*, 1021–1025. [CrossRef] [PubMed]
45. Reynolds, I.R.; Fixen, D.R.; Parnes, B.L.; Lum, H.D.; Shanbhag, P.; Church, S.; Linnebur, S.A.; Orosz, G. Characteristics and Patterns of Marijuana Use in Community-Dwelling Older Adults. *J. Am. Geriatr. Soc.* **2018**, *66*, 2167–2171. [CrossRef] [PubMed]
46. Kennedy, S.M.; Patel, R.P.; Cheh, P. Hsia, J.; Rolle, I.V. Tobacco and Marijuana Initiation Among African American and White Young Adults. *Nicotine Tobacco Res.* **2016**, *18*, 57–64. [CrossRef] [PubMed]
47. Satterlund, T.D.; Lee, J.P.; Moore, R.S. Stigma among California's Medical Marijuana Patients. *J. Psychoact. Drugs* **2015**, *47*, 10–17. [CrossRef] [PubMed]
48. Firth, C.L.; Maher, J.E.; Dilley, J.A.; Darnell, A.; Lovrich, N.P. Did marijuana legalization in Washington State reduce racial disparities in adult marijuana arrests? *Subst. Use Misuse* **2019**, *54*, 1582–1587. [CrossRef]
49. Kepple, N.J.; Freisthler, B. Place over traits? Purchasing edibles from medical marijuana dispensaries in Los Angeles, CA. *Addict. Behav.* **2017**, *73*, 1–3. [CrossRef]
50. Kogan, S.M.; Cho, J.; Oshri, A.; MacKillop, J. The Influence of Substance Use on Depressive Symptoms among Young Adult Black Men: The Sensitizing Effect of Early Adversity. *Am. J. Addict.* **2017**, *26*, 400–406. [CrossRef]

51. Pahl, K.; Brook, J.S.; Koppel, J. Trajectories of marijuana use and psychological adjustment among urban African American and Puerto Rican women. *Psychol. Med.* **2011**, *41*, 1775–1783. [CrossRef]
52. Webb, F.J.; Striley, C.W.; Cottler, L.B. Marijuana Use and Its Association with Participation, Navigation and Enrollment in Health Research among African Americans. *J. Ethn. Subst. Abus.* **2015**, *14*, 325–339. [CrossRef]

Article

Migraine Frequency Decrease Following Prolonged Medical Cannabis Treatment: A Cross-Sectional Study

Joshua Aviram [1], Yelena Vysotski [1], Paula Berman [1], Gil M. Lewitus [1], Elon Eisenberg [2,3] and David Meiri [1,*]

1 Faculty of Biology, Technion-Israel Institute of Technology, Haifa 3200003, Israel; shukiaviram@gmail.com (J.A.); lalatrololo@gmail.com (Y.V.); bermansh@gmail.com (P.B.); lgil@technion.ac.il (G.M.L.)

2 Institute of Pain Medicine, Rambam Health Care Campus, Haifa 3109601, Israel; e_eisenberg@rambam.health.gov.il

3 Rappaport Faculty of Medicine-Technion, Israel Institute of Technology, Haifa 3525433, Israel

* Correspondence: dmeiri@technion.ac.il; Tel.: +972-77-8871680 or +972-525330031

Received: 13 May 2020; Accepted: 5 June 2020; Published: 9 June 2020

Abstract: Background: Medical cannabis (MC) treatment for migraine is practically emerging, although sufficient clinical data are not available for this indication. This cross-sectional questionnaire-based study aimed to investigate the associations between phytocannabinoid treatment and migraine frequency. Methods: Participants were migraine patients licensed for MC treatment. Data included self-reported questionnaires and MC treatment features. Patients were retrospectively classified as responders vs. non-responders (≥50% vs. <50% decrease in monthly migraine attacks frequency following MC treatment initiation, respectively). Comparative statistics evaluated differences between these two subgroups. Results: A total of 145 patients (97 females, 67%) with a median MC treatment duration of three years were analyzed. Compared to non-responders, responders (n = 89, 61%) reported lower current migraine disability and lower negative impact, and lower rates of opioid and triptan consumption. Subgroup analysis demonstrated that responders consumed higher doses of the phytocannabinoid ms_373_15c and lower doses of the phytocannabinoid ms_331_18d (3.40 95% CI (1.10 to 12.00); $p < 0.01$ and 0.22 95% CI (0.05–0.72); $p < 0.05$, respectively). Conclusions: These findings indicate that MC results in long-term reduction of migraine frequency in >60% of treated patients and is associated with less disability and lower antimigraine medication intake. They also point to the MC composition, which may be potentially efficacious in migraine patients.

Keywords: cannabinoids; migraine: chronic pain; opioids; triptans; disability

1. Background

Chronic migraine constitutes a disabling neurological disorder, affecting around one to two percent of the global population worldwide [1]. Traditionally, abortive migraine treatments include triptans [2], non-steroidal anti-inflammatory drugs (NSAIDs) [3], paracetamol [4], ergots [5], opioids [6], and antiemetics [7]. Preventive treatments include antidepressants, anticonvulsants, beta-blockers, and more recently, anti-calcitonin gene-related peptide (CGRP) agents [8]. In recent years, the use of medical cannabis (MC) for the treatment of chronic pain in general has emerged, along with an increase in demand and use by migraine patients. A recent cross-sectional study found that nearly 36% of MC users reported using it to treat headache and migraine [9]. An additional survey reported about 50% reduction of migraine and headache severity following inhaled cannabis consumption [10]. Nevertheless, good clinical data supporting the beneficial effect of MC on migraine are scarce.

Both clinical and preclinical data suggest an abnormal endocannabinoid system function in migraine. In patients with chronic migraine, the cerebral spinal fluid (CSF) concentrations of anandamide (AEA) were significantly lower and the concentrations of palmitoylethanolamide (PEA) were significantly higher compared to non-migraine headache patients and controls [11]. Furthermore, reduced levels of AEA degrading enzymes were found in platelets of patients with chronic migraine [12]. In animal models of migraine, administration of AEA diminished hyperalgesic behavior [13], and the plant-derived (-)-Δ^9-*trans*-tetrahydrocannabinol (THC) showed anti-migraine effects in rats [14]. Whilst the available evidence suggests involvement of the endocannabinoid system and a potential for MC treatment to be therapeutic in migraine, more research is required to demonstrate the efficacy parameters of MC treatment for migraine. The complexity of the MC plant and how to design therapeutics from it must also be considered.

The single-compound, single-target approach in pharmaceutical science is a long-standing tradition embedded in our approach to clinical problem-solving. This is wholly different to MC treatment, which is often times multi-compound, whole-plant treatment. The cannabis plant contains hundreds of different active components, including phytocannabinoids, terpenes, and flavonoids [15]. While THC and cannabidiol (CBD) are among the most well-known phytocannabinoids, others are likely to have biological activity as well [16]. Hence, it is conceivable that various combinations of phytocannabinoids differ in their anti-migraine activity. This multi-compound effect of cannabis has been called the "entourage effect" [17], which suggests that studies examining the role of single-molecule cannabinoids in disease may not necessarily capture the synergy at play in multi-compound MC treatment. To add to the complexity of MC treatment with multiple compounds, there are hundreds of different cannabis cultivars, each with its own unique chemical composition [18]. Recently, we have developed an electrospray ionization mass spectrometry - liquid chromatography mass spectrometry (ESI-LC/MS) approach for comprehensive identification and quantification of phytocannabinoids in cannabis. We have identified over 90 phytocannabinoids, of which approximately 20 were previously unknown [19]. Quantifying the multitude of phytocannabinoids is the first step to better understanding the therapeutic potential of each cannabis cultivar, and therefore how to plan better clinical studies.

The regulations that govern cannabis use for medical purposes in Israel under the Israeli Ministry of Health (IMOH) allow only specific indications for which a patient can be issued with a MC license by their prescribing physician [20]. Whilst migraine is not an approved indication, it is sometimes comorbid with approved indications, such as gastrointestinal disease and chronic neuropathic non-cancer pain. In the case of chronic non-cancer pain, migraine is itself sometimes characterized as a chronic non-cancer pain condition, depending on its frequency and duration. In order to receive a license, the Medical Cannabis Unit (MCU) of the IMOH reviews MC license applications and provides the physician with either an approval or refusal, along with the justification for all declined applications.

Applications to the MCU include recommendations on MC routes of administration (oil extracts for sublingual use or inflorescence for inhalation and vaporization) and the starting monthly dose of 20 g (MCU approval is required for any increased dose). The physician will then recommend a specific MC cultivar or combinations of cultivars to their patients; however, the patient ultimately makes the final decision of which cultivar(s) to consume. In order for patients to determine which cultivar(s) best meets their therapeutic needs, they conduct a personal trial-and-error process. In addition, the guidelines for titration schedules, which are delivered as recommendations either by a nurse or by instructors from the companies licensed to cultivate cannabis, are not enforced. Titration scheduling covers doses per day, recommended starting dose, guidelines for increasing or decreasing the dose, and the maximum allowable dose. This means that the doses of phytocannabinoids consumed by the patient are not controlled.

The purpose of this cross-sectional study was to calculate the total dose of individual phytocannabinoids consumed by migraine patients and explore differences in dosages between subgroups of patients according to their changes in frequency of migraine attacks. Additionally, associations between changes in frequency of migraine attacks to migraine disability severity, sleep quality and timing, and migraine analgesics consumption were explored.

2. Materials and Methods

2.1. Subject

Patients were eligible to participate in this study if they were Hebrew speaking, aged ≥18 years with a standing MC license for the treatment of any approved condition, coupled with a diagnosis of comorbid migraine by their physician.

2.2. Study Procedure

Data were collected after the study was approved by the Institutional Ethics Committee of the Technion, Institute of Technology, Haifa, Israel (#011-2016). An existing database of Israeli patients with a MC license (n = 3218) was used to contact those patients who fulfilled the eligibility criteria for this cross-sectional study. Patients who had elected to disclose their email address for future studies and who also reported a diagnosis of migraine (n = 423, 13%) were invited to complete an online questionnaire after reading an explanation of the study. Prior to completing the questionnaire, the patients became participants after confirming their migraine diagnosis was received by a physician and after they signed an electronic consent form. Data were collected between August 2019 and February 2020. Participants were not offered financial compensation. While questionnaire data were being collected, the most prominent and most frequently administered cultivars from various approved cultivators in Israel were analyzed for phytocannabinoid content by ESI-LC/MS. Importantly, the chemical analyzes were performed on inflorescence cultivars, which were received from the cultivators only and not directly from the patients. Due to normal variation in chemical constituents of plant material and the expected variability between the cultivars analyzed in the lab compared to those consumed by patients, only phytocannabinoids that were consumed with minimum average concentrations of 0.1 g per month were analyzed. The individual phytocannabinoid monthly dose was calculated for each patient.

2.3. Study Questionnaires

Online questionnaire data were collected by secure survey technology Qualtrics® (Provo, Utah, version 12018) [21]. Questionnaires consisted of demographic information, including age, gender, MC treatment duration (years), and BMI. Data on migraine characteristics included the number of migraine days in the last month and the month prior to MC treatment initiation; age of migraine initiation; average current duration of a migraine attack (hours); and the presence of aura, nausea or vomiting, photo- or phonophobia, uni- or bilateral manifestation, and aggravation during physical activity of the migraine attack. Information on the analgesics and the specific abortive or preventive migraine medications was collected. Validated questionnaires included the migraine index disability scale (MIDAS) [22], the headache impact test (HIT-6) [23], and the Pittsburgh sleep quality index (PSQI) [24]. Additionally, MC treatment characteristics included administration route, cultivar name, cultivator brand, total monthly dose (grams), monthly dose of each specific cultivar (grams), and related adverse effects (AEs).

2.4. Phytocannabinoid Profiling of Cannabis Cultivars

Air-dried medical cannabis cultivars were obtained from several Israeli medical cannabis cultivators. Reagents, analytical standards, and general methodologies for phytocannabinoid extraction and analysis from cannabis were conducted according to our previously published methods [18,19].

Briefly, for phytocannabinoid extraction, 100 mg of ground cannabis inflorescences were accurately weighed and extracted with 1 mL ethanol. Samples were agitated in an orbital shaker at 25 °C for 15 min and then centrifuged at 4200 rpm. A fraction of the supernatant was collected and filtered through a 0.22 μm PTFE syringe filter and diluted at ratios of 1:9, 1:99, and 1:999 *v*/*v* cannabis extract to ethanol. Phytocannabinoid analyses were performed using a Thermo Scientific ultra-high-performance liquid chromatography (UHPLC) system coupled with a Q Exactive™ Focus Hybrid Quadrupole Orbitrap mass spectrometer (MS, Thermo Scientific, Bremen, Germany). The chromatographic conditions were as detailed in Baram et al. (2019) [18]. Identification and absolute quantification of phytocannabinoids was performed by external calibrations [19]. Compounds for which there were no analytical standards commercially available were semi-quantified [19]. For each phytocannabinoid, the concentrations of the acid and its neutral counterpart were summed and reported as the total content. For example, the concentration of total THC was calculated as Total THC = THCA × 0.877 + THC. Here, 0.877 is the molar ratio between the two compounds, which corrects for a change in the mass of (-)-Δ^9-*trans*-tetrahydrocannabinolic acid (THCA) as a result of decarboxylation. For compounds with no absolute identification, neutral or acid concentrations were utilized.

2.5. Statistical Analysis

R software (V.1.1.463) with tidyverse [25], pheatmap [26], and atable [27] packages were used to analyze differences in outcome measures by Pearson's chi-square test for categorical measures and Kruskal–Wallis rank sum test for numeric measures. For the effect size (i.e., odds ratio, OR) and confidence interval (CI), we utilized Cohen's d test. As is customary in recent migraine clinical trials [28], the primary outcome of this study was the clinically significant reduction in the monthly frequency of migraine attacks following the initiation of MC treatment (i.e., ≥50%; responders) compared to non-responders (i.e., <50%). Shapiro–Wilk test of normality demonstrated non-normal distribution for all measures; thus, data are presented as the median and lower and upper quartiles (Q1–Q3). Differences were considered significant at the $p < 0.05$ level. Incidences are presented as the number and percentage of patients.

3. Results

3.1. Subjects

We established a patient-reported outcomes database of Israeli patients with a preexisting MC license for various MCU-approved indications (n = 3218); the specific data in this database were previously described [29]. A total of 423 (13%) patients reported receiving a diagnosis of migraine in this database population. These patients' reasons for MC license approval was chronic neuropathic non-cancer pain (81%), cancer-related disorders (9%), post-traumatic stress disorders (7%), gastrointestinal disorders (2%), and neurological disorders (1%). A total of 231 (54% response rate) patients responded to participate in the current study.

A total of 145 patients reported on both the monthly frequency of migraine attacks before and after MC treatment initiation; these patients represent the sample that is analyzed and reported in this paper. The sample consisted of a majority of females (n = 97, 67%), with a median age of 45 (34–54). These patients were treated with MC for over a year (3 (2.4–4.6) years), with a range of MC treatment from one to 12 years (Table 1). Notably, no significant differences were found between responders and non-responders in the demographic and MC treatment measures.

3.2. Migraine and Sleep Features

We divided our sample into non-responders (i.e., <50%; n = 56, 39%) and responders (i.e., ≥50%; responders n = 89, 61%) based on their reduction of monthly frequency of migraine attacks from pre-MC to the current post-MC period. No significant difference was found in monthly migraine attack frequency prior to MC treatment initiation (15 (7.8–30) and 14 (8–27), respectively) (0.06 95% CI (−0.27 to 0.41); p = 0.71), strengthening the division methodology, as both subgroups started from a similar standpoint. Moreover, there were no significant differences between the subgroups in any of the current migraine features, including the age of migraine diagnosis, average duration of migraine attacks, activity-induced aggravation of migraine, unilateral migraine, bilateral migraine, presence of aura prior to migraine, nausea during migraine, or phono- or photophobia during migraine (Table 2).

We found that responders were more likely to report lower MIDAS (Figure 1A) and HIT-6 (Figure 1B) questionnaires scores (18 (5–40) and 64 (60–69), respectively) than non-responders (40 (26–80) and 68 (66–70), respectively) (0.50 95% CI (0.11 to 0.90); $p < 0.05$ and 0.66 95% CI (0.26 to 1.00); $p < 0.001$, respectively). Moreover, responders reported better sleep quality (9 (6–13)) than non-responders (11 (9–14)) (0.46 95% CI (0.03 to 0.89); $p < 0.05$)) (Figure 1C). Nonetheless, the evaluated sleep timing measures of sleep latency and sleep duration did not vary significantly between the migraine response subgroups (Table 3).

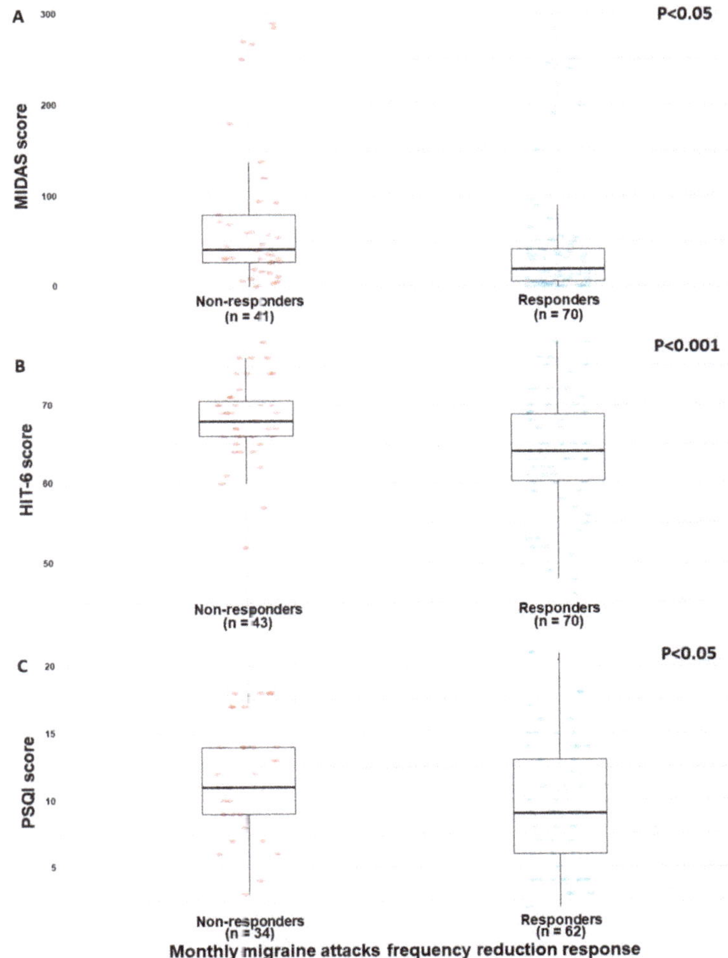

Figure 1. (**A–C**) Clinical differences between responders and non-responders. Note: MIDAS, migraine index disability scale; HIT-6, headache impact test; PSQI, Pittsburgh sleep quality index. Response refers to reduction in the monthly frequency of migraine attacks following the initiation of MC treatment (i.e., ≥50%) compared to non-responders (i.e., <50%).

Table 1. Demographic and medical cannabis (MC) treatment characteristics.

	Non-Responders N = 56	Responders N = 89	Total N = 145	Statistic (*p*)	Effect Size (CI)
Measure	**Number of patients (%)**				
Gender					
Female	35 (62)	62 (70)	97 (67)	0.51 (0.48)	0.73 (0.34–1.6)
Male	21 (38)	27 (30)	48 (33)		
Missing N	0	0	0		
	Median (IQR)				
Age (years)	46 (35–54)	44 (34–54)	45 (34–54)	0.08 (0.96)	−0.02 (−0.37–0.32)
Missing N	3	3	6		
BMI	25 (22–27)	25 (22–28)	25 (22–27)	0.13 (0.64)	0.05 (−0.28–0.39)
Missing N	0	1	1		
MC treatment duration (years)	3.5 (2.8–5.2)	3 (2–4)	3 (2.4–4.6)	0.22 (0.09)	0.46 (0.10–0.81)
Missing N	4	5	9		
Measure	**Number of patients (%)**				
MC administration route					
Inflorescence	40 (71)	72 (81)	112 (77)	2.4 (0.31)	0.13 (0–0.29)
Oil extract	7 (12)	5 (6)	12 (8)		
Combination #	7 (12)	12 (13)	19 (13)		
Missing N	2	0			
Inflorescence administration method *					
Pure MC cigarettes	22 (39)	33 (37)	55 (38)	0.04 (0.84)	1.10 (0.54–2.40)
MC cigarettes mixed with tobacco	17 (30)	30 (34)	47 (32)	0.01 (0.89)	0.89 (0.40–1.90)
Bhang	3 (5)	11 (12)	14 (10)	1.1 (0.29)	0.41 (0.07–1.70)
Electronic vaporizer	14 (25)	15 (17)	29 (20)	1.1 (0.29)	0.59 (0.24–1.50)
Manual vaporizer	5 (9)	20 (22)	25 (17)	3.3 (0.06)	2.9 (0.96–10.00)
Missing N	2	1			
Oil extract administration method *					
Sublingual	13 (23)	13 (15)	26 (18)	1.4 (0.24)	0.55 (0.21–1.40)
Swallowing	2 (4)	1 (1)	3 (2)	0.19 (0.67)	0.30 (0.005–5.90)
Missing N	2	1			

Combination refers to patients consuming MC inflorescence concomitantly with MC oil extract; * administration methods do not add up to 100% due to concomitant routes. Note: CI, confidence interval; IQR, interquartile range; BMI, body mass index; MC, medical cannabis.

Table 2. Migraine features.

	Non-Responders N = 56	Responders N = 89	Statistic (*p*)	Effect Size (CI)
Measure	**Median (IQR)**			
Age of migraine diagnosis (years)	20 (14–36)	22 (14–32)	0.07 (0.98)	0.07 (−0.27–0.42)
Missing N	1	4		
Average migraine duration (hours)	20 (5.8–35)	15 (5–48)	0.12 (0.72)	0.15 (−0.19–0.49)
Missing N	1	2		
	Number of patients (%)			
Activity induced aggravation of migraine	32 (57)	61 (69)	1.20 (0.28)	1.60 (0.73–3.3)
Missing N	1	0		
Unilateral migraine	40 (71)	59 (66)	0.39 (0.53)	0.74 (0.33–1.60)
Missing N	1	0		
Aura+	16 (29)	31 (35)	0.28 (0.60)	1.30 (0.60–2.9)
Missing N	1	0		
Nausea+	25 (45)	51 (57)	1.50 (0.23)	1.60 (0.78–3.40)
Missing N	1	0		
Phono/photo phobia+	38 (68)	60 (67)	0.00 (0.98)	0.93 (0.42–2.00)
Missing N	1	0		

Note: CI, confidence interval; IQR, interquartile range; +, positive for this manifestation.

Table 3. Sleep characteristics.

Measure	Non-Responders N = 56	Responders N = 89	Statistic (p)	Effect Size (CI)
	Median (IQR)			
Sleep quality global score (PSQI, 0–21)	11 (9–14)	9 (6–13)	0.30 (0.04)	0.46 (0.03–0.89)
Missing N	22	27		
Sleep latency (minutes)	32 (20–60)	30 (15–60)	0.09 (0.97)	−0.07 (−0.46–0.33)
Missing N	16	21		
Sleep duration (hours)	6.2 (5–7)	6 (5–7)	0.11 (0.92)	−0.09 (−0.49–0.30)
Missing N	16	20		
Subjective sleep quality *	3 (2–3)	2.5 (1–3)	0.18 (0.39)	0.42 (0.02–0.81)
Missing N	15	19		
Sleep latency *	2 (1.8–3)	2 (1–3)	0.15 (0.65)	0.2 (−0.20–0.59)
Missing N	16	21		
Sleep duration *	1 (0–2)	1 (0–2)	0.1 (0.95)	−0.02 (−0.41–0.37)
Missing N	16	20		
Habitual sleep efficiency *	1 (0–2)	0 (0–2)	0.09 (0.99)	0.08 (−0.32–0.49)
Missing N	18	22		
Sleep disturbances *	2 (2–2)	2 (1–2)	0.19 (0.33)	0.59 (0.19–0.98)
Missing N	15	19		
Use of sleeping medication *	1 (0–3)	0 (0–1.2)	0.19 (0.34)	0.35 (−0.05–0.75)
Missing N	17	21		
Daytime dysfunction *	2 (1–2)	1 (1–2)	0.18 (0.40)	0.34 (−0.06–0.74)
Missing N	17	23		

* Components of the PSQI questionnaire global score. Note: CI, confidence interval; IQR, interquartile range; PSQI, Pittsburgh sleep quality index.

3.3. MC Treatment Safety

MC-related adverse effects (AEs) were reported by 37% (n = 53) of the sample. Notably, non-responders reported higher incidences of any AEs (n = 26, 46%) than responders (n = 27, 30%) (0.46 95% CI (0.21 to 0.99), $p < 0.05$). Most of the specific AEs did not vary significantly between responders and non-responders. However, itchy and red eyes (n = 8, 9%, for both) were reported only in the responder subgroup ($\chi^2_{(1)} = 6.9$, $p < 0.01$ for both). Additionally, dry mouth was reported at higher rates among the responders (n = 9, 10%) than by non-responders (n = 2, 4%) ($\chi^2_{(1)} = 3.9$, $p < 0.05$).

In descending order of frequency, reported AEs included central nervous system AEs (n = 33, 23%), psychological AEs (n = 21, 14%), ophthalmic AEs (n = 16, 11%), gastrointestinal AEs (n = 15, 10%), musculoskeletal AEs (n = 11, 8%), cardiovascular AEs (n = 10, 7%), and auditory AEs (n = 9, 6%).

We further evaluated the associations between MC administration routes and AEs. There were no significant differences between patients reporting MC-related AEs and MC administration routes (i.e., inflorescence, oil extract, or a combination of these administration routes) (0.08 95% CI (0 to 0.25); p = 0.59). Additionally, no differences were observed between the different consumption methods (e.g., smoking, vaping, sublingual etc., $p > 0.05$).

3.4. MC Treatment Complexity

The complexity of MC treatment in Israel is due to the variety of available cultivars in Israel (about 100 different cultivars or "strains") and the options for patients to consume more than one cultivar in the same month, with varying doses of each cultivar. Consequently, the 68 patients in the current study reported consumption of 50 different MC cultivars combinations were reported in the current study by the 68 patients we had full cultivar lab information on. Notably, 46 (92%), 1 (2%) and 3 (6%) of the 50 possible combinations were compiled of cultivars that were THC-, CBD-dominant or contained equally high contents of THC:CBD, respectively. These 50 combinations comprised 38 unique cultivars. Figure 2 shows a z-score clustered heatmap of the main phytocannabinoids (presented as total concentrations in % w/w) in the 38 cultivars consumed by the sample subgroup. Based on the phytocannabinoid concentration variability, these cultivars were clustered to nine different groups. Figure 2 also shows that in the combinations of cultivars consumed, ten cultivars were consumed only by responders, eight cultivars were consumed only by non-responders, and the rest of the cultivars (n = 20) were consumed by both groups.

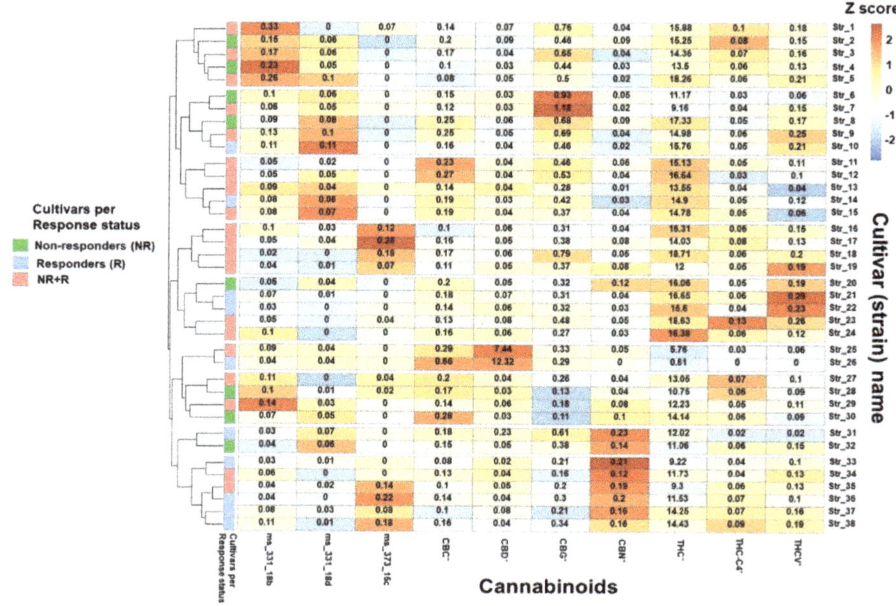

Figure 2. Relative phytocannabinoid concentrations in the most frequently consumed cultivars. Colors on the graph represent the scaled phytocannabinoid concentration variations between cultivars; the numbers in each box represent the concentration (% w/w) of the specific phytocannabinoid within each cultivar. Note: * for each phytocannabinoid, the concentrations of the acid and its neutral counterpart were summed and reported as the total content; Method used: package "pheatmap", function pheatmap, with the "Euclidean" (default) distance measure used in clustering rows, "complete" clustering method used on z-scored data scaled by row. Note: THC, (-)-Δ^9-*trans*-tetrahydrocannabinol; CBD, cannabidiol; CBC, cannabichromene; CBG, cannabigerol; CBN, cannabinol; THC-C4, (-)-Δ^9-*trans*-tetrahydrocannabinol-C4; THCV, (-)-Δ^9-*trans*-tetrahydrocannabivarin.

3.5. MC Treatment Characteristics

In this subgroup analysis we included data only from patients who smoked or vaped MC inflorescences and not those who consumed oil extracts sublingually, in order to avoid comparing between different routes of administration (different pharmacokinetics). Since the inflorescences in this study were analyzed in their natural form, monthly consumption of phytocannabinoid doses were calculated according to total phytocannabinoid concentrations rather than analyzing separate acid or neutral concentrations, in order to simulate the neutral maximum content of phytocannabinoids consumed following smoking or vaporization. This calculation corrects for any differences that may arise in phytocannabinoid profiles as a result of decarboxylation due to mishandling or storage of the MC inflorescences. Thus, the minority of patients that reported sublingual consumption of oil extract (n = 12) or combined these with inflorescences (n = 19) were not included in this subgroup analysis. Consequently, 68 (47%) patients reported exclusive MC inflorescence consumption via inhalation. Of these, 45 (66%) of them were responders and 23 (34%) were non-responders.

For the abovementioned 68 patients, we first evaluated the differences in total MC monthly dose between responders and non-responders. No significant differences were found (30 (20–40) g and 30 (20–45) g, respectively) (0.25 95% CI (−0.26 to 0.76); $p = 0.97$) (Figure 3A). Therefore, we evaluated the impact of the monthly doses of specific phytocannabinoids. As the distribution of monthly doses of specific phytocannabinoids were non-normal, we separated specific phytocannabinoids into low and high monthly dose groups, based on the distribution of consumption in our patient sample.

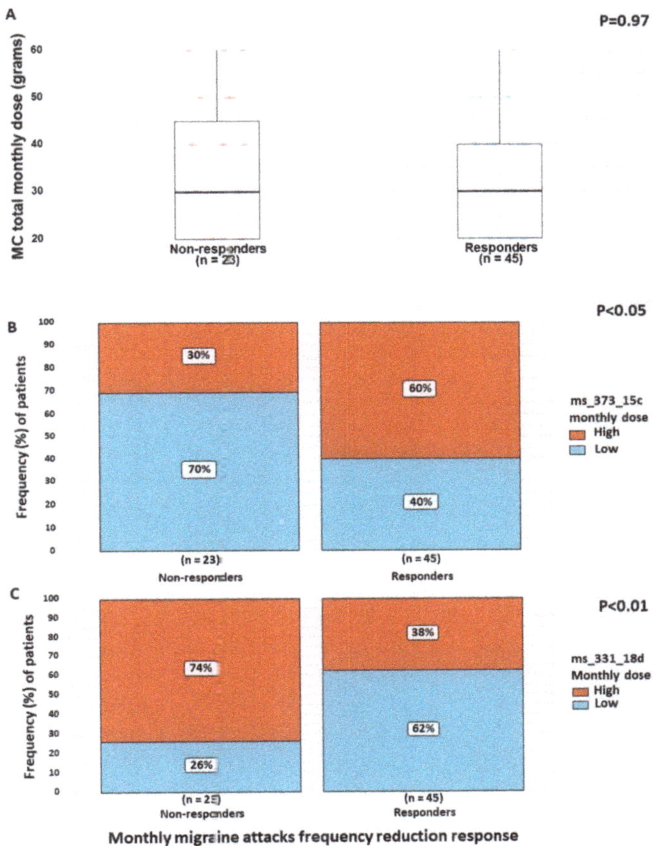

Figure 3. (**A–C**) Phytocannabinoid dose differences between responders and non-responders. Note: MC, medical cannabis. Response refers to reduction in the monthly migraine attacks frequency following the initiation of MC treatment (i.e., ≥50%) compared to non-responders (i.e., <50%).

We found that responders were more likely to consume high doses (7.9–109.5 mg per month) of the phytocannabinoid ms_373_15c ($n = 27$, 60%) and low doses (0–9.9 mg per month) of the phytocannabinoid ms_331_18d ($n = 28$, 62%) compared to non-responders, who were more likely to consume low doses (0–7.8 mg per month) of ms_373_15c ($n = 16$, 70%) and high doses (10.0–46.8 mg per month) of ms_331_18d ($n = 17$, 74%) (3.40 95% CI (1.10 to 12.00); $p < 0.05$ and 0.22 95% CI (0.05 to 0.72); $p < 0.01$, respectively) (Figure 3B,C). The other phytocannabinoids monthly doses did not vary significantly between the subgroups. Importantly, no differences were found between responders and non-responders in the daily frequency of MC consumption (5 (2.5–7) times per day and 4.5 (3–6) times per day, respectively) (0.18 95% CI (−0.34 to 0.71), $p = 0.99$). Additionally, no differences were found in the number of monthly cannabis cultivars combinations (2 (1–2) cultivars, respectively) (0.04 95% CI (−0.47 to 0.56), $p = 0.99$). Interestingly, among the 38 unique cultivars that patients consumed in their combinations, 12 contained considerable amounts of ms_373_15c and none or very low amounts of ms_331_18d. These cultivars appeared more frequently among the responders (42 appearances in cultivar combinations) than the non-responders (14 appearances in cultivar combinations).

3.6. Migraine Treatment Characteristics

A total of 65 (45%) of the patients reported any current consumption of pharmaceutical analgesic medications. Although not significant (0.51 95% CI (0.23 to 1.10), $p = 0.09$), more of the non-responders ($n = 30$, 54%) reported consumption of analgesics compared to the responders ($n = 35$, 39%). Nonetheless, there was a significant difference in the type of analgesic intake between the two groups. Non-responders consumed significantly higher rates of weak opioids ($n = 13$, 23%; e.g., tramadol hydrochloride, buprenorphine hydrochloride, etc.), strong opioids ($n = 14$, 25%; e.g., oxycodone hydrochloride, fentanyl, etc.), and triptans ($n = 9$, 16%; e.g., sumatriptan, rizatriptan, etc.) compared to responders ($n = 4$, 5%; $n = 7$, 8% and $n = 4$, 5%, respectively) (0.15 95% CI (0.03 to 0.53); $p < 0.005$, 0.25 95% CI (0.07 to 0.72); $p < 0.005$ and 0.24 95% CI (0.05 to 0.93), $p < 0.05$). No statistically significant variations were found between responders and non-responders in the consumption rates of over-the-counter analgesics, NSAIDs, anticonvulsants, antidepressants, and antiemetics.

4. Discussion

In this cross-sectional study, we evaluated patient reports on the frequency of their monthly migraine attacks, both pre- and post-MC treatment. Patients were classified as responders if they reported greater than 50% reduction in monthly migraine attacks post-MC treatment. As expected, responders reported lower current migraine disability and lower negative impact compared to non-responders.

A recent retrospective study conducted by Rhyne et al. (2016) showed that migraine patients who inhaled MC had a significant reduction in migraine frequency [30], which is in line with the results demonstrated here, and supports our finding of high rates of patient reporting of migraine frequency reduction. Migraine is classified as a pain condition. Mechanistically, endocannabinoids have been shown to have an inhibitory effect on serotonin receptors in vivo [31], which is shown to modulate pain and emetic responses. Additional in vivo data showed that THC induced an antinociception effect on the periaqueductal gray matter [32], which is believed to be involved in migraine pathophysiology [33]. Moreover, relatively low levels of the endocannabinoid anandamide (AEA) in the cerebral spinal fluid (CSF) were found to be associated with the mechanism of migraine initiation [11]. A reduction in pain in in vivo models following endocannabinoid [31] and cannabinoid [32] treatments supports our finding regarding a reduction of migraine disability in the responders group. Nonetheless, these studies still do not incorporate all the complexities of whole-plant cannabis treatment.

In this study, responders reported better migraine disability status, less negative headache impact, and better sleep quality. Whilst this result is logical, conflicting results were reported in another cross-sectional study, which demonstrated an association between improved headache disability and migraine intensity, but found no such association with headache frequency [34]. Taken together, our findings suggest that improved migraine disability status and negative impact among MC treatment responders might be attributed directly to MC treatment effects, rather than being secondary to the reduction of the frequency of migraine attacks. Here, we also reported an association between patients with poor sleep quality and less responsiveness to MC treatment in reducing the frequency of migraine attacks. A previous cross-sectional study demonstrated similar results, showing that even without MC treatment, an association was found between poor sleep quality and higher migraine attack frequency [35]. Thus, it might be suggested that migraineurs that responded to MC treatment and demonstrated a decrease in their monthly migraine frequency also had a concurrent sleep quality improvement. However, due to the current study design, we are unable to conclude whether the reported improved sleep quality can be attributed to the decrease in monthly migraine attack frequency or directly due to MC treatment effects.

There is increasing evidence that MC treatment has opioid-sparing effects [36–40]. Here, we found that responders to MC treatment also reported lower rates of consumption of opioids and triptans compared to non-responders. Both opioids and triptans are usually prescribed for migraine pain relief and not for prevention [6,41]. Thus, although we do not have information regarding the use of these medications prior to MC treatment initiation, this might be an indication that patients that responded clinically to MC treatment substituted this conventional treatment for MC.

In this study, we evaluated the differences in relative monthly dose of phytocannabinoids in each cultivar consumed, in both the responders and non-responders groups. To the best of our knowledge, this is the first study to assess the dose consumption of a wide variety of specific phytocannabinoids administered in combinations of cultivars. By doing so, we were able to elucidate associations between specific cannabinoids consumed over a monthly dose and the clinical response of migraine frequency reduction following MC treatment initiation. The most novel finding of this study was the identification that higher rates of patients that reported significant migraine frequency reduction following MC treatment also consumed higher monthly doses of ms_373_15c and lower monthly doses of ms_331_18d. Our group has previously identified these compounds in both THC- and CBD-dominant chemovars according to LC/MS/MS [18,19], however their absolute structure still needs to be elucidated. According to their MS/MS fragmentation spectra, ms_373_15c and ms_331_18d are acidic and neutral phytocannabinoids, respectively. Additionally, we identified specific cultivars that contain this favorable ratio between those compounds. However, it is important to note that we cannot attribute the anti-migraine effect of MC specifically to these phytocannabinoids, as we are yet to understand whether they are biological active. Nevertheless, we suggest using the presence of these phytocannabinoids to help in choosing specific MC chemovars for further research. Unfortunately, due to the relatively small sample size of patients in this study and a large number of cultivars with variable chemical constituents, translating these findings to the clinical setting will require a larger sample size and a more comprehensive approach. However, the work presented here could be the foundation of such a study to include these "lesser known" phytocannabinoid compounds. Currently, there are no clinical trials on migraine and MC [42]. Previous studies on migraine did not assess the phytocannabinoids mentioned in our study [43], and usually regarded "cannabis" as a single adherent medication [30], therefore disregarding the inherent complexity in MC treatment, with differences in over 90 phytocannabinoids [18] between cannabis cultivars [44].

We also found that the incidence of MC-related AEs was higher among non-responders. This may be explained by responders tolerating MC-related AEs better than non-responders. It could also be explained by the responders' success during trial-and-error to identify a specific MC chemovar that provided them relief with lower rates of AEs. Nevertheless, due to our study design, we could not corroborate these findings. Future studies should, therefore, investigate the association between MC-related AEs and treatment response a priori. Importantly, none of the patients reported aggravation of migraine AEs as a result of MC treatment.

Limitations

There are four limitations in the current study. Firstly, our results may have been biased by the small sample size; non-parametric models were used to balance this limitation. Secondly, there may be self-reporting bias. Participants were able to respond to the questionnaire under strict anonymity, ensuring there were no risks that their current treatment plan may be altered by their physician. The questionnaire has also been validated. Thirdly, since we cannot compare the initial indications for which responders and non-responders obtained their MC license, it is possible that the presented results have been biased. Nonetheless, since we identified that chronic neuropathic non-cancer pain was the predominant indication for obtaining MC license, we assumed that it is unlikely that differences between the subgroups exist. Lastly, since the frequency of migraine attacks prior to MC treatment was reported in retrospect, recall bias might have occurred.

5. Conclusions

Migraine is currently not indicated for a MC treatment license in Israel. Nevertheless, in some cases it falls under the approved chronic neuropathic non-cancer pain indication, making it possible to study migraine more extensively. In this study, we demonstrated that patients responding to MC treatment also reported less disability and lower conventional anti-migraine medications intake. Additionally, we highlighted the importance of recognizing and analyzing the doses of the pronounced MC constituents consumed by patients, which in turn allowed us to better understand MC treatment associations with reduction in migraine attacks frequency. We also identified specific cultivars that contain the favorable ratio of compounds that were associated with migraine frequency reduction. These results might shed light on the beneficial effects of MC on migraine and motivate future studies to utilize a cannabis cultivar with the specific phytocannabinoids mentioned here. This additional work could validate our results and possibly support making migraine an approved indication for MC license in Israel.

Author Contributions: Conceptualization, J.A., G.M.L., E.E., and D.M.; data curation, J.A.; formal analysis, Y.V.; funding acquisition, D.M.; investigation, J.A.; methodology, J.A., Y.V., P.B., G.M.L., E.E., and D.M.; project administration, D.M.; resources, D.M.; supervision, G.M.L. and D.M.; writing—original draft, J.A.; writing—review and editing, J.A., Y.V., P.B., G.M.L., E.E., and D.M. All authors have read and agreed to the published version of the manuscript.

Funding: The study was funded by the Evelyn Gruss Lipper Charitable Foundation, Lauren Frank Rose Donation. This sponsor had no role or influence on the study or on this submission.

Conflicts of Interest: The authors declare no conflict of interest.

References

1. Burch, R.C.; Buse, D.C.; Lipton, R.B. Migraine: Epidemiology, Burden, and Comorbidity. *Neurol. Clin.* **2019**, *37*, 631–649. [CrossRef] [PubMed]
2. Cameron, C.; Kelly, S.; Hsieh, S.C.; Murphy, M.; Chen, L.; Kotb, A.; Peterson, J.; Coyle, D.; Skidmore, B.; Gomes, T.; et al. Triptans in the Acute Treatment of Migraine: A Systematic Review and Network Meta-Analysis. *Headache* **2015**, *55*, 221–235. [CrossRef] [PubMed]
3. Pardutz, A.; Schoenen, J. NSAIDs in the acute treatment of migraine: A review of clinical and experimental data. *Pharmaceuticals* **2010**, *3*, 1966–1987. [CrossRef] [PubMed]
4. Derry, S.; Moore, R.A. Paracetamol (acetaminophen) with or without an antiemetic for acute migraine headaches in adults. *Cochrane Database Syst. Rev.* **2013**, *2013*. [CrossRef] [PubMed]
5. Tfelt-Hansen, P.; Saxena, P.R.; Dahlöf, C.; Pascual, J.; Láinez, M.; Henry, P.; Diener, H.-C.; Schoenen, J.; Ferrari, M.D.; Goadsby, P.J. Ergotamine in the acute treatment of migraine: A review and European consensus. *Brain* **2000**, *123*, 9–18. [CrossRef]
6. Kelley, N.E.; Tepper, D.E. Rescue Therapy for Acute Migraine, Part 3: Opioids, NSAIDs, Steroids, and Post-Discharge Medications. *Headache J. Head Face Pain* **2012**, *52*, 467–482. [CrossRef]
7. Becker, W.J. Acute Migraine Treatment in Adults. *Headache J. Head Face Pain* **2015**, *55*, 778–793. [CrossRef]
8. Silberstein, S.D. Preventive migraine treatment. *Contin. Lifelong Learn. Neurol.* **2015**, *21*, 973–989. [CrossRef]
9. Sexton, M.; Cuttler, C.; Finnell, J.S.; Mischley, L.K. A Cross-Sectional Survey of Medical Cannabis Users: Patterns of Use and Perceived Efficacy. *Cannabis Cannabinoid Res.* **2016**, *1*, 131–138. [CrossRef]
10. Cuttler, C.; Spradlin, A.; Cleveland, M.J.; Craft, R.M. Short- and Long-Term Effects of Cannabis on Headache and Migraine. *J. Pain* **2019**. [CrossRef]
11. Sarchielli, P.; Pini, L.A.; Coppola, F.; Rossi, C.; Baldi, A.; Mancini, M.L.; Calabresi, P. Endocannabinoids in Chronic Migraine: CSF Findings Suggest a System Failure. *Neuropsychopharmacology* **2007**, *32*, 1384–1390. [CrossRef] [PubMed]

12. Cupini, L.M.; Costa, C.; Sarchielli, P.; Bari, M.; Battista, N.; Eusebi, P.; Calabresi, P.; Maccarrone, M. Degradation of endocannabinoids in chronic migraine and medication overuse headache. *Neurobiol. Dis.* **2008**, *30*, 186–189. [CrossRef] [PubMed]
13. Greco, R.; Mangione, A.S.; Sandrini, G.; Maccarrone, M.; Nappi, G.; Tassorelli, C. Effects of anandamide in migraine: Data from an animal model. *J. Headache Pain* **2011**, *12*, 177–183. [CrossRef] [PubMed]
14. Kandasamy, R.; Dawson, C.T.; Craft, R.M.; Morgan, M.M. Anti-migraine effect of Δ9-tetrahydrocannabinol in the female rat. *Eur. J. Pharmacol.* **2018**, *818*, 271–277. [CrossRef]
15. ElSohly, M.A.; Slade, D. Chemical constituents of marijuana: The complex mixture of natural cannabinoids. *Life Sci.* **2005**, *78*, 539–548. [CrossRef]
16. Russo, E.B. Taming THC: Potential cannabis synergy and phytocannabinoid-terpenoid entourage effects. *Br. J. Pharmacol.* **2011**, *163*, 1344–1364. [CrossRef]
17. Ben-Shabat, S.; Fride, E.; Sheskin, T.; Tamiri, T.; Rhee, M.H.; Vogel, Z.; Bisogno, T.; De Petrocellis, L.; Di Marzo, V.; Mechoulam, R. An entourage effect: Inactive endogenous fatty acid glycerol esters enhance 2-arachidonoyl-glycerol cannabinoid activity. *Eur. J. Pharmacol.* **1998**, *353*, 23–31. [CrossRef]
18. Baram, L.; Peled, E.; Berman, P.; Yellin, B.; Besser, E.; Benami, M.; Louria-Hayon, I.; Lewitus, G.M.; Meiri, D. The heterogeneity and complexity of Cannabis extracts as antitumor agents. *Oncotarget* **2019**, *10*, 4091–4106. [CrossRef]
19. Berman, P.; Futoran, K.; Lewitus, G.M.; Mukha, D.; Benami, M.; Shlomi, T.; Meiri, D. A new ESI-LC/MS approach for comprehensive metabolic profiling of phytocannabinoids in Cannabis. *Sci. Rep.* **2018**, *8*, 1–15. [CrossRef]
20. Landshaft, Y.; Albo, B.; Mechoulam, R.; Afek, A. The Updated Green Book (May 2019): The Official Guide to Clinical Care in Medical Cannabis. Available online: https://www.health.gov.il/hozer/mmk154_2016.pdf (accessed on 8 June 2020).
21. Qualtrics, L.L.C. *Qualtrics (Version 12018)*; Qualtrics Labs Inc.: Provo, UT, USA, 2015. Available online: http//www.qualtrics.com (accessed on 10 May 2020).
22. Stewart, W.F.; Lipton, R.B.; Dowson, A.J.; Sawyer, J. Development and testing of the Migraine Disability Assessment (MIDAS) Questionnaire to assess headache-related disability. *Neurology* **2001**, *56*, S20–S28. [CrossRef]
23. Yang, M.; Rendas-Baum, R.; Varon, S.F.; Kosinski, M. Validation of the Headache Impact Test (HIT-6TM) across episodic and chronic migraine. *Cephalalgia* **2011**, *31*, 357–367. [CrossRef] [PubMed]
24. Shochat, T.; Tzischinsky, O.; Oksenberg, A.; Peled, R. Validation of the Pittsburgh Sleep Quality Index Hebrew translation (PSQI-H) in a sleep clinic sample. *Isr. Med. Assoc. J.* **2007**, *9*, 853–856. [PubMed]
25. Wickham, H. *Tidyverse: Easily Install and Load 'Tidyverse" Packages*; Version 1.3.0; Comprehensive R Archive Network (CRAN): Vienna, Austria, 2019.
26. Kolde, R. *R Package pheatmap: Pretty Heatmaps*; Version 1.0.8; Comprehensive R Archive Network (CRAN): Vienna, Austria, 2015.
27. Ströbel, A.; Haynes, A. *R Package a Table: Create Tables for Reporting Clinical Trials*; Version 0.1.5; Comprehensive R Archive Network (CRAN): Vienna, Austria, 2019.
28. Christensen, C.E.; Younis, S.; Deen, M.; Khan, S.; Ghanizada, H.; Ashina, M. Migraine induction with calcitonin gene-related peptide in patients from erenumab trials. *J. Headache Pain* **2018**, *19*, 105. [CrossRef] [PubMed]
29. Hergenrather, J.Y.; Aviram, J.; Vysotski, Y.; Campisi-pinto, S.; Lewitus, G.M.; Meiri, D. Cannabinoid and Terpenoid Doses are Associated with Adult ADHD Status of Medical Cannabis Patients. *Rambam Maimonides Med. J.* **2020**, *11*, 1–14. [CrossRef] [PubMed]
30. Rhyne, D.N.; Anderson, S.L.; Gedde, M.; Borgelt, L.M. Effects of Medical Marijuana on Migraine Headache Frequency in an Adult Population. *Pharmacother. J. Hum. Pharmacol. Drug Ther.* **2016**, *36*, 505–510. [CrossRef]
31. Fan, P. Cannabinoid agonists inhibit the activation of 5-HT3receptors in rat nodose ganglion neurons. *J. Neurophysiol.* **1995**, *73*, 907–910. [CrossRef]
32. Lichtman, A.H.; Cook, S.A.; Martin, B.R. Investigation of Brain Sites Mediating Cannabinoid-Induced Antinociception in Rats: Evidence Supporting Periaqueductal Gray Involvement. *J. Pharmacol. Exp. Ther.* **1996**, *276*, 585–593.
33. Goadsby, P.J.; Gundlach, A.L. Localization of 3H-Dihydroergotamine-binding sites in the cat central nervous system: Relevance to migraine. *Ann. Neurol.* **1991**, *29*, 91–94. [CrossRef]

34. Magnusson, J.E.; Becker, W.J. Migraine Frequency and Intensity: Relationship with Disability and Psychological Factors. *Headache* **2003**, *43*, 1049–1059. [CrossRef]
35. Lin, Y.K.; Lin, G.Y.; Lee, J.T.; Lee, M.S.; Tsai, C.K.; Hsu, Y.W.; Lin, Y.Z.; Tsai, Y.C.; Yang, F.C. Associations between Sleep Quality and Migraine Frequency. *Medicine USA* **2016**, *95*, 1–7. [CrossRef]
36. Bradford, A.C.; Bradford, W.D.; Abraham, A.; Adams, G.B. Association between US state medical cannabis laws and opioid prescribing in the Medicare Part D population. *JAMA Intern. Med.* **2018**, *178*, 667–672. [CrossRef] [PubMed]
37. Haroutounian, S.; Meidan, R.; Davidson, E. The Effect of Medicinal Cannabis on Pain and Quality of Life Outcomes in Chronic Pain: A Prospective Open-label Study. *Clin. J. Pain* **2016**, *32*, 1036–1043. [CrossRef] [PubMed]
38. McCarty, D. Does Medical Cannabis Reduce Use of Prescription Opioids? *Am. J. Psychiatry* **2018**, *1*, 6–7. [CrossRef] [PubMed]
39. Stith, S.S.; Vigil, J.M.; Adams, I.M.; Reeve, A.P. Effects of Legal Access to Cannabis on Scheduled II–V Drug Prescriptions. *J. Am. Med. Dir. Assoc.* **2017**, *19*, 59–64. [CrossRef]
40. Yassin, M.; Garti, A.; Robinson, D. Effect of Medicinal Cannabis Therapy (MCT) on Severity of Chronic Low Back Pain, Sciatica and Lumbar Range of Motion. *Int. J. Anesthesiol. Pain Med.* **2016**, *2*, 1–6. [CrossRef]
41. Loder, E. Triptan Therapy in Migraine. *N. Engl. J. Med.* **2010**, *363*, 63–70. [CrossRef]
42. Russo, E. Cannabis for migraine treatment: The once and future prescription? An historical and scientific review. *Pain* **1998**, *76*, 3–8. [CrossRef]
43. Baron, E.P.; Lucas, P.; Eades, J.; Hogue, O. Patterns of medicinal cannabis use, strain analysis, and substitution effect among patients with migraine, headache, arthritis, and chronic pain in a medicinal cannabis cohort. *J. Headache Pain* **2018**, *19*, 37. [CrossRef]
44. Hazekamp, A.; Fischedick, J.T. Cannabis—From cultivar to chemovar. *Drug Test. Anal.* **2012**, *4*, 660–667. [CrossRef]

Article

Efficacy of Dronabinol for Acute Pain Management in Adults with Traumatic Injury: Study Protocol of A Randomized Controlled Trial

Claire Swartwood [1], Kristin Salottolo [2], Robert Madayag [2,3] and David Bar-Or [2,*]

1 Pharmacy Department, St Anthony Hospital, Lakewood, CO 80228, USA; Claireswartwood@centura.org
2 Trauma Research Department, St Anthony Hospital, Lakewood, CO 80228, USA; Kristin.salottolo@icloud.com (K.S.); robertmadayag@centura.org (R.M.)
3 Trauma Services Department, St Anthony Hospital, Lakewood, CO 80228, USA
* Correspondence: davidbme49@gmail.com; Tel.: +1-303-788-4089

Received: 3 February 2020; Accepted: 10 March 2020; Published: 12 March 2020

Abstract: Delta-9-tetrahydrocannabinol (Δ^9-THC) and other cannabinoids present in cannabis (marijuana) have been shown to affect the normal inhibitory pathways that influence nociception in humans. The potential benefits of cannabinoids as an analgesic are likely greatest in hyperalgesic and inflammatory states, suggesting a role as a therapeutic agent for treating acute pain following injury. Dronabinol is a licensed form of Δ^9-THC. The primary objective of this single center randomized controlled trial is to evaluate the efficacy of adjunctive dronabinol versus control (systemic analgesics only, no dronabinol) for reducing opioid consumption in adults with traumatic injury. Study inclusion is based on high baseline utilization of opioids ≥50 morphine equivalents (mg) within 24 h of admission for adults aged 18–65 years with traumatic injury. There is a 48-hour screening period followed by a 48-hour treatment period after randomization. A total of 122 patients will be randomized 1:1 across 2 study arms: adjunctive dronabinol versus control (standard of care using systemic analgesics, no adjunctive dronabinol). Patients randomized to the dronabinol arm should receive their first dose within 12 h of randomization, with a dose range of 5 mg up to 30 mg daily in divided doses, in addition to systemic analgesics as needed for pain. The primary efficacy endpoint is a change in opioid consumption (morphine equivalents), assessed post-randomization (48 h after randomization) minus pre-randomization (24 h prior to randomization). This is the first randomized trial to investigate whether adjunctive dronabinol is effective in reducing opioid consumption in acute pain management of traumatic injury. Trial Registration: ClinicalTrials.gov Identifier: NCT03928015.

Keywords: Delta-9-tetrahydrocannabinol; dronabinol; marijuana; randomized controlled trial; opioids; traumatic injury

1. Introduction

Delta-9-tetrahydrocannabinol (Δ^9-THC) and other cannabinoids present in cannabis (marijuana) have been shown to affect the normal inhibitory pathways that influence nociception in humans. Cannabinoids act through the binding of two cannabinoid receptors coupled through G proteins; CB1 receptors are predominantly found at central and peripheral nerve terminals, where they mediate transmitter release, while CB2 receptors are highly expressed throughout the immune system [1].

The evidence demonstrating a therapeutic effect of THC and cannabis-based medications is still emerging but is well established for treating chronic pain based on three influential peer-reviewed publications [2–4]. These publications also provide conclusive evidence for a therapeutic effect of cannabis-based medications as anti-emetics and for multiple sclerosis symptoms. There is moderate evidence for improving sleep outcomes associated with sleep apnea, fibromyalgia, multiple sclerosis, and chronic pain. There is insufficient or low-quality evidence in all remaining conditions that have been studied. For instance, there is a dearth of research on cannabinoid use for acute pain management. A 2017 systematic review identified seven randomized controlled trials (RCT) assessing the analgesic efficacy of cannabinoid medications for acute pain [5]. Of these studies, five RCTs demonstrated that cannabinoids were equivalent to placebo, in one RCT cannabinoids were superior to placebo, and in one RCT cannabinoids were inferior to placebo. These limited and inconsistent data justify the necessity to perform additional studies on the analgesic effects of cannabinoids for acutely painful conditions.

Patients commonly experience severe, acute pain following traumatic injury that is treated with analgesics, particularly opiates. The antinociceptive properties of cannabinoids may be greatest in hyperalgesic and inflammatory states, suggesting a therapeutic role for treating pain following injury [6]. Moreover, pre-clinical studies support a potential role of Δ^9-THC and cannabinoids as an adjunctive agent to opioids in painful conditions, via synergistic enhancement of mu opioid antinociception as well as the prevention of tolerance to and withdrawal from opiates [7–9].

Recently published preliminary clinical research from our group examined the effect of adjunctive dronabinol for acute pain management among 66 trauma patients [10]. Cases demonstrated a significant reduction in opioid consumption (morphine equivalents) from baseline with adjunctive dronabinol (−79 mg, $p < 0.001$), while the change in opioid consumption for matched controls was unchanged from baseline (−9 mg, $p = 0.63$), resulting in a nine-fold greater reduction in opioid consumption for cases versus controls that was significantly different between pairs (difference: −70 mg, $p = 0.02$). There were no differences in secondary outcomes. These results suggest that adjunctive dronabinol used as part of a multimodal analgesia regimen may result in a marked reduction in opioid consumption

Two subset analyses of this matched cohort study provide mixed evidence that the opioid sparing effect of dronabinol may be greater in patients who are marijuana users. Among the subset of 19 cases who were marijuana users, opioid consumption was significantly reduced with adjunctive dronabinol (−97 mg, $p < 0.001$) versus no change in opioid consumption in 19 matched controls (1 mg, $p = 0.70$), with a difference between pairs that was significant: −108 mg, $p = 0.01$) [10]. However, when examining the subset of patients who received dronabinol, there were no differences in the change in opioid consumption for patients who were marijuana users ($n = 21$, −97 mg reduction with dronabinol) compared to non-marijuana users ($n = 15$, −64 mg reduction with dronabinol), $p = 0.41$ (unpublished).

We are recruiting patients in a RCT to evaluate the efficacy of adjunctive dronabinol on opioid utilization for acute pain management. The primary trial objective is to evaluate the efficacy of adjunctive dronabinol versus control (systemic analgesics only, no dronabinol) for reduction in opioid consumption in adults with traumatic injury. Dronabinol is a licensed form of Δ^9-THC. Dronabinol is not FDA approved for acute pain management; however, it has been in use at our level I trauma center system formulary without restriction since 2015.

2. Materials and Methods

2.1. Study Design and Setting

This is an open label RCT being performed at a single level I trauma center: St. Anthony Hospital in Lakewood, CO. This RCT was designed primarily to determine whether adjunctive dronabinol reduces opioid consumption compared to control. The study was designed with a stratified randomization by baseline marijuana use, which is intended to determine whether the treatment effect of dronabinol is greater in chronic marijuana users compared to recreational or non-marijuana users. This stratified randomization design was incorporated based on the gestalt that cannabis-based medication has a greater benefit for marijuana users.

There is a 48-hour screening/randomization window, a 48-hour treatment window, and a total participation period extending through the acute hospitalization. A description of the clinical trial is posted at ClinicalTrials.gov.

2.2. Study Subjects

Patients are being recruited from the participating trauma center to which they are acutely presenting. A total of 122 adult trauma patients will be randomized 1:1 across 2 study arms: adjunctive dronabinol or control (systemic analgesics only), as shown in Figure 1.

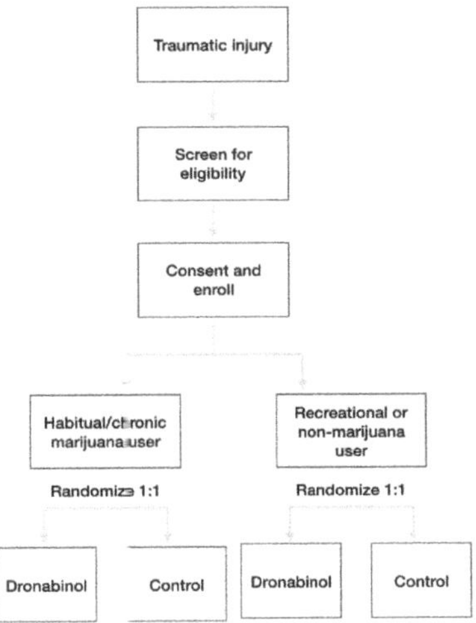

Figure 1. Subject Disposition.

Patients should fulfill all of the following inclusion criteria:

- Male or female, 18 years to 65 years old (inclusive).
- Diagnosis of traumatic injury based on ICD10 diagnosis of S00-T14, which covers injuries to any region of the body.
- Moderately high initial morphine equivalent use ≥50 mg within a 24-hour window during the screening period. Opioids will be converted to morphine equivalents (mg) using an Equianalgesic conversion chart, Table 1 [11].
- Willing to disclose current marijuana status (current user (habitual/chronic or recreational), former user, never user).

Table 1. Oral morphine milligram equivalents (MME) conversion factors.

Opioid (mg, Except Where Noted)	Oral MME Conversion Factor [1]
Buprenorphine	N/A
Codeine	0.15
Fentanyl, intravenous (mcg)	0.3
Hydrocodone	1
Hydromorphone	4
Meperidine	0.1
Methadone	3
Morphine, oral	1
Morphine, intravenous	3
Oxycodone	1.5
Tramadol	0.1

[1] Formula: Strength per Unit X (Number of Units/Days Supply) X MME conversion factor = MME/Day.

Patients fulfilling one or more of the following criteria may not be enrolled in the study:

- Patients on a pain management agreement
- Patients who are nil per os (NPO) at the time of randomization or are expected to be NPO within the next 48 h, with the exception for a brief NPO period for surgical procedures
- Patients who have received or are expected to receive neuraxial/locoregional blocks for pain within the next 48 hours
- Patients with a known allergy or previous hypersensitivity reaction to dronabinol or sesame oil
- Patients prescribed dronabinol between arrival and prior to randomization
- Pregnancy or breast feeding
- Incarceration (presumed; patients are not arraigned until after hospital discharge).

2.3. Study Visits

The following procedures will be performed at screening, within 48 h of hospital admission: ensure patient meets inclusion and exclusion criteria; record 24-hour total morphine equivalents; record habitual marijuana usage; obtain informed consent via patient or proxy.

Once patients are confirmed to meet all criteria and have signed an informed consent, they will be randomized 1:1 across the two study arms (dronabinol or control, Figure 1). The following procedures will be performed during randomization: randomize the patient using the Microsoft Excel blinded randomization schema; record pain using the patient self-reported pain numeric rating scale (NRS, 0–10 scale).

The following procedures will be performed during the acute hospitalization, post-randomization: if randomized to the dronabinol arm, administer the first dose within the first scheduled dose window and within 12 h of randomization. Record all doses received, including date, time and dose; record all opioid and non-opioid systemic analgesics received, including route/dose/frequency; record all non-analgesic concomitant medications; record pain NRS scores at the following time points: once admitted in hospital bed, preoperatively in the OR prior to anesthesia, one hour post operatively; record all analgesic complications; record all documented drug use from patient self-report and urine drug screening results. Detailed information on the regularity of marijuana use will also be recorded.

The following procedures will be performed at hospital discharge: record discharge pain NRS score; record discharge location; re-consent, if necessary.

2.4. Randomization and Blinding

Patients will be assigned to treatment by a randomization schedule developed and maintained by an independent statistician. The randomization allocation sequence was computer generated and is blinded, with allocation hidden until a patient has met all inclusion and exclusion criteria and provided informed consent. Randomization will occur in a 1:1 fashion in blocks of 2 and 4 and stratified by habitual marijuana user (yes/no).

The assessor, participants, treatment team, and statisticians are unblinded. All assessments are standard and routinely collected by the assessors (ICU and general ward nursing staff), including pain NRS scores and analgesia administration.

2.5. Intervention

The study drug is dronabinol (Marinol®, AbbVie, Inc; Chicago, IL, USA). Eligible patients will receive adjunctive dronabinol vs. control (no dronabinol, systemic analgesia only). Patients will be allocated to a treatment in accordance with the randomization schedule following confirmation of eligibility.

Patients who consent to participate in the study will have an order in the electronic medical record that will be used to assist with treatment compliance and for dispensing dronabinol, when applicable. Patients randomized to the dronabinol arm should receive their first dose within 12 h of randomization. The initial dosing and any changes in dosing will be determined by the prescribing/treating clinician.

The target dose is 5 mg twice daily; the dose may be adjusted to within 5 mg to 30 mg daily in divided doses (e.g., 2.5 mg twice daily–10 mg three times daily). Patients who are randomized to the control arm will have an order set that specifies no administration of dronabinol for 48 h.

Patients in both arms will receive as needed (pro re nata, PRN) non-opioid/opioid analgesia as determined by the care team; patients who are not randomized to the dronabinol arm will receive these analgesics only, while patients randomized to the dronabinol arm will receive dronabinol in addition to PRN non-opioid/opioid analgesia. A target pain numeric rating score for trauma patients is 4 or less on a 0–10 scale. Higher pain scores ≥5 typically warrant analgesia, as determined by the attending physician and care team for the patient's specific needs. These established guidelines will ensure patients are receiving analgesia based on self-reported pain, independent of treatment arm.

After the 48-hour treatment window post-randomization, the use of adjunctive dronabinol for the remaining acute hospitalization will be at the patient's and physician's discretion. Except for the analgesia protocol, all other interventions will follow techniques used in the context of everyday clinical practice, and thus will be identical for participants in both arms. The following medications are discouraged: neuraxial and locoregional nerve blocks.

2.6. Outcome Measures

Patients will be followed to hospital discharge for outcomes of morphine equivalent use, length of stay, pain NRS scores, hospital complications, and analgesic complications.

The primary outcome is morphine equivalents. All opioids consumed will be converted to morphine equivalents, as shown in Figure 1 [11]. The clinical effects of treatment arm on morphine equivalents will be evaluated at 48 h after randomization.

Secondary outcomes include the following:

- Morphine equivalents: overall (hospital admission through discharge or death)
- Non-opioid analgesics: overall doses received (admission through acute hospitalization discharge or death), and examined by non-opioid drug
- Acute hospitalization length of stay
- Pain NRS scores: in ED prior to randomization, once admitted in hospital bed, preoperatively in the OR prior to anesthesia, one-hour post operatively, at hospital discharge
- Time (hours) to transition to non-opioid analgesia
- Incidence of hospital complications
- Safety (Incidence of analgesic complications)

Analgesic complications will be recorded irrespective of the presence or absence of a causal relationship, and include the including:

- Allergic reaction
- Nausea and vomiting
- Respiratory depression (hypoxia and hypopnea)
- Hypotension
- Urinary retention
- Constipation/ileus
- Abdominal pain
- Dizziness
- Euphoria
- Paranoid reaction
- Somnolence
- Delirium
- Over-sedation.

2.7. Statistics

Significance is set at an alpha value of 0.05. SAS (Cary, NC) software will be used for statistical analysis. All efficacy analyses will be performed in the intent-to-treat population, defined as all patients who are randomized. Subset analyses will be performed by habitual marijuana use.

The primary endpoint is the change in morphine equivalents and will be assessed as: post-randomization (48 h after randomization) minus pre-randomization (24 h prior to randomization). No imputation will occur for the primary endpoint. The change in morphine equivalents (mg) will be analyzed with an analysis of covariance (ANCOVA) model to examine the effect of treatment arm, adjusted for age, gender, injury severity score, and clinical characteristics that differ between groups with $p < 0.15$. Of note, our study inclusion criteria allow for patients to present with polytrauma. We anticipate the majority of patients will have injuries to the thorax and extremities, with few patients presenting with severe TBI because administration of opioids and other drugs that alter a neurological assessment tend to be used sparingly. Should there be differences in injury patterns, despite the 1:1 randomization procedure, these differences will be adjusted for in the primary ANCOVA analysis.

Secondary efficacy analyses include the difference between treatment groups in: hospital disposition, hospital complications, and analgesic complications, reported as proportion (%) and analyzed with chi-square tests; morphine equivalents over the hospitalization, hospital length of stay (days), time (h) to transition to non-opioid analgesia, pain NRS scores at all specified time points, reported as median (IQR) and analyzed with a Wilcoxon rank-sum test. Analgesic complications will be described by severity as mild, moderate and severe.

2.8. Sample Size

The planned enrollment is 122 patients total randomized 1:1 across two study arms: dronabinol or no dronabinol (systemic analgesia only). The sample size is based on a 38% reduction in morphine equivalents with adjunctive dronabinol vs. an 8% reduction in morphine equivalents for systemic analgesics only, with a pooled standard deviation of 58. The analysis was performed using two sample mean tests with normal approximation and equal weights. These estimates were derived via bootstrapping of the final matched study sample of 66 patients. The power to demonstrate the main effect of dronabinol over systemic analgesics is 80% using a 2-tailed alpha of 0.05.

2.9. Ethical Considerations

The study was approved from the Institutional Review Board for St. Anthony Hospital (Catholic Health Initiatives). There may be patients incurring cognitive impairment (due to head injury or acuity of illness). The study coordinator will discuss with the treating team and will directly assess the consenting capacity of the patient. The study nurse follows the current hospital protocol regarding the use of consent by a legally authorized representative. In these clinical situations where the patient's representative initially consents, the patient will be "re-consented" when able to assure that they want to continue in the study.

Safety outcomes will be reported to the head of the medical executive committee at an ongoing basis. If/when the rate and/or severity of the monitored safety events becomes unacceptable, the medical executive committee has procedures in place to protect research subjects.

An interim analysis will be performed when >50% ($n = 62$) of patients have been enrolled and discharged from the hospital to determine clinical equipoise. A stopping guideline of $p < 0.001$ will be used for the primary end point.

3. Discussion

This is the first randomized trial to investigate whether the addition of dronabinol is effective at reducing opioid consumption for acute pain management of traumatic injury. There are numerous strengths of this study. This clinical trial improves upon our previously published matched cohort study and removes many of the limitations of that study: patients will now be matched by self-reported marijuana use; the pre-treatment period for the controls will be identical to cases rather than being estimated based on the median time from admission to first administration of dronabinol among cases; we will know why controls were not prescribed dronabinol; there are complications and adverse effects that are associated with both systemic analgesics and dronabinol, which will be recorded and analyzed in this trial by treatment arm and by severity. Additional strengths of this study are that it is investigator-initiated and independent from pharmaceutical or other industry interests, and the findings (whether positive or negative) will be submitted to a peer reviewed scientific journal for publication.

Another benefit of this study is the stratified randomization by chronic marijuana usage. Earlier work by the study investigators suggests that pre-injury marijuana use results in increased consumption of opioid analgesics and greater self-reported pain following traumatic injury compared to trauma patients who are marijuana naïve [12]. If the randomized trial demonstrates a greater treatment effect in the subset of chronic marijuana users, this will have wide-ranging clinical implications for acute pain management, because trauma patients have a high prevalence of marijuana use and other substance abuse issues, reported in 40–50% of patients [13,14] that appears to be increasing over time [15]. Thus, if marijuana use significantly affects acute pain management then chronic marijuana users will merit special consideration during acute pain management.

While there are now 11 states that have legalized recreational marijuana, we believe Colorado is uniquely able to study this issue because of the high utilization in our state. Colorado was the first state to legalize and commercialize recreational marijuana, with retail shops opening on January 1st, 2014. A recent study identified that commercialization of recreational marijuana in Colorado was associated with an increased use of marijuana or an increased risk of traumatic injury while using marijuana [15].

Opioids are established and effective analgesics for managing pain in the traumatic and critical care setting due to their proven efficacy in treating moderate to severe acute pain [16]. The Center for Disease Control and Prevention (CDC) estimates that approximately 130 Americans are dying each day from opioid overdose, resulting in an opioid epidemic. We believe the use of dronabinol as a tool in the clinician's tool kit to decrease reliance on opioids is an appealing option. Some possible benefits of this study include better pain control and a lower need for opiates for participants. Use of dronabinol to reduce or maintain the opioid regimen, rather than increasing narcotic dosages to detrimentally high levels, may also reduce the negative effects of opioids on vascular neurologic response and respiratory depression.

One of the primary limitations of this trial is that the study is open label. Patients are still prospectively randomized to active treatment vs. control and all assessments are standard and routinely collected by the assessors (ICU and general ward nursing staff), including pain scores and analgesia administration. However, we are unable to blind patients because there are no orally administered placebo pills that are on hospital formulary to be used for this study (unavoidable blinding). We did not blind clinicians because the dosing of dronabinol may need to be modified and is allowed within the range of 5 mg to 30 mg daily in divided doses. Although a blinded study would be preferred to reduce knowledge bias, the study design is compatible with real-world situations and increases the external validity of the study.

Additional limitations are as follows. First, our preliminary study was conducted in 2017, around the peak of the opioid epidemic [17,18]. Since that time, there have been enterprise-wide initiatives to use alternatives to opioids [19,20], which could impact our enrollment criteria. However, our study has potentially greater implications in the current setting where opioid alternatives are sought. Second, and related, the data used to power the RCT were recorded in 2017, and it is possible that opioid consumption will be less in both groups (dronabinol and control), but whether this translates to a different treatment effect with dronabinol remains to be seen. Third, marijuana use is based on self-reporting because admission urine toxicology testing is only utilized in about 50% of patients, with a bias towards screening younger patients. Unlike blood alcohol tests, urine toxicology testing seldom results in a change in care and thus are not routinely ordered following traumatic injury. We will not be requiring a change in practice for ordering urine toxicology testing as part of our study. However, our unpublished research demonstrates the percent agreement between urine toxicology findings and patient self-report is 81% for cannabis. The negative predictive value of 95% demonstrates that a negative self-report correctly identifies 95% of patients who test negative for cannabis, while the specificity provides an 85% chance that a patient will not test positive for cannabis if the patient denies use. Fourth, the results of this study are only be applicable to dronabinol and not to other cannabinoids, such as the recently trending cannabidiol (CBD). Finally, the study is currently approved as a single-center RCT, which limits its generalizability. The authors are amenable to adding additional sites which use dronabinol on formulary without restrictions.

There are two additional risks to the patient that need to be mentioned. First, this study involves an experimental (investigational) drug that has not been approved by the U.S. Food and Drug Administration (FDA) for the specific indication of acute pain management. Dronabinol is only FDA approved for loss of appetite due to HIV and chemotherapy-induced nausea and vomiting. This study is not intended to result in an FDA Investigational New Drug Application. Second, dronabinol is a synthetic version of THC. There is a risk that the study medication will result in a positive urine drug screen test for cannabis for two weeks or more in patients who are not a current user of marijuana products. In most cases, if an employee has a recent prescription for dronabinol, that is sufficient to report the result to the employer as a negative.

Trial Status

The trial has been recruiting patients since October 2019 and will continue until 122 patients have been randomized. Protocol version 1.2. Two amendments have occurred since trial commencement. First, the inclusion criteria of a minimum baseline pain score ≥5 was removed. The second amendment modified the sample size calculation to incorporate the full preliminary study findings, rather than a smaller pilot population.

Author Contributions: Conceptualization, C.S.; methodology, K.S. and C.S.; software, K.S.; formal analysis, K.S.; investigation, C.S. and R.M.; resources, D.B.-O.; writing—original draft preparation, K.S.; writing—review and editing, C.S., R.M., D.B.-O.; supervision, D.B.-O.; project administration, C.S. and D.B.-O. All authors have read and agreed to the published version of the manuscript.

Funding: This research received no external funding.

Conflicts of Interest: The authors declare no conflict of interest.

References

1. Pertwee, R.G. Cannabinoid pharmacology: The first 66 years. *Br. J. Pharmacol.* **2006**, *147* (Suppl. 1), S163–S171. [CrossRef] [PubMed]
2. National Academies of Sciences Engineering, and Medicine. *The Health Effects of Cannabis and Cannabinoids: The Current State of Evidence and Recommendations for Research*; The National Academies Press: Washington, DC, USA, 2017.
3. Koppel, B.S.; Brust, J.C.; Fife, T.; Bronstein, J.; Youssof, S.; Gronseth, G.; Gloss, D. Systematic review: Efficacy and safety of medical marijuana in selected neurologic disorders: Report of the Guideline Development Subcommittee of the American Academy of Neurology. *Neurology* **2014**, *82*, 1556–1563. [CrossRef] [PubMed]
4. Whiting, P.F.; Wolff, R.F.; Deshpande, S.; Di Nisio, M.; Duffy, S.; Hernandez, A.V.; Keurentjes, J.C.; Lang, S.; Misso, K.; Ryder, S.; et al. Cannabinoids for Medical Use: A Systematic Review and Meta-analysis. *JAMA* **2015**, *313*, 2456–2473. [CrossRef] [PubMed]
5. Stevens, A.J.; Higgins, M.D. A systematic review of the analgesic efficacy of cannabinoid medications in the management of acute pain. *Acta Anaesthesiol. Scand.* **2017**, *61*, 268–280. [CrossRef] [PubMed]
6. Iversen, L.; Chapman, V. Cannabinoids: A real prospect for pain relief? *Curr. Opin. Pharmacol.* **2002**, *2*, 50–55. [CrossRef]

7. Cichewicz, D.L.; Martin, Z.L.; Smith, F.L.; Welch, S.P. Enhancement mu opioid antinociception by oral delta9-tetrahydrocannabinol: Dose-response analysis and receptor identification. *J. Pharmacol. Exp. Ther.* **1999**, *289*, 859–867. [PubMed]
8. Cichewicz, D.L.; McCarthy, E.A. Antinociceptive synergy between delta(9)-tetrahydrocannabinol and opioids after oral administration. *J. Pharmacol. Exp. Ther.* **2003**, *304*, 1010–1015. [CrossRef] [PubMed]
9. Cichewicz, D.L.; Welch, S.P. Modulation of oral morphine antinociceptive tolerance and naloxone-precipitated withdrawal signs by oral Delta 9-tetrahydrocannabinol. *J. Pharmacol. Exp. Ther.* **2003**, *305*, 812–817. [CrossRef] [PubMed]
10. Schneider-Smith, E.; Salottolo, K.; Swartwood, C.; Melvin, C.; Madayag, R.; Bar-Or, D. A Matched Pilot Study Examining Cannabis-Based Dronabinol for Acute Pain Following Traumatic Injury. *Trauma Surg. Acute Care Open* **2020**, *0*, e000391. [CrossRef] [PubMed]
11. National Center for Injury Prevention and Control. *CDC Compilation of Benzodiazepines mr, Stimulants, Zolpidem, and Opioid Analgesics with Oral Morphine Milligram Equivalent Conversion Factors*; 2018 version; Centers for Disease Control and Prevention: Atlanta, GA, USA, 2018. Available online: https://www.cdc.gov/drugoverdose/resources/data.html (accessed on 18 December 2017).
12. Salottolo, K.; Peck, L.; Tanner, A.; Carrick, M.M.; Madayag, R.; McGuire, E.; Bar-Or, D. The grass is not always greener: A multi-institutional pilot study of marijuana use and acute pain management following traumatic injury. *Patient Saf. Surg.* **2018**, in press. [CrossRef] [PubMed]
13. Cowperthwaite, M.C.; Burnett, M.G. Treatment course and outcomes following drug and alcohol-related traumatic injuries. *J. Trauma Manag. Outcomes* **2011**, *5*, 3. [CrossRef] [PubMed]
14. Rivara, F.P.; Jurkovich, G.J.; Gurney, J.G.; Seguin, D.; Fligner, C.L.; Ries, R.; Raisys, V.A.; Copass, M. The magnitude of acute and chronic alcohol abuse in trauma patients. *Arch. Surg.* **1993**, *128*, 907–912. [CrossRef] [PubMed]
15. Chung, C.; Salottolo, K.; Tanner, A., 2nd; Carrick, M.M.; Madayag, R.; Berg, G.; Lieser, M.; Bar-Or, D. The impact of recreational marijuana commercialization on traumatic injury. *Inj. Epidemiol.* **2019**, *6*, 3. [CrossRef] [PubMed]
16. Minkowitz, H.S.; Rathmell, J.P.; Vallow, S.; Gargiulo, K.; Damaraju, C.V.; Hewitt, D.J. Efficacy and safety of the fentanyl iontophoretic transdermal system (ITS) and intravenous patient-controlled analgesia (IV PCA) with morphine for pain management following abdominal or pelvic surgery. *Pain Med.* **2007**, *8*, 657–668. [CrossRef] [PubMed]
17. Ahmad, F.B.; Rossen, L.M.; Spencer, M.R.; Warner, M.; Sutton, P. *Provisional Drug Overdose Death Counts*; National Center for Health Statistics: Hyattsville, MD, USA, 2018.
18. Tully, A.; Anderson, L.; Adams, W.; Mosier, M.J. Opioid creep in burn center discharge regimens: Doubled amounts and complexity of narcotic prescriptions over seven years. *Burns* **2019**, *45*, 328–334. [CrossRef] [PubMed]
19. Baird, J.; Faul, M.; Green, T.C.; Howland, J.; Adams, C.A., Jr.; Hodne, M.J.; Bohlen, N.; Mello, M.J. Evaluation of a Safer Opioid Prescribing Protocol (SOPP) for Patients Being Discharged From a Trauma Service. *J. Trauma Nurs.* **2019**, *26*, 113–120. [CrossRef] [PubMed]
20. Urman, R.D.; Boing, E.A.; Khangulov, V.; Fain, R.; Nathanson, B.H.; Wan, G.J.; Lovelace, B.; Pham, A.T.; Cirillo, J. Analysis of predictors of opioid-free analgesia for management of acute post-surgical pain in the United States. *Curr. Med. Res. Opin.* **2019**, *35*, 283–289. [CrossRef] [PubMed]

MDPI
St. Alban-Anlage 66
4052 Basel
Switzerland
Tel. +41 61 683 77 34
Fax +41 61 302 89 18
www.mdpi.com

Brain Sciences Editorial Office
E-mail: brainsci@mdpi.com
www.mdpi.com/journal/brainsci

www.ingramcontent.com/pod-product-compliance
Lightning Source LLC
LaVergne TN
LVHW071005170726
843515LV00006B/1076
9783039439959